AF389087

Rapid Detection of Mycotoxin Contamination

Rapid Detection of Mycotoxin Contamination

Editor

András Székács

MDPI • Basel • Beijing • Wuhan • Barcelona • Belgrade • Manchester • Tokyo • Cluj • Tianjin

Editor
András Székács
Agro-Environmental
Research Centre, Institute of
Environmental Sciences,
Hungarian University of
Agriculture and Life Sciences
Budapest
Hungary

Editorial Office
MDPI
St. Alban-Anlage 66
4052 Basel, Switzerland

This is a reprint of articles from the Special Issue published online in the open access journal *Toxins* (ISSN 2072-6651) (available at: www.mdpi.com/journal/toxins/special_issues/Detection_Mycotoxin_Contamination).

For citation purposes, cite each article independently as indicated on the article page online and as indicated below:

LastName, A.A.; LastName, B.B.; LastName, C.C. Article Title. *Journal Name* **Year**, *Volume Number*, Page Range.

ISBN 978-3-0365-1694-3 (Hbk)
ISBN 978-3-0365-1693-6 (PDF)

Contents

About the Editor

András Székács

András Székács, PhD in chemistry, Doctor of the Hungarian Academy of Sciences, is a professor at the Hungarian University of Agriculture and Life Sciences (MATE), Deputy Director of the Insitute of Environmental Sciences, and Head of the Agro-Environmental Research Centre and of the Department of Environmental Analysis and Technology. He is also a honorary professor at the Budapest University of Technology and Economics (BME). His main fields of research are chemical and environmental safety of agricultural and chemical technologies (including mycotoxins, pesticides, genetically modified crops, surfactants, anthropogenic contaminants, etc.), as well as instrumental and immunoanalysis of organic micropollutants. Currently, he is a member of the Management Board of the European Food Safety Authority, and he is a member of the Scientific Advisory Body of the OECD Co-operative Research Programme. The number of his peer reviewed publications: 208, cumulated impact factor 288.1, h-index: 25.

Preface to "Rapid Detection of Mycotoxin Contamination"

Mycotoxin occurrence in crops and subsequent contamination in food and feed is currently a major concern in environmental and food safety, affecting both crop production and animal husbandry. In turn, rapid detection of mycotoxin levels in food and feed, as well as in other biological and environmental matrices, is of key importance both in mycotoxin monitoring and exposure assessment. Recent developments, utilization, evaluation, and possible improvements of methods that allow rapid, sensitive, and accurate detection of various mycotoxins have been chosen to be the topic of this Book of Toxins, comprising 12 original research articles and a review. Overall, 56 authors have contributed from 10 countries (China, France, Germany, Hungary, Iran, Kenya, Poland, Sweden, UK, US), discussing various aspects of mycotoxin research, with mycotoxin analysis involved, classical instrumental analytical or biosensor method development, sample preparation and handling to support method accuracy, as well as applications in routine monitoring or decontamination assessment. Target analyte mycotoxins included aflatoxins, deoxynivalenol, diacetoxyscirpenol, fumonisins, fusarenone-X, HT-2 toxins, nivalenol, ochratoxins, sterigmatocystin, T-2 toxin, and zearalenone.

In discussing rapid detection, method development is an obvious focus area aiming for improved analytical characteristics (analytical sensitivity, accuracy, precision, linearity, robustness, and ruggedness, limits of detection and quantification, applicability). Thus, the development of classical instrumental analytical methods (e.g., liquid chromatography-mass spectrometry), immunoanalytical methods (e.g., a magnetic particle-based immunoassay, a fluorescent immunoassay, and several immunosensors), as well as a fluorescent sensor format utilising selective aptamers as recognition elements, are reported. Efforts to improve method applicability by enhanced sample preparation as well as sample selection are also described. Last, but not least, successful application of various rapid analysis methods in different commodity and environmental matrices is documented. Target matrices included traditional ones, e.g., laboratory fungal cultures, cereal, and feed samples, but in addition, surface water is also discussed as a novel environmental matrix of mycotoxins as emerging surface water contaminants. An additional emerging mycotoxin contamination problem is the spread of toxicogenic fungi with climate change tendencies. A particular corresponding issue is decontamination of mycotoxins in contaminated commodities, which is exemplified in this book by effective decomposition of aflatoxin and sterigmatocystin in maize by Lactobacilli and non-Lactobacillus lactic acid bacterial strains.

Due to the diverse topics covered, the book can account for the interest of a wide range of readers, from researchers to experts in practical applications.

András Székács

Editor

Editorial

Mycotoxins as Emerging Contaminants. Introduction to the Special Issue "Rapid Detection of Mycotoxin Contamination"

András Székács

Agro-Environmental Research Centre, Institute of Environmental Sciences, Hungarian University of Agriculture and Life Sciences, Herman O. út 15, H-1022 Budapest, Hungary; szekacs.andras@uni-mate.hu

Citation: Székács, A. Mycotoxins as Emerging Contaminants. Introduction to the Special Issue "Rapid Detection of Mycotoxin Contamination". *Toxins* **2021**, *13*, 475. https://doi.org/10.3390/toxins 13070475

Received: 8 June 2021
Accepted: 6 July 2021
Published: 9 July 2021

Publisher's Note: MDPI stays neutral with regard to jurisdictional claims in published maps and institutional affiliations.

Concerns for human and environmental health regarding mycotoxins are predominantly raised in connection with their occurrence in food and feed (especially in grains) [1,2]. Thus, mycotoxin contamination is an emerging problem in agriculture. These toxic secondary metabolites produced by some fungal species belong to chemically diverse groups of low molecular weight fungal metabolites with a range of toxic effects including genotoxicity and endocrine disruption [3–6]. In addition, mycotoxins have been identified recently as emerging contaminants in aqueous environments as well [7,8]. In turn, rapid detection of mycotoxins became an essential requirement in both food/feed and environmental monitoring that also triggered method development [9,10].

Recent developments, utilization, evaluation and possible improvements of methods that allow rapid, sensitive, and accurate detection of various mycotoxins have been chosen to be the topic of this Special Issue. Overall, 56 authors contributed 13 articles (12 original research articles and a review) discussing various aspects of mycotoxin research, but with mycotoxin analysis involved in all cases. Thus, through a compilation of current progress in the field, the Special Issue focuses on several aspects of mycotoxin analysis. Its scope encompasses classical instrumental analytical or biosensoric method development, sample preparation and handling to support method accuracy, as well as applications in routine monitoring or in decontamination assessment.

1. Method Development

Panasiuk et al. [11] developed an ultra-performance liquid chromatography-tandem mass spectrometry (UHPLC-MS/MS) method for a range of target mycotoxins including deoxynivalenol (DON), 3- and 15-acetyl-DON, DON-3-glucoside, nivalenol (NIV), and fusarenone-X. Sample preparation for the method included solid–liquid extraction, dispersive solid-phase extraction (QuEChERS), solid-phase extraction with hydrophilic-lipophilic balance column, and several immunoaffinity columns; the highest efficacy being achieved with the last. However, of the six immunoaffinity columns tested, none showed cross-reactivity to all of the mycotoxins, therefore no single immunoaffinity separation can be advised. The optimized method using a Mycosep 225 Trich column clean-up was validated with a large number of feedstuff samples including wheat, maize, and animal feeds. A similar LC-MS/MS-based procedure is reported by Nakhjavan et al. [12] for multi-mycotoxin analyses. The method employing immunoaffinity clean-up, solid-phase extraction, or QuEChERS sample preparation was optimized for simultaneous quantitation of aflatoxins (aflatoxins B1, B2, G1 and G2, AFB1, AFB2, AFG1, and AFG2), ochratoxin A (OTA), zearalenone (ZEN), deoxynivalenol (DON), NIV, diacetoxyscirpenol, fumonisins (fumonisins B1, B2 and B3, FB1, FB2, and FB3), T-2 toxin and HT-2 toxin in feed, and it allows limits of detection (LODs) ranging between 0.0003 and 0.05 µg/mL for the various mycotoxins tested.

Majdinasab et al. [13] reviewed colorimetric methods for industrial monitoring of mycotoxins in food and feed e.g., grains and cereals, grape juice, or red wine, and discuss the advantages and disadvantages for each method. Colorimetric strategies for various mycotoxins including T-2, DON, OTA, aflatoxins, ZEN, or FB1 (but not to FB2 or FB3) consist of enzyme-linked assays, lateral flow assays, microfluidic devices, and homogenous in-solution strategies that can utilize various (bio)receptors such as antibodies or aptamers.

The development of several immunoanalytical methods for mycotoxin detection is presented in the Special Issue. A competitive nanoparticle-based magnetic immunodetection assay for the detection and quantification of AFB1 with a LOD of 1.1 ng/mL is reported by Pietschmann et al. [14]. The method is based on magnetic separation of streptavidin-labeled magnetic particles, using an immobilized AFB1 antigen and biotinylated monoclonal AFB1-specific antibodies. The binding of antibodies to the immobilized antigen is competed by the free analyte (AFB1) in the solution (sample). Bound (i.e., uninhibited) antibodies on the solid surface are detected by frequency mixing magnetic detection. The LOD of the method is 1.1 ng/mL, comparable to a laboratory-based enzyme-linked immunosorbent assay (ELISA) method with a LOD of 0.29–0.39 ng/mL. The development of a portable instrument for ZEN by enzyme-linked fluorescent immunoassay (ELFIA) is reported by Gémes et al. [15], but as this instrument is a novel application for detection of mycotoxins as emerging water contaminants, it is discussed among the applications of routine monitoring (see Section 3. Applications in routine monitoring).

Several immunosensors on the basis of the same ZEN-specific polyclonal antibody are presented in the Special Issue for the detection of ZEN. An immobilized antibody-based competitive optical planar waveguide-based immunosensor by Nabok et al. [16] allowed a concentration-dependent detection of ZEN in the 0.01–1000 ng/mL concentration range. The optimized experimental benchtop planar waveguide setup is planned to be further developed into a portable hand-held biosensor including the signal processing electronics, suitable for in-field use. Using a similar sensor design but utilizing both immobilized antibody- (direct) and immobilized antigen-based (competitive) architectures, novel optical waveguide light mode spectroscopy (OWLS)-based immunosensors are reported by Székács et al. [17]. Covalent immobilization on the sensor surface was devised by epoxy-, amino-, and carboxyl-functionalization, and standard sigmoid curves in the optimized sensor formats allowed an outstanding LOD of 0.002 pg/mL for ZEN in the competitive immunosensor setup with a dynamic detection range of 0.01–1 pg/mL ZEN concentrations. The OWLS format represents five orders of magnitude improvement in LOD compared to the corresponding competitive ELISA, and the selectivity of the immunosensor for ZEN is outstanding on the basis of cross-reactivities determined for structurally related and unrelated compounds. The method was shown applicable in maize extract.

In addition to immunoanalytical (antibody-based) setups, the development of a label-free aptamer-based fluorescent sensor is reported by Qian et al. [18] for the detection of OTA. The aptasensor utilizing a nucleotide recombination hybridization chain reaction amplification element allows high selectivity for OTA with a LOD of 2.0 pg/mL (4.9 pM). The elegant aptamer setup utilizes two hairpin nucleotide probes (H1 and H2). H1 contains a central loop portion capable of specific complex formation with OTA and two 6-nucleotide long terminal sequences complementary with each other. H2 is similar in structure, where the central loop is a G-quadruplex sequence capable to bind with N-methyl-mesoporphyrin IX and thus, forms a complex with enhanced fluorescent excitability. In the system, complex formation between OTA and H1 initiates repeated recombination-driven binding of numerous H2 probes, each incorporating N-methyl-mesoporphyrin IX molecules into the elongating H2 chain and resulting in amplification of the fluorescent signal. Other mycotoxins (ochratoxin B, AFB1) do not cross-react with the detection system and do not disturb the binding of OTA either. The detection method was demonstrated to be effective in wheat flour and red wine as commodity matrices.

2. Sample Preparation and Handling to Support Method Accuracy

As seen also from studies on method development in this Special Issue [11,12], sample preparation is a step of key importance in the chemical analysis process; not only due to its required features of applicability and recovery, but also because novel standardized methods, such as the QuEChERS dispersive solid-phase extraction protocol can facilitate standardization of the analytical procedure improving inter-laboratory standard errors. The work of Kibugu et al. [19] clearly illustrates the importance of appropriate sample selection and preparation methods to maintain analysis performance quality descriptors including accuracy, precision, linearity, robustness, and ruggedness, as well as limits of detection and quantification. Their detailed statistical analysis of the determination of AFB1 content in chicken feed, using hierarchical sampling (from primary to quaternary with gradually decreasing sample sizes), wet milling with solvent extraction, and AFB1 quantification by a commercial ELISA kit, indicates accurate, precise, stable, reliable, and cost-effective analysis with improved inherent variability, which allows the processing of a lowered recommended test portion sample size of 50 g, and is suitable for laboratories not equipped with automated sample-splitting equipment.

3. Applications in Routine Monitoring

Analytical approaches utilized in practical applications may not have to be entirely based on novel principles—application of traditional detection methods can be devised for given tasks. The work reported by Alshannaq et al. [20] adapts a high-performance liquid chromatography method coupled with diode array (DAD) and fluorescence (FLD) detectors to screen for the possible presence of aflatoxins (AFB1, AFB2, AFG1, AFG2) in aflatoxigenic and non-toxigenic laboratory fungal cultures of *Aspergilli*, including *Aspergillus flavus*, *A. oryzae*, and *A. parasiticus*. In their method, readily available and easily applied in most mycology laboratories, the limit of quantification (LOQ) for AFs was found to be 2.5 to 5.0 ng/mL with DAD and 0.025 to 2.5 ng/mL with FLD with medium recoveries of 76–88%. Hong et al. [21] apply an immunochromatographic assay based on digital detection using colloidal gold nanoparticles labeled to monoclonal antibodies to detect ZEN in authentic cereal (corn, wheat, wheat flour, cereal product) and feed samples within a monitoring campaign carried out in China in 2019. Their survey included 187 cereal and cereal product samples and allowed a LOD of 0.25 ng/mL and recoveries between 87 and 117%.

A possible route for mycotoxin exposure has been linked to mycotoxins as surface water contaminants [7,8,22,23]. The occasional occurrence of mycotoxins in surface and drinking water is not a newly identified phenomenon, but its particular significance has been emphasized lately [7,8,22,24–26], classifying mycotoxins and their metabolites as emerging surface water contaminants [7,27], and assessing their routes of occurrence [28,29]. Gémes et al. [15] report the development of an ELFIA method as a module of a portable, in situ fluorimeter instrument installed in a mobile laboratory vehicle to detect ZEN in water with a LOD of 0.09 ng/mL. This LOD appears to be quite favorable compared to reported ELISAs, but a major advantage of the ELFIA method lies in its combined in situ applicability in the determination of important water quality parameters detectable by induced fluorimetry—e.g., total organic carbon content, algal density or the level of other organic micropollutants. The immunofluorescence module also appears to be flexible; with the use of other expedient antibodies it can be expanded to other target analytes.

Mycotoxins are also emerging contaminants in traditional matrices (commodities, feedstuff) in previously atypical geographical areas due to pathogen migration caused by climate change [30–33]. In consequence, decontamination by the use of suitably isolated metabolic enzymes capable to decompose, preferably selectively, certain mycotoxins is of great interest both from the aspects of fundamental research and technology development. Thus, enzymatic decomposition [34] and surface binding on microbial cell walls [35,36] of mycotoxins have been extensively studied, and two studies have been devoted to this topic in this Special Issue by Kosztik et al. [37] and Bata-Vidács et al. [38]. By their cell wall polysaccharides binding various mycotoxins, certain microbes are capable of absorbing, or in rare cases degrading, these substances. Thus, these microorganisms can be utilized in the biological detoxification of given mycotoxins. Such binding potential of *Lactobacilli* [37] and non-*Lactobacillus* lactic acid bacteria [38] towards AFB1 and sterigmatocystin (ST) is reported in this Special Issue, as the first report on microbial ST binding. Among

105 phylogenetically characterized *Lactobacillus* strains, 14 strains were able to bind AFB1 above 5%, 58 strains showed minor (below 3%) binding capacity, and 33 strains could not bind the mycotoxin. The highest AFB1 binding capacities (8–12%) were obtained for a strain of *L. pentosus* and three strains of *L. plantarum*. In addition, among 49 lactic acid bacteria other than lactobacilli, three strains of *Pediococcus acidilactici*, as well as one strain of *Enterococcus hirae*, and one of *E. lactis* had higher AFB1 binding ability (7.6%, 4.6%, 4.6%, 4.6%, 3.5%, respectively). Among 39 similarly phylogenetically characterized *Lactobacillus* strains, 27 and 12 strains were able to bind ST above 5% and between 0.8% and 5%, respectively. The highest ST binding capacities (above 20%) were obtained for five strains of *L. plantarum*, a strain of *L. paracasei*, and a strain of *L. pentosus*. In addition, the ST binding ability of strains belonging to the genus *Pediococcus* was found to be 2–3 times higher than the AFB1 binding capacities. The best AFB1 binding *Pediococcus* strain was also the best ST binding. This can be explained by the fact that the two structurally similar mycotoxins bind to the same cell wall polysaccharide receptor of the bacterium.

Author Contributions: A.S. is Guest Editor for the Special Issue "Rapid Detection of Mycotoxin Contamination" and is the sole author of this Editorial. The author has read and agreed to the published version of the manuscript.

Funding: The work was funded by the Hungarian National Research, Development and Innovation Office, project NVKP_16-1-2016-0049.

Institutional Review Board Statement: Not applicable.

Informed Consent Statement: Not applicable.

Conflicts of Interest: The author declare no conflict of interest.

References

1. Lee, H.J.; Ryu, D. Worldwide occurrence of mycotoxins in cereals and cereal-derived food products: Public health perspectives of their co-occurrence. *J. Agric. Food Chem.* **2017**, *65*, 7034–7051. [CrossRef] [PubMed]
2. Khaneghah, A.M.; Fakhri, Y.; Gahruie, H.H.; Niakousari, M.; Sant'Ana, A.S. Mycotoxins in cereal-based products during 24 years (1983–2017): A global systematic review. *Trends Food Sci. Technol.* **2019**, *91*, 95–105. [CrossRef]
3. Marin, S.; Ramos, A.J.; Cano-Sancho, G.; Sanchis, V. Mycotoxins: Occurrence, toxicology, and exposure assessment. *Food Chem. Toxicol.* **2013**, *60*, 218–237. [CrossRef] [PubMed]
4. Rocha, E.B.; Freire, F.C.O.; Maia, E.F.; Guedes, I.F.; Rondina, D. Mycotoxins and their effects on human and animal health. *Food Control* **2014**, *36*, 159–165. [CrossRef]
5. Ostry, V.; Malir, F.; Toman, J.; Grosse, Y. Mycotoxins as human carcinogens—The IARC Monographs classification. *Mycotoxin Res.* **2017**, *33*, 65–73. [CrossRef]
6. Cimbalo, A.; Alonso-Garrido, M.; Font, G.; Manyes, L. Toxicity of mycotoxins in vivo on vertebrate organisms: A review. *Food Chem. Toxicol.* **2020**, *137*, 111161. [CrossRef]
7. Gromadzka, K.; Wa´skiewicz, A.; Goliński, P.; Swietlik, J. Occurrence of estrogenic mycotoxin—Zearalenone in aqueous environmental samples with various NOM content. *Water Res.* **2009**, *43*, 1051–1059. [CrossRef] [PubMed]
8. Picardo, M.; Filatova, D.; Nuñez, O.; Farré, M. Recent advances in the detection of natural toxins in freshwater environments. *TrAC Trends Anal. Chem.* **2019**, *112*, 75–86. [CrossRef]
9. Nolan, P.; Auer, S.; Spehar, A.; Elliott, C.T.; Campbell, K. Current trends in rapid tests for mycotoxins. *Food Addit. Contam. Part A Chem. Anal. Control. Expo. Risk Assess.* **2019**, *36*, 800–814. [CrossRef] [PubMed]
10. Yang, Y.; Li, G.; Wu, D.; Liu, J.; Li, X.; Luo, P.; Hu, N.; Wang, H.; Wu, Y. Recent advances on toxicity and determination methods of mycotoxins in foodstuffs. *Trends Food Sci. Technol.* **2020**, *96*, 233–252. [CrossRef]
11. Panasiuk, L.; Jedziniak, P.; Pietruszka, K.; Posyniak, A. Simultaneous determination of deoxynivalenol, its modified forms, nivalenol and fusarenone-X in feedstuffs by the liquid chromatography–tandem mass spectrometry method. *Toxins* **2020**, *12*, 362. [CrossRef]
12. Nakhjavan, B.; Ahmed, N.S.; Khosravifard, M. Development of an improved method of sample extraction and quantitation of multi-mycotoxin in feed by LC-MS/MS. *Toxins* **2020**, *12*, 462. [CrossRef] [PubMed]
13. Majdinasab, M.; Aissa, S.B.; Marty, J.L. Advances in colorimetric strategies for mycotoxins detection: Toward rapid industrial monitoring. *Toxins* **2021**, *13*, 13. [CrossRef] [PubMed]
14. Pietschmann, J.; Spiegel, H.; Krause, H.-J.; Schillberg, S.; Schröper, F. Sensitive aflatoxin B1 detection using nanoparticle-based competitive magnetic immunodetection. *Toxins* **2020**, *12*, 337. [CrossRef]

15. Gémes, B.; Takács, E.; Gádoros, P.; Barócsi, A.; Kocsányi, L.; Lenk, S.; Csákányi, A.; Kautny, S.; Domján, L.; Szarvas, G.; et al. Development of an immunofluorescence assay module for determination of the mycotoxin zearalenone in water. *Toxins* **2021**, *13*, 182. [CrossRef]
16. Nabok, A.; Al-Jawdah, A.M.; Gémes, B.; Takács, E.; Székács, A. An optical planar waveguide-based immunosensors for determination of *Fusarium* mycotoxin zearalenone. *Toxins* **2021**, *13*, 89. [CrossRef]
17. Székács, I.; Adányi, A.; Szendrő, I.; Székács, A. Direct and competitive optical grating immunosensors for determination of *Fusarium* mycotoxin zearalenone. *Toxins* **2021**, *13*, 43. [CrossRef]
18. Qian, M.; Hu, W.; Wang, L.; Wang, Y.; Dong, Y. A Non-enzyme and non-label sensitive fluorescent aptasensor based on simulation-assisted and target-triggered hairpin probe self-assembly for ochratoxin a detection. *Toxins* **2020**, *12*, 376. [CrossRef]
19. Kibugu, J.; Mdachi, R.; Munga, R.; Mburu, D.; Whitaker, T.; Huynth, T.P.; Grace, D.; Lindahl, J.F. Improved sample selection and preparation methods for sampling plans used to facilitate rapid and reliable estimation of aflatoxin in chicken feed. *Toxins* **2021**, *13*, 216. [CrossRef]
20. Alshannaq, A.F.; Yu, J.-H. A Liquid chromatographic method for rapid and sensitive analysis of aflatoxins in laboratory fungal cultures. *Toxins* **2020**, *12*, 93. [CrossRef]
21. Hong, X.; Mao, Y.; Yang, C.; Liu, Z.; Li, M.; Du, D. Contamination of zearalenone from China in 2019 by a visual and digitized immunochromatographic assay. *Toxins* **2020**, *12*, 521. [CrossRef]
22. Bucheli, T.D.; Wettstein, F.E.; Hartmann, N.; Erbs, M.; Vogelgsang, S.; Forrer, H.-R.; Schwarzenbach, R.P. *Fusarium* mycotoxins: Overlooked aquatic micropollutants? *J. Agric. Food Chem.* **2008**, *56*, 1029–1034. [CrossRef] [PubMed]
23. Waśkiewicz, A.; Gromadzka, K.; Bocianowski, J.; Pluta, P.; Goliński, P. Zearalenone contamination of the aquatic environment as a result of its presence in crops/Pojava mikotoksina u vodenom okolišu zbog njihove prisutnosti u usjevima. *Arch. Ind. Hyg. Toxicol.* **2012**, *63*, 429–435. [CrossRef] [PubMed]
24. Olivera, B.R.; Mata, A.T.; Ferreire, J.P.; Crespo, M.T.B.; Pereira, V.J.; Bronze, M.R. Production of mycotoxins by filamentous fungi in untreated surface water. *Environ. Sci. Pollut. Res. Int.* **2018**, *25*, 17519–17528. [CrossRef]
25. Al-Gabr, H.M.; Zheng, T.; Yu, X. Fungi contamination of drinking water. *Rev. Environ. Contam. Toxicol.* **2013**, *228*, 121–139. [CrossRef]
26. Picardo, M.; Sanchís, J.; Núñez, O.; Farré, M. Suspect screening of natural toxins in surface and drinking water by high performance liquid chromatography and high-resolution mass spectrometry. *Chemosphere* **2020**, *261*, 127888. [CrossRef] [PubMed]
27. Kolpin, D.W.; Schenzel, J.; Meyer, M.T.; Phillips, P.J.; Hubbard, L.E.; Scott, T.-M.; Bucheli, T.D. Mycotoxins: Diffuse and point source contributions of natural contaminants of emerging concern to streams. *Sci. Total Environ.* **2014**, *470–471*, 669–676. [CrossRef] [PubMed]
28. Jarošová, B.; Javůrek, J.; Adamovský, O.; Hilscherová, K. Phytoestrogens and mycoestrogens in surface waters—Their sources, occurrence, and potential contribution to estrogenic activity. *Environ. Int.* **2015**, *81*, 26–44. [CrossRef] [PubMed]
29. Babič, M.N.; Gunde-Cimerman, N.; Vargha, M.; Tischner, Z.; Magyar, D.; Veríssimo, C.; Sabino, R.; Viegas, C.; Meyer, W.; Brandão, J. Fungal contaminants in drinking water regulation? A tale of ecology, exposure, purification and clinical relevance. *Int. J. Environ. Res. Public Health* **2017**, *14*, 636. [CrossRef]
30. Dobolyi, C.; Sebők, F.; Varga, J.; Kocsubé, S.; Szigeti, G.; Baranyi, N.; Szécsi, Á.; Tóth, B.; Varga, M.; Kriszt, B.; et al. Occurrence of aflatoxin producing *Aspergillus flavus* isolates in maize kernel in Hungary. *Acta Aliment.* **2013**, *42*, 451–459. [CrossRef]
31. Medina, A.; Akbar, A.; Baazeem, A.; Rodriguez, A.; Magan, N. Climate change, food security and mycotoxins: Do we know enough? *Fung. Biol. Rev.* **2017**, *31*, 143–154. [CrossRef]
32. Leggieri, M.C.; Toscano, P.; Battilani, P. Predicted aflatoxin B1 increase in Europe due to climate change: Actions and reactions at global level. *Toxins* **2021**, *13*, 292. [CrossRef] [PubMed]
33. Liu, C.; Van der Klerx, H.J. Quantitative modeling of climate change impacts on mycotoxins in cereals: A review. *Toxins* **2021**, *13*, 276. [CrossRef] [PubMed]
34. Dohnal, V.; Wu, Q.; Kuca, K. Metabolism of aflatoxins: Key enzymes and interindividual as well as interspecies differences. *Arch. Toxicol.* **2014**, *88*, 1635–1644. [CrossRef]
35. Lahtinen, S.J.; Haskard, C.A.; Ouwehand, A.C.; Salminen, S.J.; Ahokas, J.T. Binding of aflatoxin B1 to cell wall components of *Lactobacillus rhamnosus* strain GG. *Food Addit. Contam.* **2004**, *21*, 158–164. [CrossRef] [PubMed]
36. Chapot-Chartier, M.P.; Vinogradov, E.; Sadovskaya, I.; Andre, G.; Mistou, M.Y.; Trieu-Cuot, P.; Furlan, S.; Bidnenko, E.; Courtin, P.; Péchoux, C.; et al. The cell surface of *Lactococcus lactis* is covered by a protective polysaccharide pellicle. *J. Biol. Chem.* **2010**, *285*, 10464–10471. [CrossRef]
37. Kosztik, J.; Mörtl, M.; Székács, A.; Kukolya, J.; Bata-Vidács, I. Aflatoxin B1 and sterigmatocystin binding potential of Lactobacilli. *Toxins* **2020**, *12*, 756. [CrossRef]
38. Bata-Vidács, I.; Kosztik, J.; Mörtl, M.; Székács, A.; Kukolya, J. Aflatoxin B1 and sterigmatocystin binding potential of non-Lactobacillus LAB strains. *Toxins* **2020**, *12*, 799. [CrossRef]

Article

Simultaneous Determination of Deoxynivalenol, Its Modified Forms, Nivalenol and Fusarenone-X in Feedstuffs by the Liquid Chromatography–Tandem Mass Spectrometry Method

Łukasz Panasiuk *, Piotr Jedziniak, Katarzyna Pietruszka and Andrzej Posyniak

Department of Pharmacology and Toxicology, National Veterinary Research Institute, Al. Partyzantów 57, 24-100 Puławy, Poland; piotr.jedziniak@piwet.pulawy.pl (P.J.); katarzyna.pietruszka@piwet.pulawy.pl (K.P.); aposyn@piwet.pulawy.pl (A.P.)
* Correspondence: lukasz.panasiuk@piwet.pulawy.pl; Tel.: +48-81-886-3177

Received: 27 April 2020; Accepted: 29 May 2020; Published: 1 June 2020

Abstract: A liquid chromatography-tandem mass spectrometry method was developed for simultaneous determination of deoxynivalenol (DON), 3-acetyldeoxynivalenol (3Ac-DON), 15-acetyldeoxynivalenol (15Ac-DON), DON-3-glucoside (DON-3Glc) nivalenol and fusarenone-X in feedstuffs. Different techniques of sample preparation were tested: solid-liquid-extraction, QuEChERS, solid phase extraction with OASIS HLB columns or immunoaffinity columns and a Mycosep 225 Trich column. None of the six immunoaffinity columns tested showed cross-reactivity to all of the mycotoxins. Surprisingly, the results show that if the immunoaffinity columns bound 3Ac-DON, then they did not bind 15Ac-DON. The most efficient sample preparation was achieved with a Mycosep 225 Trich column clean-up. The chromatography was optimised to obtain full separation of all analytes (including 3Ac-DON and 15Ac-DON isomeric form). The validation results show the relative standard deviations for repeatability and reproducibility varied from 4% to 24%. The apparent recovery ranged between 92% and 97%, and the limit of quantification described a 1.30 to 50 µg/kg range. The method trueness was satisfactory, as assessed by a proficiency test and analysis of reference material. A total of 99 feed samples were analysed by the developed method, revealing the presence of DON and DON-3Glc in 85% and 86% of examined animal feeds, respectively at concentrations between 1.70 and 1709 µg/kg. The ratios DON-3Glc to DON in the surveyed feedstuffs were from a low of 3% to high of 59%.

Keywords: type B trichothecenes; modified mycotoxins; isomer separation; method validation

Key Contribution: Full separation of all compounds was obtained including the isomeric forms. ^{13}C-labeled internal standards were used. The proficiency of the method was successfully demonstrated.

---

1. Introduction

Type B trichothecenes are a group of mycotoxins produced by *Fusarium* genera (*F. graminearum* and *F. culmorum*) and are currently some of the most prevalent and important contaminants of cereals in the field. So far, the best-known toxins in feedstuffs in this group are deoxynivalenol (DON), nivalenol (NIV) and fusarenone-X (FUS-X) [1]. These toxins are resistant to milling, processing and heat and, therefore, it is very hard to eliminate them from the feed chain [2]. Among type B trichothecenes, DON is the most prevalent and hazardous mycotoxin and its occurrence can cause many adverse health effects in animals, such as feed refusal, emesis, suboptimal weight gain and diarrhoea, which can lead to economic losses. All animal species evaluated to date are susceptible to DON in this order of vulnerability: pigs > mice > rats > poultry ≈ ruminants [3].

Moreover, in recent years, the occurrence of so-called "modified mycotoxins" is an increasing concern in feed and food safety, since they remain undetected in testing for their parent mycotoxin [4]. Modified forms of DON can be formed by fungi as the acetylated derivatives: 3-acetyldeoxynivalenol (3Ac-DON) and 15-acetyldeoxynivalenol (15Ac-DON) or as a form of the parent toxins conjugated with glucose (DON-3glucoside; DON-3Glc). DON-3Glc is produced as part of the defense system of a plant infected by toxigenic fungi. These toxic compounds can transform into their parent toxin by hydrolysis in the mammalian digestive system [5]. DON-3Glc has lower toxicity than its precursor, while both acetylated forms possess equivalent or much stronger toxicity to animals, and their conversion into their native form also cannot be excluded [6]. Moreover, recently published papers show that 15Ac-DON has a higher toxicity then 3Ac-DON [7].

At the time of writing, the guidance values for the native form of DON in feedstuffs are set down in Commission Recommendation 2006/576/EC with amending Recommendation 2016/1319 [8,9], but other toxins (DON-3Glc, 3Ac-DON, 15Ac-DON, NIV and FUS-X) are not included. Consequently, the occurrence of DON modified forms would imply an underestimation of the level of DON contamination in feedstuffs. Nevertheless, in recent guidelines the European Food Safety Authority (EFSA) has launched calls for data on occurrence in food and feed of DON, NIV and modified DON mycotoxins to enable drafting of a scientific opinion on mycotoxins with respect to food and feed safety [10–12].

The important issue is the simultaneous determination of DON and its modified forms. Various methods for their analysis in cereals and feedstuffs have been reported, such as liquid chromatography (LC) coupled with tandem mass spectrometry (MS/MS) [13–16], fluorescence detection (FLD) [17], photodiode-array detection (PDA) [18–20], and ultra-high-performance supercritical fluid chromatography-tandem mass spectrometry (UHPSFC-MS/MS) [21]. Moreover, in recent years the introduction of high-resolution mass spectrometers (HRMS) has allowed screening of non-target compounds, novel compound identification and retrospective data analysis [22]. In previously published studies for chromatographic separation, authors used mostly C18 columns [13,16,19,23].

Different sample preparation techniques and clean-up approaches have been used in feedstuffs: solid-liquid-extraction (SLE) without clean-up [16,23], the quick, easy, cheap, effective, rugged and safe (QuEChERS) technique [24], solid-phase-extraction (SPE) [14] and immunoaffinity columns (IAC) [17,25,26]. While SLE is frequently applied for multi-mycotoxin analysis in feedstuffs, the inclusion of clean-up strategies in the procedure could increase the sensitivity of the method, as well as decrease high matrix effect for the difficult matrix.

One of the challenges in this analysis is the chromatographic separation of the 3Ac-DON and 15Ac-DON isomers, which only differ in structure by the position of the acetyl group. They have the same daughter ions, so it is crucial to fully separate them by LC-MS/MS for accurate quantification. Therefore it is important to derive validated methods for accurate assessment of exposure to DON and its metabolites by determining their levels in feedstuffs (due to their different toxicities).

The paper describes the development of a method for determination of DON, its modified forms, NIV and FUS-X with particular reference to the following aspects: full separation of all analytes (including isomeric forms of acetylated DON), comparison of different strategies for sample preparation and clean-up (SLE, QuEChERS, SPE with OASIS HLB columns or IACs and a Mycosep 225 Trich column) as well as verification of the method in a proficiency test (PT) and with naturally contaminated feed samples.

2. Results and Discussion

2.1. LC-MS/MS Optimisation

The optimisation of LC-MS/MS parameters was accomplished by directly applying tuning solutions of the selected mycotoxins at concentrations of 1 µg/mL each, using 0.1% CH_3COOH and MeOH as a mobile phase. The MRM mode was used and the analytes were ionized in both positive (ESI^+) and negative modes (ESI^-) (Table S1). In the case of positional isomers, proper quantification of 3Ac-DON and 15Ac-DON cannot be achieved through the difference of the product ion in MRM mode. In positive mode (339 *m/z*) the most abundant parent ion was the same for both analytes, but different product ions were chosen (339/231.2 and 339/279.1 for 3Ac-DON and 339/321.2 and 339/261.2 for 15Ac-DON) [27,28]. However, other studies show higher ionisation in positive mode with $NH4^+$ adducts for 3Ac-DON and 15-AcDON [13,23] or in negative modes [14,16]. If two different daughter ions for both isomeric forms are be chosen, positive false results could sometimes occur. For proper qualification and quantification it is therefore essential to have successful baseline separation of analytes.

2.2. Chromatographic Separation

Due to the different polarities of the analysed toxins, several chromatographic columns (Table S2) and mobile phases were tested. Based on the literature data, hydrophilic interaction liquid chromatography (HILIC) was tested as a first choice. Satisfactory separation on DON and DON-3Glc was achieved with Luna HILIC column (Phenomenex, Torrance, CA, USA), For other toxins no satisfactory retention times were obtained, and poor separation of all analytes was observed (Figure 1A). Kinetex C18 and Kinetex Biphenyl columns (Phenomenex) were also compared (Figure 1B,C). Although, these columns enable DON, DON-3Glc, NIV and FUS-X to be separated, 3Ac- and 15Ac-DON were not. In our study, the best results were achieved using a Luna Omega Polar C18 column (Phenomenex; 100×2.1; 1.6 µm) which is designed for polar compounds. Application of this column with MeOH as the organic mobile phase allowed all analytes to be separated except 3Ac and 15Ac-DON, their peaks still broadening (Figure 2A), and co-eluted.

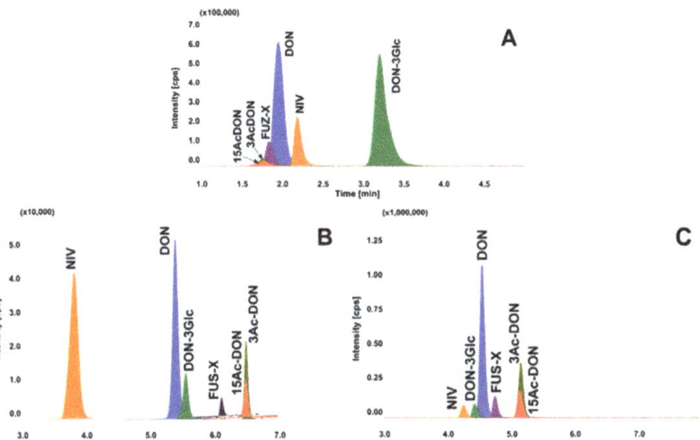

Figure 1. LC–MS/MS chromatograms of analysed toxins, tested in the same mobile phase but at different chromatographic columns: (**A**) Phenomenex Luna HILIC (**B**) Phenomenex Kinetex Biphenyl 100×2.1mm, 1.7 µm; (**C**) Phenomenex Kinetex C18 100×2.6 mm, 2.1 µm.

To obtain full separation of peaks, ACN was used as an organic mobile phase with specific gradient mode (2–6 min with an almost isocratic gradient from 15–18% of ACN) (Figure 2B). Moreover, the choice of ACN shortened retention time for all analytes, as well as the time of analysis. To date a lot of studies have described the chromatographic separation of DON, its modified mycotoxins, and other type B Trichothecenes in different biological matrixes [13,14,16,18,27–34]. Nevertheless, most of the authors did not achieve baseline separation of acetylated forms of DON. Only a few studies described the full chromatographic separation of these compounds [17,35,36]. Yoshinari et al. (2013) obtained retention times for 3Ac-DON and 15Ac-DON of 5.48 and 5.60 min, respectively. In a different study Goncalves et al. [17] achieved partial separation of the isomeric form (both peaks co-eluted). Contrary results were demonstrated by Slododchikova et al. [37] who concluded that the best for separation of the isomeric forms of acetylated DON was a pentafluorophenyl column and MeOH as the mobile phase.

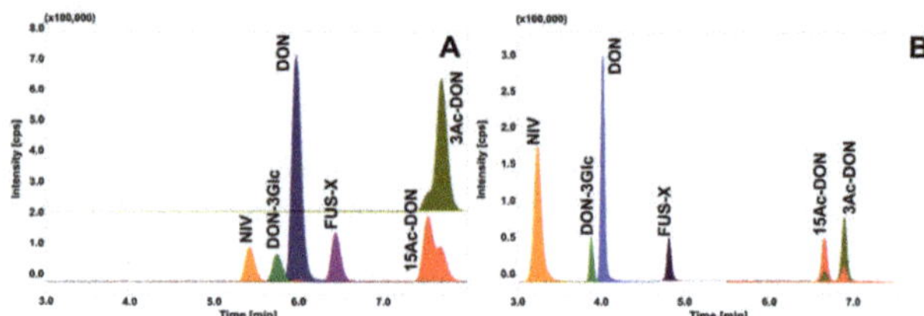

Figure 2. Chromatographic separation of tested compounds on the same column (Phenomenex Luna ® Omega C18 100 × 2.1, 1.6 µm) with different organic mobile phase: (**A**) MeOH; (**B**) ACN.

In conclusion, the usage of a Phenomenex Luna Omega Polar C18 column (100 × 2.1; 1.6 µm) column with the combination of 0.2% CH_3COOH with ACN as the mobile phase in a specific gradient mode achieves full separation of all tested compounds with a total run time of 12 min (Figure 2B).

2.3. Sample Preparation

2.3.1. IAC Testing

Six commercially available IACs were tested (DONTest, DZT MS-PREP, DON PREP, B-TeZ IAC, DONStar, and DONaok). Cross reactivity with the modified mycotoxins depended on the immobilized antibody [38]. While all the IACs showed excellent recovery for DON (Figure 3), none of them bound all DON metabolites and other toxins. For DON-3Glc, DONTEST, DZT MS-PREP, DON PREP and DONaok cross-reacted which is in line with other researchers' results [17,20,39]. Contrary to findings in other papers, none of IACs tested retained 3Ac-DON and 15Ac-DON simultaneously. Our results showed that if the antibodies bound 3Ac-DON, they did not bind 15Ac-DON. These results are in disagreement with those of another study, where a DONTEST column was used to determine DON derivatives with good recoveries (over 80%), although with HPLC post-column derivatisation and fluorescence detection [17]. In turn, Versilovskis et al. [39] demonstrated that DZT MS-PREP, DONPREP and DONaok cross-reacted with 15Ac-DON, although the recoveries obtained were low, and did not exceed 25%. However, the chromatographic method did not separate 3Ac- and 15Ac-DON and the LC-MS/MS method was based on the MRM of the $[M^+H^+]$ ions for both isomers, which can lead to false positive results. Lack of fully separated analytes could be a reason why other authors achieved discrepancy results. Moreover, we also checked possible cross-reactivity IACs with others B-trichothecene: NIV and FUS-X. From all tested IACs DONTEST, DONSTAR and B-TEZ bound NIV. Our results are in agreement with Uhlig et al. [40] where the authors highlighted that DONTEST retained NIV. For FUS-X, only DONTEST IAC bound toxins (25%), which was not tested in any previous papers. Because the DONTEST IAC showed best results of all the tested columns (but not full satisfactory), it was chosen for further evaluation.

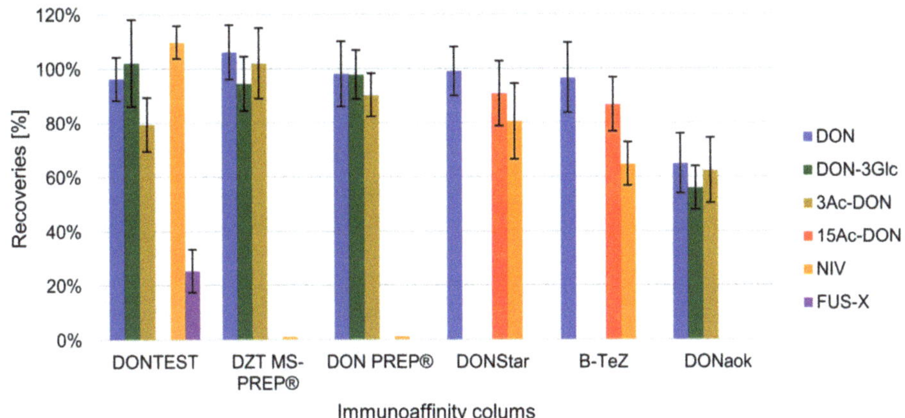

Figure 3. Mycotoxins recoveries obtained with IACs columns available on the market, obtained from different suppliers.

2.3.2. Comparison of Different Strategies for Sample Preparation and Clean-Up

The suitability for extraction and clean-up of DONTEST IAC, OASIS HLB and Mycosep 225 was tested by using them according to the manufacturer's instruction. The QuEChERS technique and SLE were prepared based on our previous experience [41,42]. As is shown in Figure 4 the best results were obtained for Mycosep 225 columns which are dedicated products for Trichothecene analysis. In our study, obtained ER were in the range of 86–94%, except for DON-3Glc where 30% recovery was achieved. However, application of matrix-matched calibration curves could effectively compensate for recovery losses (see the Method Validation section) [43]. Lower recovery of DON-3Glc using a Mycosep 225 column was previously reported [30].

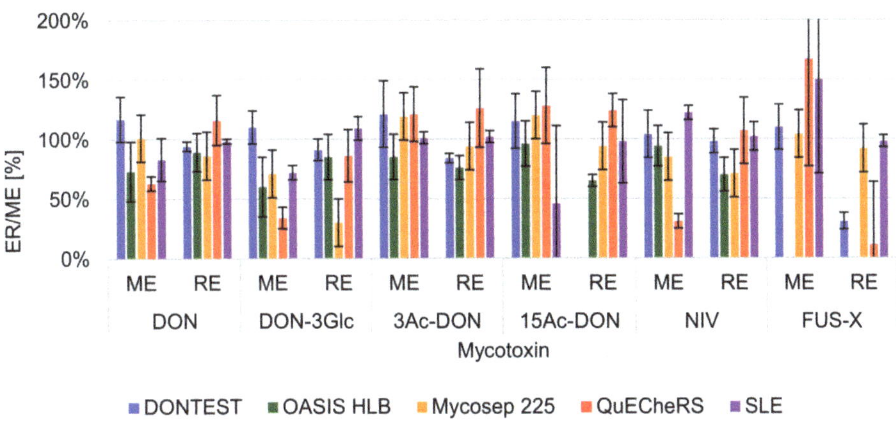

Figure 4. Extraction recovery (ER) and matrix effect (ME) of the tested methods for sample preparation.

A significant advantage of these columns was the lowest ME for all compounds (71–120%). "Push-through" columns were previously used with grain extracts [30], where the authors reported recovery for NIV, DON and FUS-X in the range of 75–85%. DONTEST IAC and SPE OASIS HLB cartridges were not suitable for the current study, because no recovery was observed by the former of 15Ac-DON or by the latter of FUS-X. QuEChERS and SLE show high ion suppression, e.g., 34% with DON-3Glc and 46% with 15Ac-DON, respectively. Consequently, the final procedure included clean-up with Mycosep 225 columns. Compared to other authors [15,39] the showed method allows for determination wider range of B-trichotecenes e.g., 15Ac-DON, NIV or FUS-X. Moreover, application for clean-up of sample Mycosep 225 is not as expensive as selective clean-up with IAC, which is frequently used for determination of DON and its metabolites in feeds [17,25,26].

2.4. Method Validation

The method was successfully validated for all tested mycotoxins in feedstuffs (Table 1). Good specificity of the methods was confirmed by analysis of 20 pseudo-blank samples. No interference peaks (S/N > 3) were detected in the retention time (±2.5%) for targeted analytes. The determined LOD and LOQ for all analytes were in the 1.78–15.0 and 5.87–49.5 µg/kg ranges, respectively. These low LOD and LOQ for DON and its modified mycotoxins were comparable with those disclosed in other publications [23,44], or even lower [45]. The calibration curves were linear over the calibration range for all compounds, resulting in R^2 values between 0.998 and 0.999. The REC was determined for each toxin based on a sample fortified at three VL (0.5 × VL, 1.0 × VL, and 1.5 × VL) and was above 90%, which fulfilled established criteria [46]. All values for repeatability and within-laboratory reproducibility were in the range of 4–24%, showing good precision for all toxins. Moreover, RSDr was between 4% and 22%, indicating that this method could be adopted for a wide range of feedstuffs. The expanded measurement uncertainty U (%) was satisfactory at below 35% for all tested toxins, showing acceptable method performance. Due to the large variability and complexity of the analysed samples, significant ME was expected [47]. In our study, ME values ranged from 61% to 120%. To compensate for this enhanced or suppressed signal, matrix-matched calibration curves were used as well as IS for DON and DON-3Glc. The MIX IS was added after extraction only to compensate for possible ME, to limit use of the expensive IS. This result shows that the developed procedure can be applied as a confirmatory method for determination of DON, its metabolites and others type B trichothecenes in feedstuffs.

2.5. Method Trueness, PT

Method trueness was evaluated by analysing three RMs (Table 2) and comparing them with the reference values. For each matrix, DON concentrations were in the uncertainty range of each sample. It is worth noting, that RMs were naturally contaminated samples, so other mycotoxins were also found. For example in the RM M15362D maize sample (Chiron) DON-3Glc, 3Ac-DON, 15Ac-DON and NIV were quantified at levels of 431, 19.0, 202, 203 µg/kg, respectively. Moreover, the molar ratios between DON-3Glc and 15-Ac to DON were high (43% and 20%, respectively). These results also indicate that if we analyse RM contaminated in real circumstances it is highly probably that metabolites of DON can be found and these data could be used for evaluation of method trueness. In the case of a PT organised by the European Union Reference Laboratory mycotoxins and plant toxins, z-score results obtained for samples (wheat and maize) were in a tolerable range $-2 \leq z \leq 2$ (−0.48 and 1.43). Thus, the developed method can produce accurate results in accordance with the reference values and could be applied in future to real contaminated feedstuff samples. It is worth to mention that in this PT acetylated forms of DON and DON-3Glc were covered only by less than half and one third of the laboratories, respectively. Also, false positive and negative results were reported related to 15Ac-DON, what indicate that determination of isomeric form is a challenge to analytical researchers.

Table 1. Validation parameters for the analysed mycotoxins in feedstuffs.

| | REC (%) | | | ME (%) | | | Calibration Range(µg/kg) | R^2 | Precision (CV, %) | | | | | | LOD (µg/kg) | LOQ (µg/kg) | U (%) |
| | | | | | | | | | RSD | | | RSD$_r$ | | | | | |
	0.5 × VL	1.0 × VL	1.5 × VL	0.5 × VL	1.0 × VL	1.5 × VL			0.5 × VL	1.0 × VL	1.5 × VL	0.5 × VL	1.0 × VL	1.5 × VL			
DON	104 ± 4	93 ± 6	99 ± 6	82 ± 12	101 ± 17%	87 ± 8	90–1800	0.999	9	9	11	4	7	6	10.1	33.3	26.0
DON-3Glc	93 ± 10	92 ± 12	96 ± 10	78 ± 12	61 ± 15%	74 ± 14	10–200	0.999	20	8	13	10	15	10	1.78	5.87	12.0
3Ac-DON	105 ± 12	94 ± 6	104 ± 10	109 ± 7	119 ± 20%	113 ± 7	10–200	0.999	9	9	9	12	7	10	2.43	8.02	26.0
15Ac-DON	97 ± 12	92 ± 16	90 ± 18	115 ± 15	120 ± 16%	111 ± 12	10–200	0.999	4	9	13	22	19	21	5.65	18.6	25.0
NIV	106 ± 14	97 ± 15	94 ± 15	96 ± 6	85 ± 7%	89 ± 6	90–1800	0.999	16	13	24	13	19	17	3.30	10.9	25.0
FUS-X	101 ± 12	93 ± 12	96 ± 11	93 ± 6	104 ± 15%	88 ± 8	50–200	0.998	6	9	10	8	16	14	15.0	49.5	35.0

Table 2. Trueness of results obtained by the developed method; mycotoxins concentration determined in proficiency testing materials (EURL) and RMs.

Reference Sample	Matrix	Analyte	Reference Concentration (µg/kg)	Concentration Uncertainty (µg/kg)	Measured Concentration (µg/kg) [a]	Calculated z-Score
Proficiency Test EURLPT-MP01	wheat	DON	570	±14.8	502	−0.48
		DON-3Glc	215	±22.2	217	−0.04
		3Ac-DON	35.2	±2.40	32.0	−0.37
	maize	DON	751	±20.4	729	−0.12
		DON-3Glc	35.0	±1.48	47.5	1.43
		3Ac-DON	92.8	±3.71	90.2	−0.11
		15Ac-DON	152	±11.5	159	0.18
RM M15362D (Chiron)	maize	DON	1077	±73.0	993	—
		DON-3Glc	—	—	431	—
		3Ac-DON	—	—	19.0	—
		15Ac-DON	—	—	202	—
		NIV	—	—	203	—
RM TET007RM (Fapas)	wheat	DON	1810	±311	1956	—
		DON-3Glc	—	—	125	—
RM TET030RM (Fapas)	maize	DON	1208	±117	1077	—
		DON-3Glc	—	—	190	—
		15Ac-DON	—	—	121	—

[a] measured in duplicate

2.6. Application to Real Contaminated Feedstuffs

The validated method was applied to the analysis of 99 feedstuff samples (Figure 5). DON and DON-3Glc were detected in 85% and 86% of surveyed samples and the mean concentrations were 511 µg/kg and 94.0 µg/kg, respectively (concentration ranging for DON between 10.1 and 1709 µg/kg and for DON-3Glc 1.70 and 385 µg/kg). Moreover, the ratio of the DON-3Glc concentration to that of DON ranged from 3% to 59% with a mean of 19%. The ratios of DON-3Glc/DON obtained in our study coincide with previously reported investigation [19,33]. The incidences of other toxins (3Ac-DON, 15Ac-DON, and NIV) were 35%, 26% and 23%, respectively.

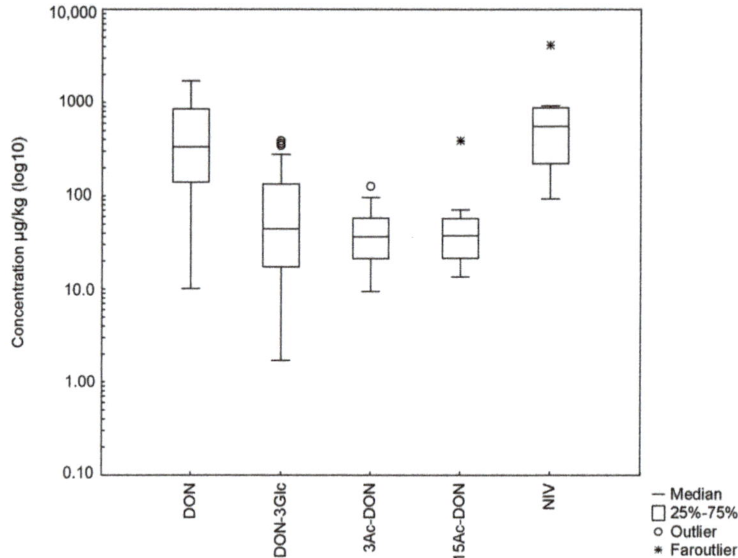

Figure 5. The mycotoxins concentrations in contaminated feedstuffs (n = 99).

3. Conclusions

A UHPLC-MS/MS method for the determination of DON, its metabolites and other type B trichothecenes in feedstuffs was successfully developed and validated. The main novelty of this method is that full separation of all compounds was achieved, including the isomeric forms 3Ac-DON and 15Ac-DON and that a DON-3Glc IS was used as the internal standard for quantification of DON-3Glc. In case of the IACs testing for they cross-reactivity features for DON modified forms none of them bound all derivatives and other toxins. The use of the commercially available Mycosep 225 columns allowed for quick and easy sample preparation. The results of RM analysis and the PT confirm the trueness of the method. Application of the validated method on feedstuffs revealed occurrence of DON and DON-3Glc in over 80% of positive samples. The developed method can be a tool for accurate qualification and quantification of mycotoxins and could be adopted as a confirmatory method for determination of DON and its modified mycotoxins NIV and FUS-X in a wide range of feedstuffs.

4. Materials and Methods

4.1. Chemicals and Standards

Six brands of IAC were compared for their cross-reactivity features: DONTest WB from Vicam, (Milford, MA, USA), DZT MS-PREP and DON PREP from R- Biopharm Rhone Ltd. (Glasgow, UK), B-TeZ IAC Deoxynivalenol from BioTeZ Berlin Buch GmbH (Berlin, Germany), DONStar from Romer

Labs Diagnostic GmbH (Tulln, Austria) and aokinImmunoClean DON (DONaok) from Aokin AG (Berlin, Germany). DON PREP, B-TeZ IAC Deoxynivalenol and DONStar—were kindly provided free of charge by suppliers for testing purposes. Mycosep 225 Trich columns were purchased from Romer Labs Diagnostic GmbH. Oasis HLB cartridges were obtained from Waters (Milford, MA, USA). Acetonitrile (analytical and LC-MS grade; ACN), methanol (LC-MS grade; MeOH), acetic acid and C18 bulk sorbent were sourced from J.T. Baker of Avantor Performance Materials (Deventer, The Netherlands). Magnesium sulphate ($MgSO_4$) was from Chempur (Piekary Śląskie, Poland) and water was prepared using a Milli-Q apparatus (MerckMillipore, Burlington, MA, USA) to attain purity of 18.2 MΩ. Mycotoxin standards of DON, U-[13C15] DON (DON IS), 3Ac-DON, 15Ac-DON, NIV and FUS-X were obtained from Sigma Aldrich (St. Louis, MO, USA). DON-3Glc and U-[13C21] DON-3G (DON-3Glc IS) were purchased from Romer Labs. The primary standard stock solutions were prepared in ACN. All standards were stored according to their manufacturer's recommendations. The chloride and pottassioum chloride used to make phosphate buffered saline (PBS) were sourced from POCh (Gliwice, Poland) and the sodium hydrophosphate dehydrate from Chempur. PBS was made as follows: 8 g of sodium chloride, 3.6 g of sodium hydrophosphate dihydrate and 0.2 g of potassium chloride were dissolved in 1L of deionized water.

4.2. Mixed Working Solution

A mixed working solution (MIX6) was prepared in ACN from the individual stock of six mycotoxins at a concentration of 9 µg/mL for DON and NIV and 1 µg/mL for 3Ac-DON, 15Ac-DON, DON-3Glc and FUS-X. The internal standards solution (MIX IS) was mixed in ACN to achieve concentrations of 1 µg/mL and 0.5 µg/mL for DON IS and DON-3Glc IS, respectively. All working standard solutions were stored at 2–8 °C.

4.3. Samples and Reference Materials

Poultry and swine feedstuff samples (total *n* = 99) were collected in 2017 and 2018 by Veterinary Inspectorate officers working with feed manufacturers, in accordance to Commission Regulation (EC) NO. 2009/152 [48]. Delivered samples were milled using a ZM 200 ultra-centrifugal high-speed instrument with 1 mm sieve (Retsch GmbH, Haan, Germany;) and stored in a dark place at room temperature until analysis. A validation study was conducted using low contamination feed samples (pseudo-blanks) [49] with DON concentration of 50 ± 13 µg/kg (Figure 6).

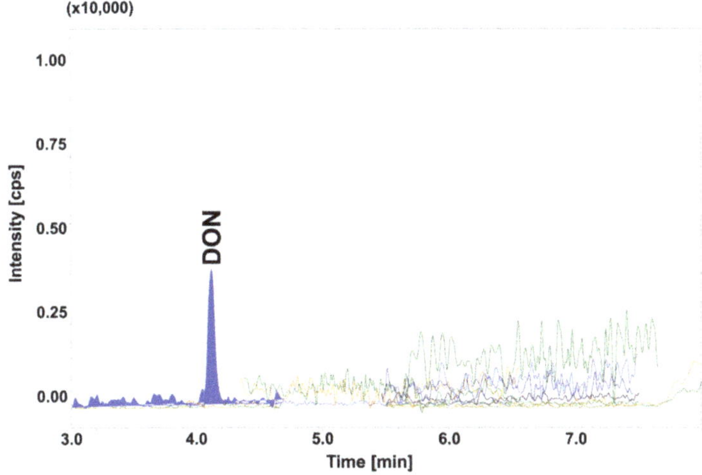

Figure 6. Chromatogram of pseudo-blank sample with DON concentration of 50 ± 13 µg/kg.

For confirmation of the method trueness, three reference materials (RMs) were tested: maize (TET030RM; Fapas, Fera Science, York, UK), maize (12199.15-G; Chiron AS, Trondheim, Norway) and wheat (TET007RM; Fapas) (Table 2). Moreover, the method was verified in the "Deoxynivalenol and related compounds in food and feed matrices"; EURLPT-MP01 proficiency test on DON, 3Ac-DON, 15Ac-DON and DON-3Glc in wheat and maize carried out by the European Union Reference Laboratory mycotoxins and plant toxins.

4.4. IAC Testing

The cross-reactivity of six selected IAC was evaluated by loading 4 mL of water or PBS (according to the manufacturer's recommendation) spiked with the MIX 6 mixed mycotoxin solution (at the level at which the analytes concerned could be extracted from samples). Next, the IAC was flushed with 4 mL of water and analytes were eluted with two portions of 1.5 mL of MeOH. The solvent was evaporated to dryness under a nitrogen stream at 40 °C and the dry residue was dissolved in 200 µL of 0.2% acetic acid and transferred to an autosampler vial. Each IAC was tested in triplicate. For comparison of their suitability for detection of the relevant mycotoxins, recovery was calculated as the ratio of the measured concentration of toxins to neat solvent.

4.5. Compared Strategies for Sample Preparation and Clean-Up

In addition to IACs, several other techniques of sample preparation have been tested: SLE, QuEChERS, SPE with OASIS HLB cartridges and Mycosep 225 Trich column (contain a mixture of adsorbent materials). The protocol for sample preparation for SLE and QuEChERS was based on our previously described methods [41,42].

The OASIS HLB column was tested according to the manufacturer's protocol. Briefly, 1 g of sample was weighed into a 50 mL plastic tube and extracted for 30 min with 8 mL of H_2O. Next, the sample was centrifuged and 2 mL of extract was transferred into OASIS HLB cartridges previously conditioned with 3 mL of MeOH and 3 mL of H_2O. Subsequently, the column was washed with 5 mL of 5% MeOH and eluted with 3 mL of MeOH. All sample were collected and evaporated to dryness under a gentle nitrogen stream at 40 °C. The residue was dissolved in 200 µL of 0.2% CH_3COOH and transferred to an autosampler vial.

For IAC DONTEST 1 g of sample was weighed into a 50 mL plastic tube and extracted for 30 min with 8 mL of H_2O. Next, the sample was centrifuged and 2 mL of extract was passed onto DONTEST. Subsequently, the column was washed with 10 mL water and finally eluted with 2 mL methanol and collected to glass tube. After evaporation (N_2, 40 °C) sample was dissolved in 200 µL of 0.2% CH_3COOH and transferred to an autosampler vial.

For the sample preparation using for clean-up Mycosep 225 Trich column (based on the manufacturer's recommendation with slight modification) 1 g of previously milled sample was weighed into a 50 mL centrifuge tube. Next, 8 mL of $ACN:H_2O$ (84:16; *v/v*) mixture was added and was extracted in a rotary shaker for 30 min, followed by centrifugation for 15 min at 4000 rpm. Subsequently, 6 mL of supernatant was transferred to a glass tube and pushed through a Mycosep 225 Trich column. Purified extract (2 mL) was collected, and the sample was evaporated to dryness in a gentle nitrogen stream at 40 °C. The dry residue was reconstituted in 200 µL of 0.2% CH3COOH and transferred to an autosampler vial. For final optimized sample preparation additionally, 10 µL of MIX IS was added before sample evaporation. Extraction recovery (ER) and matrix effect (ME) were calculated to find an appropriate method for sample preparation. In this case, ER was calculated as the ratio of the area of the analyte(s) recorded for the sample spiked with the target compound(s) before extraction to the area for the spiked sample after extraction.

4.6. LC-MS/MS Analysis

The analysis was performed with a Nexera X2 system with an LCMS-8050 triple-quadrupole mass spectrometer (Shimadzu, Kioto, Japan). LabSolutions software (version 5.60 SP2, Shimadzu,

Kioto, Japan) was used for data acquisition and processing. Chromatographic separation was tested using four chromatographic columns with the different stationary and mobile phase composition (Table S2). Column and autosampler temperatures were set at 45 °C and 4 °C, respectively. The final optimised mobile phase A consisted of 0.2% CH3COOH in water/ACN (95:5; *v/v*) (eluent A) and ACN/0.2% CH3COOH in water (95:5; *v/v*) (eluent B). A gradient elution was used as follows: 0–2 min 15% B, 2.1–6 min 18% B, 6.1–9 min isocratic step at 100% B, and 9.1–12 min 0% B. The total run time was 12 min at a flow rate of 0.3 mL/min and the injection volume was 5 µL.

The mass spectrometry detection was carried out using multiple reaction monitoring (MRM) with positive and negative electrospray ionization (ESI+/−) (Table S2). The following ion-source settings were used: nebulising gas flow: 2 L/ min, heating gas flow: 10 L/min, drying gas flow: 10 L/min, interface temperature: 300 °C, desolvation line temperature: 250 °C, heat block temperature: 400 °C, and Q1 and Q3 resolution: Unit.

The identification of the analyte was performed according to the SANTE/12089 /2016 Guidance document on identification of mycotoxins in food and feed [50]. Four identification criteria were used: comparison of peak retention time in test samples with retention times of calibration standards; the retention time of the internal standard being within the tolerance range ±0.05 min relative to the appropriate standard; selection of at least two characteristic fragmentary ions and calculation of their ion ratio (within ±30% (relative) of average of calibration standards from the same sequence); and the peaks having an S/N ratio of at least 3.

4.7. Method Validation

The method was validated for feedstuffs and the following parameters were verified and calculated: specificity, limit of detection (LOD) and quantification (LOQ), linearity, apparent recovery (REC, %), precision as repeatability (RSD, CV%) and within-laboratory reproducibility (RSDr, CV%), trueness and matrix effect (ME, %) [46,50]. All validation parameters were calculated using the relative peak area with respect to DON IS and DON-3Glc IS. The specificity was checked by analysing 20 different pseudo-blank feed samples to evaluate possible interferences. LOD and LOQ were also calculated based on Guidance Document on the Estimation of LOD and LOQ for Measurements in the Field of Contaminants in Feed and Food using paired observation approach [27]. The three validation levels (VL) 0.5 × VL, 1.0 × VL, and 1.5 × VL used for DON and NIV were: 450, 900, and 1350 µg/kg, and for DON-3Glc, 3Ac-DON, 15Ac-DON and FUS-X they were: 50, 100, and 150 µg/kg. The decisions on fortification levels for each analyte were made on basis of the lowest guidance value EU in feedstuffs (only for DON, 900 µg/kg) [8]. For other toxins—based on concentration data in feedstuffs reported by others authors [13,18,35] and LC-MS/MS detection. The linearity was determined using a matrix-matched calibration plot. The concentration ranges of the five-point calibration curves were 90–1800 µg/kg for DON and NIV and 10–200 µg/kg for DON-3Glc 3Ac-DON, 15Ac-DON and 50–200 µg/kg for FUS-X. The REC (apparent recovery) was calculated by quantifying the mycotoxins using matrix-matched calibration curves at the three VL. For the repeatability study, one kind of pseudo-blank feedstuff was spiked at three levels in six repetitions. The within-laboratory reproducibility was assessed by analysis of six different feedstuffs for animals (wheat, maize, feedstuffs for pigs, poultry, fish and mixed cereals) at the three VL on different days by two operators. Trueness was evaluated by analysing three RM in duplicate (Table 2).

To evaluate the ME, six different pseudo-blank samples were extracted with the proposed procedure and spiked after extraction with pure standards of all tested mycotoxins at the same level as in the within-laboratory reproducibility study. The responses of the mycotoxins were compared to a neat standard solution. With such a calculation, the ratio of 1:1 (100%) would mean no matrix effects, ion enhancement would result in a ratio above 100%, and ion suppression in a ratio below 100%. To compensate for possible losses caused by ME, DON IS was used for quantification of DON, NIV, 3-AcDON, 15Ac-DON, FUS-X and DON-3Glc IS for DON-3Glc. Additionally, the expanded measurement uncertainty (U) was calculated with MUkit software (Envical SYKE, Helsinki, Finland) for the 1.0 × VL spiking level using the Nordtest approach [51].

Supplementary Materials: The following are available online at http://www.mdpi.com/2072-6651/12/6/362/s1, Table S1: LC-MS/MS parameters for detection of mixed mycotoxins by the mass spectrometer; Table S2: LC column used during the chromatographic set up.

Author Contributions: Investigation, Ł.P.; Methodology, Ł.P. and K.P.; Writing—original draft, Ł.P.; Writing—review & editing, P.J. and A.P. All authors have read and agreed to the published version of the manuscript.

Funding: Funded by the KNOW (Leading National Research Centre) Scientific Consortium 'Healthy Animal–Safe Food', a decision of the Ministry of Science and Higher Education No. 05-1/KNOW2/2015.

Conflicts of Interest: The authors declare no conflict of interest.

References

1. Streit, E.; Schatzmayr, G.; Tassis, P.; Tzika, E.; Marin, D.; Taranu, I.; Tabuc, C.; Nicolau, A.; Aprodu, I.; Puel, O.; et al. Current situation of mycotoxin contamination and co-occurrence in animal feed focus on Europe. *Toxins* **2012**, *4*, 788–809. [CrossRef] [PubMed]

2. Sugita-Konishi, Y.; Park, B.J.; Kobayashi-Hattori, K.; Tanaka, T.; Chonan, T.; Yoshikava, K.; Kumagai, S. Effect of Cooking Process on the Deoxynivalenol Content and Its Subsequent Cytotoxicity in Wheat Products. *Biosci. Biotechnol. Biochem.* **2006**, *70*, 1764–1768. [CrossRef] [PubMed]

3. Pestka, J.J. Deoxynivalenol: Toxicity, mechanisms and animal health risks. *Anim. Feed Sci. Technol.* **2007**, *137*, 283–298. [CrossRef]

4. Rychlik, M.; Humpf, H.U.; Marko, D.; Dänicke, S.; Mally, A.; Berthiller, F.; Klaffke, H.; Lorenz, N. Proposal of a comprehensive definition of modified and other forms of mycotoxins including "masked" mycotoxins. *Mycotoxin Res.* **2014**, *30*, 197–205. [CrossRef] [PubMed]

5. Berthiller, F.; Crews, C.; Dall'Asta, C.; de Saeger, S.; Haesaert, G.; Karlovsky, P.; Oswald, I.P.; Seefelder, W.; Speijers, G.; Stroka, J. Masked mycotoxins: A review. *Mol. Nutr. Food Res.* **2013**, *57*, 165–186. [CrossRef]

6. Freire, L.; Sant'Ana, A.S. Modified mycotoxins: An updated review on their formation, detection, occurrence, and toxic effects. *Food Chem. Toxicol.* **2018**, *111*, 189–205. [CrossRef]

7. Broekaert, N.; Devreese, M.; Demeyere, K.; Berthiller, F.; Michlmayr, H.; Varga, E.; Adam, G.; Meyer, E.; Croubels, S. Comparative in vitro cytotoxicity of modified deoxynivalenol on porcine intestinal epithelial cells. *Food Chem. Toxicol.* **2016**, *95*, 103–109. [CrossRef]

8. European Commission. *Commission Recommendation 2006/576/EC of 17 August 2006 on the Presence of Deoxynivalenol, Zearalenone, Ochratoxin A, T-2 and HT-2 and Fumonisins in Products Intended for Animal Feeding*; Office of the European Union: Brussels, Belgium, 2006; Volume 229, pp. 7–9.

9. European Commission. *Commision Recommendation 2016/1319 of 29 July 2016—Amending Recommendation 2006/576/EC as Regards Deoxynivalenol, Zearalenone and Ochratoxin A in Pet Food*; Office of the European Union: Brussels, Belgium, 2016; Volume 73, pp. 58–60. Available online: http://eur-lex.europa.eu/legal-content/EN/TXT/PDF/?uri=CELEX:32016H1319&from=EN (accessed on 29 May 2020).

10. European Food Safety Authority (EFSA). Scientific Opinion on risks for animal and public health related to the presence of nivalenol in food and feed. *EFSA J.* **2013**, *11*, 3262. [CrossRef]

11. European Food Safety Authority (EFSA). Risks to human and animal health related to the presence of deoxynivalenol and its acetylated and modified forms in food and feed. *EFSA J.* **2017**, *15*. [CrossRef]

12. European Food Safety Authority (EFSA). Scientific Opinion on the risks for human and animal health related to the presence of modified forms of certain mycotoxins in food and feed. *EFSA J.* **2014**, *12*. [CrossRef]

13. Zhao, Z.; Rao, Q.; Song, S.; Liu, N.; Han, Z.; Hou, J.; Wu, A. Simultaneous determination of major type B trichothecenes and deoxynivalenol-3-glucoside in animal feed and raw materials using improved DSPE combined with LC-MS/MS. *J. Chromatogr. B Anal. Technol. Biomed. Life Sci.* **2014**, *963*, 75–82. [CrossRef] [PubMed]

14. Chen, D.; Cao, X.; Tao, Y.; Wu, Q.; Pan, Y.; Peng, D.; Liu, Z.; Huang, L.; Wang, Y.; Wang, X.; et al. Development of a liquid chromatography—Tandem mass spectrometry with ultrasound-assisted extraction and auto solid-phase clean-up method for the determination of Fusarium toxins in animal derived foods. *J. Chromatogr. A* **2013**, *1311*, 21–29. [CrossRef] [PubMed]

15. Yoshinari, T.; Tanaka, T.; Ishikuro, E.; Horie, M.; Nagayama, T.; Nakajima, M.; Naito, S.; Ohnishi, T.; Sugita-Konishi, Y. Inter-laboratory Study of an LC-MS/MS Method for Simultaneous Determination of Deoxynivalenol and Its Acetylated Derivatives, 3-Acetyl-deoxynivalenol and 15-Acetyl-deoxynivalenol in Wheat. *Shokuhin Eiseigaku Zasshi* **2013**, *54*, 75–82. [CrossRef]

16. Vendl, O.; Berthiller, F.; Crews, C.; Krska, R. Simultaneous determination of deoxynivalenol, zearalenone, and their major masked metabolites in cereal-based food by LC-MS-MS. *Anal. Bioanal. Chem.* **2009**, *395*, 1347–1354. [CrossRef] [PubMed]

17. Gonçalves, C.; Stroka, J. Cross-reactivity features of deoxynivalenol (DON)-targeted immunoaffinity columns aiming to achieve simultaneous analysis of DON and major conjugates in cereal samples. *Food Addit. Contam. Part A* **2016**, *33*, 1053–1062. [CrossRef] [PubMed]

18. Pascale, M.; Panzarini, G.; Powers, S.; Visconti, A. Determination of Deoxynivalenol and Nivalenol in Wheat by Ultra-Performance Liquid Chromatography/Photodiode-Array Detector and Immunoaffinity Column Cleanup. *Food Anal. Methods* **2014**, *7*, 555–562. [CrossRef]

19. Bryła, M.; Ksieniewicz-Woźniak, E.; Waśkiewicz, A.; Szymczyk, K.; Jędrzejczak, R. Natural Occurrence of Nivalenol, Deoxynivalenol, and Deoxynivalenol-3-Glucoside in Polish Winter Wheat. *Toxins* **2018**, *10*, 81. [CrossRef]

20. Trombete, F.; Barros, A.; Vieira, M.; Saldanha, T.; Venâncio, A.; Fraga, M. Simultaneous Determination of Deoxynivalenol, Deoxynivalenol-3-Glucoside and Nivalenol in Wheat Grains by HPLC-PDA with Immunoaffinity Column Cleanup. *Food Anal. Methods* **2016**, *9*, 2579–2586. [CrossRef]

21. De Boevre, M.; van Poucke, C.; Ediage, E.N.; Vanderputten, D.; van Landschoot, A.; de Saeger, S. Ultra-High-Performance Supercritical Fluid Chromatography as a Separation Tool for *Fusarium* Mycotoxins and Their Modified Forms. *J. AOAC Int.* **2018**, *101*, 627–632. [CrossRef]

22. Righetti, L.; Paglia, G.; Galaverna, G.; Dall'Asta, C. Recent Advances and Future Challenges in Modified Mycotoxin Analysis: Why HRMS Has Become a Key Instrument in Food Contaminant Research. *Toxins* **2016**, *8*, 361. [CrossRef]

23. De Boevre, M.; di Mavungu, J.D.; Maene, P.; Audenaert, K.; Deforce, D.; Haesaert, G.; Eeckhout, M.; Callebaut, A.; Berthiller, F.; van Peteghem, C.; et al. Development and validation of an LC-MS/MS method for the simultaneous determination of deoxynivalenol, zearalenone, T-2-toxin and some masked metabolites in different cereals and cereal-derived food. *Food Addit. Contam. Part A* **2012**, *29*, 819–835. [CrossRef] [PubMed]

24. Dzuman, Z.; Zachariasova, M.; Lacina, O.; Veprikova, Z.; Slavikova, P.; Hajslova, J. A rugged high-throughput analytical approach for the determination and quantification of multiple mycotoxins in complex feed matrices. *Talanta* **2014**, *121*, 263–272. [CrossRef] [PubMed]

25. Li, Y.; Wang, Z.; de Saeger, S.; Shi, W.; Li, C.; Zhang, S.; Cao, X.; Shen, J. Determination of deoxynivalenol in cereals by immunoaffinity clean-up and ultra-high performance liquid chromatography tandem mass spectrometry. *Methods* **2012**, *56*, 192–197. [CrossRef] [PubMed]

26. Suman, M.; Bergamini, E.; Catellani, D.; Manzitti, A. Development and validation of a liquid chromatography/linear ion trap mass spectrometry method for the quantitative determination of deoxynivalenol-3-glucoside in processed cereal-derived products. *Food Chem.* **2013**, *136*, 1568–1576. [CrossRef] [PubMed]

27. Veršilovskis, A.; Geys, J.; Huybrechts, B.; Goossens, E.; de Saeger, S.; Callebaut, A. Simultaneous determination of masked forms of deoxynivalenol and zearalenone after oral dosing in rats by LC-MS/MS. *World Mycotoxin J.* **2012**, *5*, 303–318. [CrossRef]

28. Andrade, P.D.; Dantas, R.R.; de Moura-Alves, T.L.D.; Caldas, E.D. Determination of multi-mycotoxins in cereals and of total fumonisins in maize products using isotope labeled internal standard and liquid chromatography/tandem mass spectrometry with positive ionization. *J. Chromatogr. A* **2017**, *1490*, 138–147. [CrossRef] [PubMed]

29. Fan, Z.; Bai, B.; Jin, P.; Fan, K.; Guo, W.; Zhao, Z.; Han, Z. Development and Validation of an Ultra-High Performance Liquid Chromatography-Tandem Mass Spectrometry Method for Simultaneous Determination of Four Type B Trichothecenes and Masked Deoxynivalenol in Various Feed Products. *Molecules* **2016**, *21*, 747. [CrossRef] [PubMed]

30. Radová, Z.; Holadová, K.; Hajšlová, J. Comparison of two clean-up principles for determination of trichothecenes in grain extract. *J. Chromatogr. A* **1998**, *829*, 259–267. [CrossRef]

31. Lattanzio, V.M.T.; Ciasca, B.; Powers, S.; Visconti, A. Improved method for the simultaneous determination of aflatoxins, ochratoxin A and Fusarium toxins in cereals and derived products by liquid chromatography-tandem mass spectrometry after multi-toxin immunoaffinity clean up. *J. Chromatogr. A* **2014**, *1354*, 139–143. [CrossRef]

32. Nathanail, A.V.; Sarikaya, E.; Jestoi, M.; Godula, M.; Peltonen, K. Determination of deoxynivalenol and deoxynivalenol-3-glucoside in wheat and barley using liquid chromatography coupled to mass spectrometry: On-line clean-up versus conventional sample preparation techniques. *J. Chromatogr. A* **2014**, *1374*, 31–39. [CrossRef]

33. Berthiller, F.; Dalla'sta, C.; Corradini, R.; Marchelli, R.; Sulyok, M.; Krska, R.; Adam, A.; Schuhmacher, R. Occurrence of deoxynivalenol and its 3-ß-D-glucoside in wheat and maize. *Food Addit. Contam.* **2009**, *26*, 507–511. [CrossRef] [PubMed]

34. Broekaert, N.; Devreese, M.; de Mil, T.; Fraeyman, S.; de Baere, S.; de Saeger, S.; de Backer, P.; Croubels, S. Development and validation of an LC – MS / MS method for the toxicokinetic study of deoxynivalenol and its acetylated derivatives in chicken and pig plasma. *J. Chromatogr. B* **2014**, *971*, 43–51. [CrossRef] [PubMed]

35. Yoshinari, T.; Ohnishi, T.; Kadota, T.; Sugita-Konishi, Y. Development of a Purification Method for Simultaneous Determination of Deoxynivalenol and Its Acetylated and Glycosylated Derivatives in Corn Grits and Corn Flour by Liquid Chromatography–Tandem Mass Spectrometry. *J. Food Prot.* **2012**, *75*, 1355–1358. [CrossRef] [PubMed]

36. Neuhof, T.; Ganzauer, N.; Koch, M.; Nehls, I. A Comparison of Chromatographic Methods for the Determination of Deoxynivalenol in Wheat. *Chromatographia* **2009**, *69*, 1457–1462. [CrossRef]

37. Slobodchikova, I.; Vuckovic, D. Liquid chromatography—High resolution mass spectrometry method for monitoring of 17 mycotoxins in human plasma for exposure studies. *J. Chromatogr. A* **2018**, *1548*, 51–63. [CrossRef]

38. Goryacheva, I.Y.; de Saeger, S. Immunochemical detection of masked mycotoxins: A short review. *World Mycotoxin J.* **2012**, *5*, 281–287. [CrossRef]

39. Veršilovskis, A.; Huybrecht, B.; Tangni, E.K.; Pussemier, L.; de Saeger, S.; Callebaut, A. Cross-reactivity of some commercially available deoxynivalenol (DON) and zearalenone (ZEN) immunoaffinity columns to DON- and ZEN-conjugated forms and metabolites. *Food Addit. Contam. Part A Chem. Anal. Control. Exp. Risk Assess.* **2011**, *28*, 1687–1693. [CrossRef]

40. Uhlig, S.; Stanic, A.; Hussain, F.; Miles, C.O. Selectivity of commercial immunoaffinity columns for modified forms of the mycotoxin 4-deoxynivalenol (DON). *J. Chromatogr. B Anal. Technol. Biomed. Life Sci.* **2017**, *1061*, 322–326. [CrossRef]

41. Panasiuk, L.; Jedziniak, P.; Pietruszka, K.; Piatkowska, M.; Bocian, L. Frequency and levels of regulated and emerging mycotoxins in silage in Poland. *Mycotoxin Res.* **2019**, *35*, 17–25. [CrossRef]

42. Jedziniak, P.; Panasiuk, Ł.; Pietruszka, K.; Posyniak, A. Multiple mycotoxins analysis in animal feed with LC-MS/MS: Comparison of extract dilution and immunoaffinity clean-up. *J. Sep. Sci.* **2019**, *42*, 1240–1247. [CrossRef]

43. Malachova, A.; Sulyok, M.; Schuhmacher, R.; Berthiller, F.; Hajslova, J.; Veprikova, Z.; Zachariasova, M.; Lattanzio, V.M.T.; de Saeger, S.; di Mavungu, J.D.; et al. Collaborative investigation of matrix effects in mycotoxin determination by high performance liquid chromatography coupled to mass spectrometry. *Qual. Assur. Saf. Crops Foods* **2013**, *5*, 91–103. [CrossRef]

44. Geng, Z.; Yang, D.; Zhou, M.; Zhang, P.; Wang, D.; Liu, F.; Zhu, Y.; Zhang, M. Determination of Deoxynivalenol-3-Glucoside in Cereals by Hydrophilic Interaction Chromatography with Ultraviolet Detection. *Food Anal. Methods* **2013**, *7*, 1139–1146. [CrossRef]

45. Xu, J.J.; Zhou, J.; Huang, B.F.; Cai, Z.X.; Xu, X.M.; Ren, Y.P. Simultaneous and rapid determination of deoxynivalenol and its acetylate derivatives in wheat flour and rice by ultra high performance liquid chromatography with photo diode array detection. *J. Sep. Sci.* **2016**, *39*, 2028–2035. [CrossRef]

46. European Commission. *Commission Regulation (EC) No 401/2006 of 23 February 2006 Laying down the Methods of Sampling and Analysis for the Official Control of the Levels of Mycotoxins in Foodstuffs*; Office of the European Union: Brussels, Belgium, 2006; Volume 2006, pp. 12–34.

47. Matuszewski, B.K.; Constanzer, M.L.; Chavez-Eng, C.M. Strategies for the Assessment of Matrix Effect in Quantitative Bioanalytical Methods Based on HPLC–MS/MS. *Anal. Chem.* **2003**, *75*, 3019–3030. [CrossRef] [PubMed]

48. European Commission. *Commission regulation (EC) No 152/2009 of 27 January 2009 Laying down the Methods of Sampling and Analysis for the Official Control of Feed*; Office of the European Union: Brussels, Belgium, 2009; Volume 36, pp. 1–130.

49. Haedrich, J.; Wenzl, T.; Schaechtele, A.; Robouch, P.; Stroka, J. *Guidance Document on the Estimation of LOD and LOQ for Measurements in the Field of Contaminants in Feed and Food*; Publications Office of the European Union: Luxembourg, 2016. [CrossRef]

50. European Commission (EC). *Guidance Document on Identification of Mycotoxins in Food and Feed*; Office of the European Union: Brussels, Belgium, 2017; pp. 1–4.

51. Magnusson, B.; Näykki, T.; Hovind, H.; Krysell, M.; Sahlin, E. *Nordtest, Handbook for Calculation of Measurement Uncertainty in Environmental Laboratories*, 4th ed.; NT TR 537; Nordtest: Taastrup, Denmark, 2017; Volume 4.

Article

Development of an Improved Method of Sample Extraction and Quantitation of Multi-Mycotoxin in Feed by LC-MS/MS

Bahar Nakhjavan *, Nighat Sami Ahmed and Maryam Khosravifard

Center for Analytical Chemistry, California Department of Food and Agriculture, Sacramento, CA 95832, USA; Nighat.ahmed@cdfa.ca.gov (N.S.A.); Maryam.khosravifard@cdfa.ca.gov (M.K.)
* Correspondence: bahar.nakhjavan@cdfa.ca.gov; Tel.: +1-916-228-6846

Received: 16 June 2020; Accepted: 17 July 2020; Published: 19 July 2020

Abstract: A multi-mycotoxin chromatographic method was developed and validated for the simultaneous quantitation of aflatoxins (AFB1, AFB2, AFG1 and AFG2), ochratoxin A (OTA), zearalenone (ZON), deoxynivalenol (DON), nivalenol (NIV), diacetoxyscirpenol (DAS), fumonisins (FB1, FB2 and FB3), T-2 toxin (T-2) and HT-2 toxin (HT-2) in feed. The three most popular sample preparation techniques for determination of mycotoxins have been evaluated, and the method with highest recoveries was selected and optimized. This modified QuEChERS (quick, easy, cheap, effective, rugged and safe) approach was based on the extraction with acetonitrile, salting-out and cleanup with lipid removal. A reconstitution process in methanol/water was used to improve the MS responses and then the extracts were analyzed by LC-MS/MS. In this method, the recovery range is 70–100% for DON, DAS, FB1, FB2, FB3, HT-2, T-2, OTA, ZON, AFG1, AFG2, AFB1 and AFB2 and 55% for NIV in the spike range of 2–80 µg/kg. Method robustness was determined with acceptable z-scores in proficiency tests and validation experiments.

Keywords: mycotoxins; feed; modified QuEChERS; LC-MS/MS

Key Contribution: This study describes an improved analytical method for quantitation of common mycotoxins with acceptable recoveries in different feed products.

---

1. Introduction

Mycotoxins are the most common contaminants in agricultural crops produced by several species of mold and fungi. During growth, maturity, harvest, storage and processing of food and animal feed products, the fungus produces mycotoxins and other secondary metabolites [1]. These mycotoxin-contaminated food and feed threaten human and animal health even at very low concentration [2]. Various degrees of toxicity in food commodities can cause acute or chronic diseases such as immune suppression, cancer, pathological lesions and growth problems [3,4]. Moreover, the presence of mycotoxins in consuming animal products such as milk and meat are a significant safety concern as well [5,6]. Hence, creating an accurate and fast analytical method to quantify the contamination levels of mycotoxins plays a vital role in food and feed safety assessment risks.

Among hundreds of mycotoxins, a few have been recognized as a food safety concern including aflatoxins, fumonisins, ochratoxin A, zearalenone, deoxynivalenol, nivalenol, diacetoxyscirpenol, T-2 and HT-2. After two decades of research, simultaneous quantitative determination of mycotoxins and their derivatives in one analysis is challenging due to the wide polarity, solubility and physicochemical properties of these compounds.

Food crops and feed materials can be easily exposed to moisture that is needed for the growth of molds and fungi. High level of contamination by these mycotoxins has been frequently reported in food and feed. This attracted much attention in recent years due to high risk of contamination and consumption of these commodities all over the world [7,8].

Among a wide variety of mycotoxin sample preparations, accelerated solvent extraction [9], ultrasonic extraction [10], liquid-liquid extraction [11], immunoaffinity column [12] and solid-phase extraction [13] have been intensively researched. Most of the existing methods suffer from poor recovery, insufficient sensitivity and non-reproducibility. This makes these methods unsuitable for simultaneous determination of multi-mycotoxin. Chemical diversity, polarity and solubility of the mycotoxins are important characteristics that can significantly affect extraction efficiency.

Despite all these significant features, co-eluting of matrix components is the most challenging issue. Co-eluting matrix components creates enhancement or suppression of an analyte ionization affecting quantitation [14,15]. Matrix-matched calibration, standard addition and isotopic internal standards are the common solutions for compensation of this problem [16]. Matrix-matched calibration standards were used to reduce the interferences from the extraction process and improve the quantitation results. Therefore, developing an extraction technique to overcome all these challenges is daunting.

A variety of analytical instrumentations have been reported for mycotoxins such as TLC, GC-MS, LC-MS, HPLC-FLD, HPLC-UV and LC-MS/MS [17–21]. Liquid chromatography coupled with triple mass spectroscopy is the most recognized analytical instrumentation for the wide range of chemical contaminants such as mycotoxins in agricultural commodities. LC-MS/MS is known as a sensitive, selective, specific and efficient technique because of its versatility and reliability [22].

The objective of this work was to develop a robust and reliable extraction and clean up technique for multi-mycotoxin in a wide range of agricultural commodities by using LC-MS/MS and validate it through different approaches. The results illustrated here show that this sample preparation method can be applied in many laboratories analyzing feed materials.

2. Results and Discussion

2.1. Optimization of Extraction Process

Multi-mycotoxin extractions were prepared by using three different sample preparation techniques in corn. Immunoaffinity (method A) [23], solid-phase extraction (method B) [24] and QuEChERS (quick, easy, cheap, effective, rugged and safe) (method C) [25,26] methods applied for this study are shown in Table 1. They were evaluated based on analyte recovery (Rec.) and relative standard deviation (RSD).

The number of mycotoxins giving acceptable recoveries is seven in method A, eight in method B and nine in method C with significant improvement for DAS, FB1 and OTA in respect to the total number of 14 analyzed mycotoxins.

As shown in Table 1, extraction efficiency using acetonitrile-water in method C was increased for FB1 while there was no change for FB2 and FB3 due to their chemical structure. FB1 with two hydroxyl groups has more solubility than FB2 and FB3 with one hydroxyl group in acetonitrile-water as an extraction solution. Aqueous acetonitrile in method C in comparison with aqueous methanol and acetonitrile:methanol in methods A and B provided better recovery for FB1, DAS and OTA. In method C, cleaning of the sample extract was carried out by the EMR lipid removal in order to minimize ion source contamination. Thus, the QuEChERS method was further modified and used as a reference method. In method D, the extraction efficiency is affected by changing the pH because the addition of 0.3% formic acid in the extraction mixture changes the state of ionization for FB1, FB2, FB3, DAS and OTA as acidic and NIV as high polar compounds. It promotes the extraction of the neutral form of acidic mycotoxins into the organic phase due to their ionization constant. Moreover, this degree of acidity prevents the retention of acidic mycotoxins such as OTA in the cleanup process. Acetonitrile and methanol cannot extract NIV lonely, and the presence of low amounts of acid is necessary during extraction due to the incomplete partitioning. However, nivalenol as the most polar mycotoxin cannot meet the satisfactory level of recovery, 55% is a remarkable change in this report. The reconstitution step at the end of the process plays an important role in improving the signal responses and better peak shapes.

Figure 1 shows extracted ion chromatogram (XIC) and total ion chromatogram (TIC) for all mycotoxins with 50 ng/mL AFB1, AFB2, AFG1 and AFG2; 1000 ng/mL FB1, FB2, FB3, T-2, HT-2, DON and ZON; 2000 ng/mL NIV; 500 ng/mL DAS and OTA in corn matrix blank. XIC is a chromatogram created by taking intensity values at a single, discrete mass value, or a mass range, from a series of mass spectral scans. TIC is a chromatogram created by summing up intensities of all mass spectral peaks belonging to the same scan. The co-elution of some mycotoxin compounds was acceptable because these related compounds illustrate different MRM transitions in LC-MS/MS.

Table 1. Recovery (Rec.) and Relative Standard Deviation (RSD) values (n = 5) of 14 mycotoxins using different extraction techniques.

Analyte	Spike Level μg/kg	Method A Rec. (%) and RSD (%)	Method B Rec. (%) and RSD (%)	Method C Rec. (%) and RSD (%)	Method D Rec. (%) and RSD (%)
DON	40	94.4 and 7.1	100.2 and 8.4	83.6 and 1.9	85.0 and 4.2
DAS	20	10.6 and 9.9	12.1 and 18.5	82.6 and 5.5	96.1 and 3.7
FB1	40	31.7 and 19.6	38.9 and 19.9	62.8 and 6.3	75.7 and 4.2
FB2	40	55.5 and 9.5	58.5 and 9.0	54.7 and 3.0	78.9 and 5.3
FB3	40	58.7 and 7.7	60.5 and 4.5	67.6 and 7.1	76.9 and 4.7
HT-2	40	98.4 and 6.5	121.6 and 4.0	93.5 and 0.9	95.9 and 4.0
T-2	40	94.9 and 7.1	113.4 and 6.5	87.3 and 4.1	99.0 and 2.4
OTA	20	7.9 and 8.5	9.3 and 7.2	58.9 and 3.2	86.0 and 1.3
ZON	40	62.5 and 6.6	103.8 and 2.9	96.7 and 2.9	81.8 and 6.0
AFG1	2.0	108.4 and 3.5	109.5 and 4.2	87.5 and 4.2	82.9 and 2.8
AFG2	2.0	111.3 and 4.4	112.0 and 1.9	91.6 and 3.4	88.8 and 5.4
AFB1	2.0	81.8 and 4.3	86.7 and 6.6	78.5 and 3.4	79.2 and 0.94
AFB2	2.0	83.6 and 5.7	94.3 and 5.9	84.4 and 2.9	84.5 and 5.2
NIV	80	ND	ND	ND	58.8 and 3.7

Method A (Vicam), Method B (solid-phase extraction (SPE)), Method C (QuEChERS (quick, easy, cheap, effective, rugged and safe)) and Method D (Modified QuEChERS).

Solvent Mixture and Mobile Phase

Among different solvents, acetonitrile provides better recovery for a wide range of mycotoxins and matrices. However, there are reports of using methanol as a suitable co-extraction solvent in this field [27]. Different proportions of acetonitrile, methanol and water have been studied to discover an efficient extraction procedure with high recoveries for all analytes. In addition, to obtain the best chromatogram with the lowest noise signal, different mobile phases were used.

To optimize the extraction step, different proportions of acetonitrile and water were applied. Three volume ratios, acetonitrile:water (80:20, v:v), (60:40, v;v) and (50:50, v:v) were employed and acetonitrile:water (50:50, v:v) was confirmed as the most appropriate volume ratio. Various proportions of acidic water (0.1%, 0.2%, 0.3%, 0.5% and 1%) were tested and the extraction with equal proportion of acetonitrile and water containing 0.3% formic acid was suitable for all acidic and neutral mycotoxins.

Different various mobile phases and additives were evaluated, including acetonitrile and methanol (mobile phase) and formic acid, ammonium formate, ammonium fluoride, acetic acid and ammonium acetate (additive). Methanol containing formic acid, ammonium formate, and ammonium fluoride was found to give superior peak shape compared to acetonitrile. The reconstitution mixture was added at the end of the sample preparation step because the compatibility of mobile phases with final extraction solvent leads to improved signal response and prevents peak fronting and unstable retention time compared with no reconstitution. Since the nature of polar compounds makes them prone to dissolve in more polar solutions, reconstitution of samples with polar mobile phases is strongly recommended.

In this research, the combination of acetonitrile and acidic water as an extraction solvent at the beginning of the sample preparation and reconstitution with methanol:water (50:50, v:v) provides higher recoveries and better resolution for all 14 mycotoxins.

2.2. Validation

Animal feeds contain a mixture of crop ingredients, which makes them complex for mycotoxin analysis. Among the feedstuffs, corn has attracted the most attention in mycotoxin analysis due to its high production, consumption and contamination. This protein rich grain expedites the animal growth to prepare them for market weight quickly at a low cost. Corn is a good source of proteins, carbohydrates, vitamins, unsaturated lipids and minerals. It has a high level of matrix effect that makes it a great candidate for research. Therefore, the proposed extraction method was initially developed and validated in corn and further evaluated using different feed matrices.

As a part of validation study, this method was used in 5 different feed matrices (sheep food, dried distiller grain, dairy food, fish food and goat starter) provided from Association of American Feed Control Officials. AAFCO program materials were prepared from different commercial feedstuffs purchased from US marketplaces in order to monitor the use and performance of methods in analytical laboratories. Participating in these proficiency tests is a valid source of method evaluation because several laboratories take part simultaneously and determine 12 mycotoxins in various animal feeds with different level of contaminations. These samples are delivered to the laboratories as powdered samples, and there is not usually enough information about their ingredients and components. They are mixtures of various crops and agricultural products. The matrices considered in this validation were selected among 2019–2010 AAFCO proficiency tests. Sheep food, dried distiller grain, dairy food, fish food and goat starter represent a wide range of feedstuffs with diverse physicochemical properties. As shown in Table 2, the applicability of the method was confirmed by satisfactory results for all mycotoxins ($z \leq \pm2$), although it is worth noting that sheep food, dried distillers' grain and fish food with lower z-scores were affected by the presence of co-eluting matrix interferences leading to signal suppression. The Chromatogram of the dried distiller grain is presented in Figure 2. It depicts the separation of DON, AFB2, AFB1, HT-2, FB1, T-2, FB3, OTA, ZON and FB2 at different retention times.

In addition, a FERN comparison exercise was conducted for Aflatoxin B1 in 12 dog foods, which was successfully reported. All these matrices were spiked and analyzed for quality control purpose.

Researches indicate that matrix effect is the most important factor in the extraction process for mycotoxin analysis in feed. To eliminate matrix effects, matrix-matched calibration must be in the same or similar matrices that are being studied. Therefore, it is possible that the matrix-matched calibration cannot completely account for different degrees of incurred mycotoxins, and matrix interferences lead to signal suppression and enhancement. As shown in the accuracy and precision section, the recoveries for the in-house validation were obtained using the same matrices for calibration standard and matrix spike samples while there was no such condition for samples in proficiency tests. Recoveries higher than 70% were obtained for DON, DAS, FB1, FB2, FB3, HT-2, T-2, OTA, ZON, AFG1, AFG2, AFB1 and AFB2 and 55% for NIV with relative standard deviation lesser than 12%.

The presented analytical procedure was used for the analysis of more than 50 routine feed samples. The applicability of this method was confirmed by comparison with our old laboratory methods using water and acetonitrile/0.5% acetic acid for extraction followed by salting out reagent and hexane. Additionally, the method validation was performed with three different levels of spike and five replicates.

Chromatograms of all analyzed mycotoxins in the lowest matrix-matched calibration standard are presented in Figure 3. Fragmentation reactions were carried out in Multiple Reaction Monitoring mode and two product ions, a quantifier ion and a qualifier ion, were measured for 14 mycotoxins. During the evaluation, it was proved that the sensitivity of the MRM transitions is related to the quality and freshness of the used solvents.

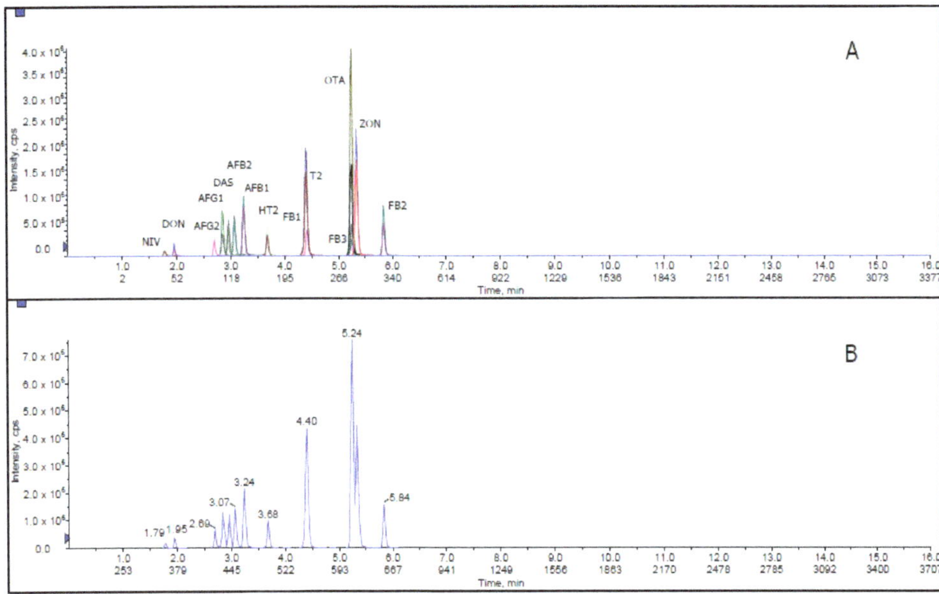

Figure 1. (**A**) Extracted ion chromatogram (XIC) and (**B**) total ion chromatogram (TIC) of mycotoxins at 50 ng/mL AFB1, AFB2, AFG1 and AFG2; 1000 ng/mL FB1, FB2, FB3, T-2, HT-2, DON and ZON; 2000 ng/mL NIV; 500 ng/mL DAS and OTA in corn matrix blank.

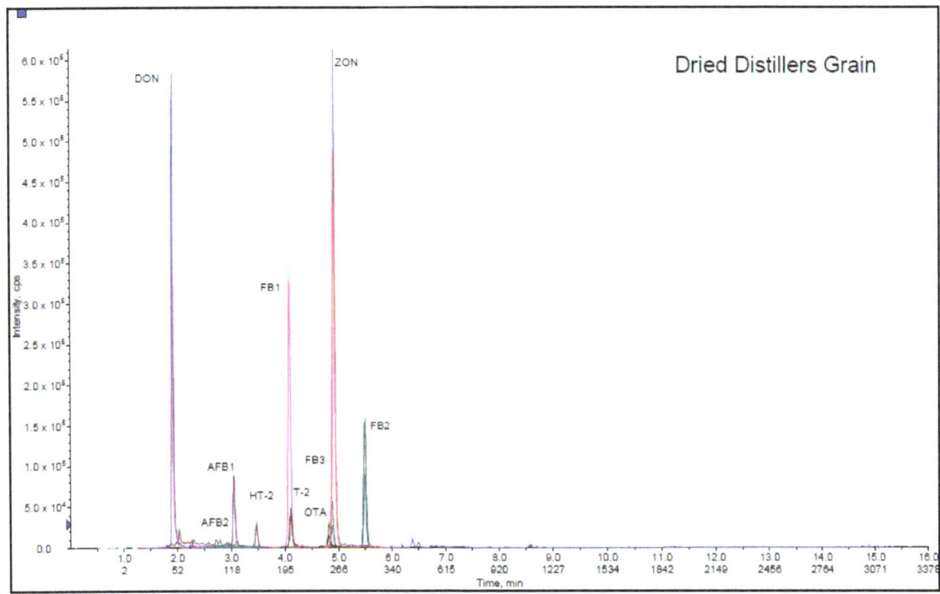

Figure 2. Extracted ion chromatogram of dried distillers' grain.

Table 2. Results of mycotoxin analysis in Association of American Feed Control Officials (AAFCO) proficiency tests.

Sample Type	Sheep Food			Dried Distillers Grain			Dairy Food			Fish Food			Goat Starter		
Toxin	Assessed Value, µg/kg	Lab Value, µg/kg	z-Score	Assessed Value, µg/kg	Lab Value, µg/kg	z-Score	Assessed Value, µg/kg	Lab Value, µg/kg	z-Score	Assessed Value, µg/kg	Lab Value, µg/kg	z-Score	Assessed Value, µg/kg	Lab Value, µg/kg	z-Score
Don	2181	993	−1.8	9673	4559	−1.8	5354	4036	−0.8	2366	988	−1.9	2221	1615	−0.91
FB1	4432	1770	−2.0	1306	560	−1.8	2905	2016	−1.0	2470	991	−2.0	3190	3321	0.14
FB2	1332	460	−2.1	373	155	−1.8	844	698	−0.6	581	233	−1.9	933	875	−0.20
FB3	595	234	−1.9	189	100	−1.4	357	380	0.2	337	164	−1.6	ND	ND	ND
HT-2	96.1	57.7	−1.2	76.0	58.4	−0.7	74	97	−1.0	37.2	27	−0.8	234	288	0.72
T-2	91.5	30.0	−2.0	55.0	25.5	−1.6	97	60.5	−1.2	69.5	28	−1.8	262	214	-0.57
OTA	107	63.8	1.2	7.6	6.3	−0.5	151	200	1.0	186.7	103	−1.4	316	364	0.48
ZON	335	126	−1.9	574	231.1	−1.9	885	730	−0.6	368.3	144	−1.9	377	342	-0.30
AFG1	0.59	0.62	0.14	ND	ND	ND	ND	ND	ND	ND	ND	ND	1.37	1.6	0.46
AFG2	ND	ND	ND	ND	ND	ND	ND	ND	ND	ND	ND	ND	ND	ND	ND
AFB1	24.3	14.1	−1.2	15.1	4.4	−2.1	38.5	28.4	−0.8	53.7	17.6	−2.0	85.8	89.5	0.13
AFB2	1.5	0.6	−1.7	1.3	1.0	−0.6	7.8	6.2	−0.6	3.5	1.1	−1.9	16	23	1.3

% satisfactory z-scores (z ≤ ±2), questionable z-scores (z ≤ ±3) and unsatisfactory z-scores (z ≥ ±3).

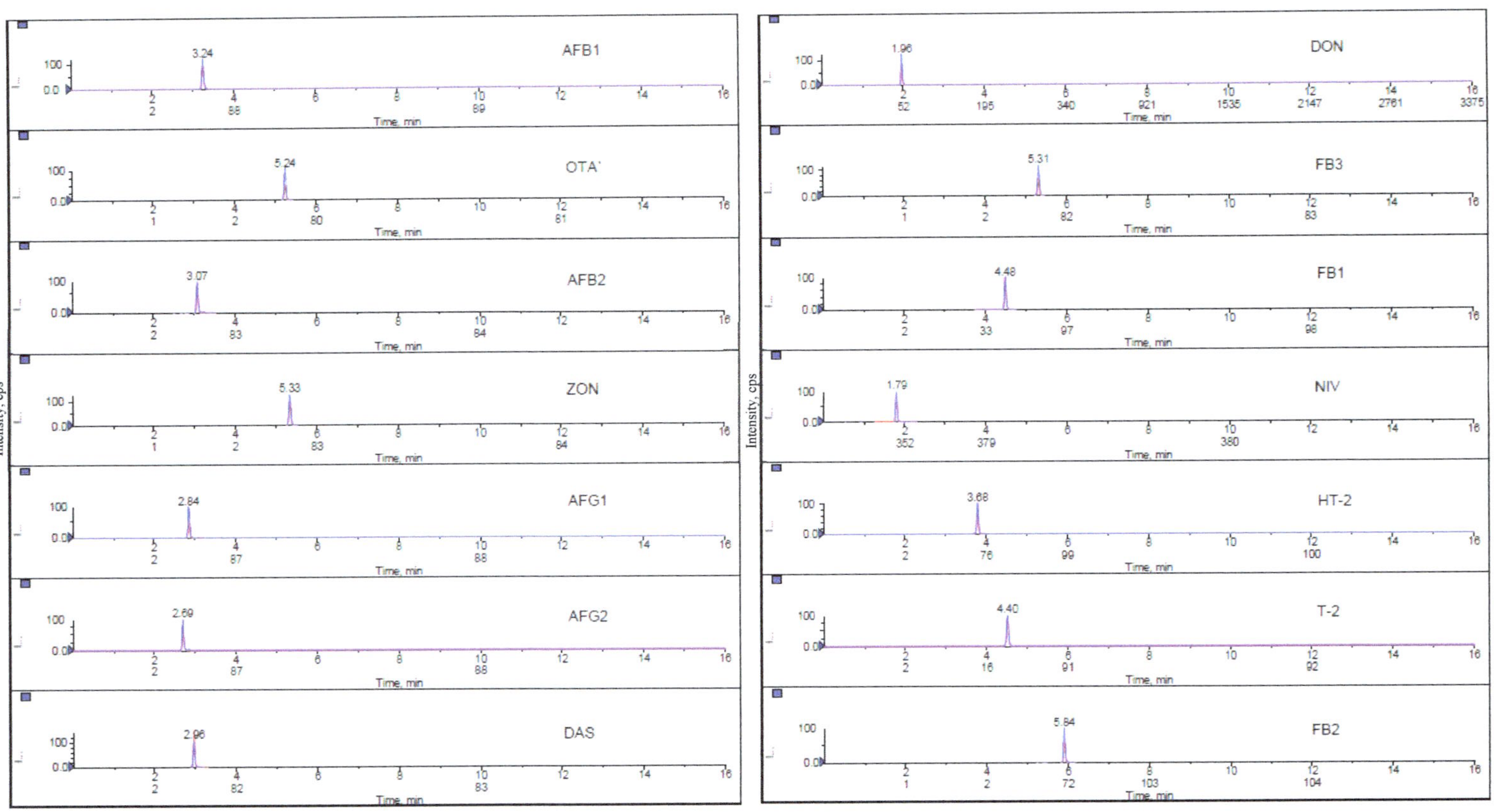

Figure 3. Multiple reaction monitoring (MRM) chromatograms of 14 mycotoxins at the lowest level of calibration standard.

3. Conclusions

The wide range of agricultural products, contaminant ranges and different distribution ways make mycotoxin an important issue in the world. Concerns with potential human health impact associated with consumption of dairy products drive research in this field. Due to the complexity of animal feed products, developing a methodology for extraction and clean-up processes covering recognized mycotoxins is necessary. For this purpose, three analytical sample preparation techniques, immunoaffinity, solid-phase extraction and QuEChERS, were compared, and the best was optimized. Corn is considered to be a complex matrix with severe matrix interferences, and matrix-matched calibration was used to reduce ion source contamination and decrease matrix effect because of co-eluting matrix components. This LC-MS/MS method was designed to create a robust and reliable approach for simultaneous analysis of 14 mycotoxins in various feeds. These selected compounds with different degrees of toxicity are representative of an important group of mycotoxins. Recovery values ranged from 70–100% for DON, DAS, FB1, FB2, FB3, HT-2, T-2, OTA, ZON, AFG1, AFG2, AFB1 and AFB2 and 55% for NIV. The results from 5 AAFCO proficiency tests have been reported on various matrices along with in-house validation.

4. Materials and Methods

4.1. Chemicals and Reagents

HPLC grade acetonitrile, methanol, water and formic acid were purchased from Fisher Scientific. Japanese aflatoxin mixture (25 µg/mL) from Sigma-Aldrich; deoxynivalenol (25 µg/mL), nivalenol (25 µg/mL), zearalenone (10 µg/mL), ochratoxin A (10 µg/mL) from Romer Labs; diacetoxyscirpenol (100 µg/mL), T-2 toxin (100 µg/mL), HT-2 toxin (100 µg/mL), fumonisin B1 (100 µg/mL), fumonisin B2 (100 µg/mL), and fumonisin B3 (100 µg/mL) from Trilogy. QuEChERS EN extraction salts and Captiva EMR-Lipid cartridge were acquired from Agilent (CA, USA).

4.2. Sample Preservation and Storage

Samples were ground finely with an Ultra Centrifugal Retsch Mill ZM200 (Retsch GmbH, Haan, NRW., Germany) to pass a 500 µm sieve. After grinding, they were homogenized to a powder-like consistency. All samples were stored in freezer before and after extraction process. Each sample was mixed carefully before weighing.

4.3. Sample Preparation

Sample preparation and cleanup process are necessary steps in various matrices for mycotoxin determination. Without sample cleanup, analysis can damage the chromatography column and Mass spectrometer. Immunoaffinity (IMA), solid-phase extraction (SPE), QuEChERS and modified QuEChERS methods were described in this section.

4.3.1. Immunoaffinity Column Method (Method A)

A corn sample (10 g) was extracted with 50 mL Phosphate Buffered Saline (PBS). After shaking for 60 min and centrifugation for 10 min, 35 mL of supernatant (extract A) was filtered through a glass microfiber filter. Subsequently, 35 mL methanol was added to the 15 mL remaining supernatant and the sample was shaken and centrifuged the same as extract A. After these two steps, 10 mL extract was diluted with 90 mL PBS and filtered through a glass microfiber filter (extract B). Extract B (50 mL) was passed over Myco6in1$^+$ (multiantibody IMA column, VICAM, Watertown, MA, USA) and washed with 20 mL PBS. At this step, 5 mL extract A was passed through the same Myco6in1+ column and washed with 10 mL water. The extract was eluted two times with 1.5 mL methanol:water (80:20, v:v) containing 0.5% acetic acid. The last eluted step was repeated and collected in another vial. Then, the extract was evaporated to dryness at 40–50 °C under a gentle stream of nitrogen. Both tubes were reconstituted with methanol:water (80:20, v:v) containing 0.5% acetic acid and combined into one vial.

4.3.2. Solid-Phase Extraction (Method B)

A corn sample (2 g) was weighed into a 50 mL centrifuge tube with 2 mL of water. 10 mL acetonitrile:water (50:50, v:v) was added and shaken for 15 min in Geno/Grinder (SPEX Sample Prep, Metuchen, NJ, USA). After centrifugation, the diluted supernatant was passed through the SPE cartridge (Oasis HLB, 500mg, 6 mL). Volumes of 3 mL water and 3 mL hexane were added into the cartridge and eluted by 2 mL acetonitrile and 2 mL methanol. The extract was evaporated and reconstituted with methanol:water (50:50, v:v).

4.3.3. QuEChERS Method (Method C)

In a 50 mL centrifuge tube, 5 g of a corn sample was extracted with 10 mL of water and 10 mL of acetonitrile. After a brief vortex, the mix was placed in Geno/Grinder for 20 min to homogenize well. Anhydrous $MgSO_4$ (4.0 g) and 0.5 g NaCl were added and shaken for 2 more minutes on a Geno/Grinder homogenizer and then centrifuged for 7 min. Subsequently, 1.6 mL of extract was diluted with 0.4 mL of water. The diluted sample was passed through a lipid removal column under constant vacuum and evaporated with nitrogen stream to the final volume of 1mL.

4.3.4. Modified QuEChERS Method (method D)

A corn sample (5 g) was weighed into a 50 mL centrifuge tube with 10 mL of water containing 0.3% formic acid and 10 mL of acetonitrile. After a brief vortex, the mix was placed in Geno/Grinder for 20 min to homogenize well. QuEChERS salts were added and shaken for 2 more minutes on a Geno/Grinder homogenizer and then centrifuged for 7 min. Subsequently, 1.6 mL of the extract was diluted with 0.4 mL of water. The diluted sample was passed through a lipid removal under constant vacuum. The extract was evaporated to dryness at 40–50 °C under a gentle stream of nitrogen and reconstituted in mobile phase A/mobile phase B (50/50, v/v).

This modified QuEChERS method uses an acidic condition to increase the extraction efficiency by using water with 0.3% formic acid. The polarity of compounds can be changed by using different pH levels [28]. It also offers evaporation/reconstitution for improving the MS responses and peak broadening. However, nivalenol cannot be completely extracted in acetonitrile; acidic condition and compatibility with mobile phases are essential in this procedure. The presence of the 0.3% formic acid improves the analyte partitioning into the organic phase and its elution as the first compound in the chromatographic run. Moreover, it prevents the retention of Ochratoxin A on the lipid removal cartridge and provides better results. Finally, the salting-out step with anhydrous $MgSO_4$ and NaCl followed by lipid removal and reconstitution steps achieved highest recoveries.

4.4. Matrix-Matched Calibration Preparation

All mycotoxin standard solutions were purchased from ISO 17034 accredited vendors. Aflatoxin mix (aflatoxin G1, G2, B1 and B2), and respectively, 25 µg/mL, DON 25 µg/mL, NIV 25 µg/mL, ZON 10 µg/mL, OTA 10 µg/mL, DAS 100 µg/mL, T-2 100 µg/mL, HT-2 100 µg/mL, FB1 100 µg/mL, FB2 100 µg/mL and FB3 100 µg/mL were used to prepare a combination standard. A mixture of 50 ng/mL AFB1, AFB2, AFG1 and AFG2; 1000 ng/mL FB1, FB2, FB3, T-2, HT-2, DON and ZON; 2000 ng/mL NIV, 500 ng/mL DAS and OTA in matrix blank was prepared for the highest level of calibration curve, and a mixture of 0.25 ng/mL AFB1, AFB2, AFG1 and AFG2; 5 ng/mL FB1, FB2, FB3, T-2, HT-2, DON and ZON; 10 ng/mL NIV, 2.5 DAS and OTA in matrix blank was prepared for the lowest level of calibration curve.

To create a matrix-matched calibration, eight point of standards must be prepared in the same or similar matrix group that was chosen for analysis. They were prepared in the same way as actual samples with 0.2% formic acid. All calibrant solutions for matrix-matched calibration were made by using clean matrix extracts. Mycotoxin standard solutions were stored in a refrigerator at 4 ± 4 °C.

4.5. Equipment Conditions

The liquid chromatography and MS/MS optimization were studied in this research to find the most appropriate operating conditions by individual injection of each mycotoxin standard.

4.5.1. Liquid Chromatography Separation Conditions

Shimadzu liquid chromatography equipped with Kinetex C18 column (2.6 µm particle size, 100 × 3.00 mm, Phenomenex, Torrance, CA, USA) was used. The mobile phases were mobile phase A, 0.5 mM ammonium fluoride, 5 mM ammonium formate and 0.1% formic acid in water; mobile phase B, 0.5 mM ammonium fluoride, 5 mM ammonium formate and 0.1% formic acid in methanol. Samples were eluted using a gradient at a flow rate of 0.45 mL/min throughout the 16 min run-time at 40 °C with injection volume of 10 µL. The gradient conditions were optimized as follows: 10% B from 0.01 to 0.5 min, 10–50% B from 0.5 to 10 min, 100% from 10–11 min, 10% 11–13.5 and 0% B from 13.5 to 16 min.

4.5.2. Mass Spectroscopy Conditions

To achieve a mass spectrum of the mycotoxins, a Triple Quad 5500 ABSciex mass spectrometer (AB SCIEX instruments, Foster, CA, USA) with a positive ESI interface is used. Mass spectrometer operating parameters are summarized as follows: curtain gas: 20 psig, ion spray voltage, 4500; temperature, 400 °C; ion source gas 1, 20; ion source gas 2, 30; collision gas, 8; MRM detection window, 60 sec and target scan time, 1. The mass spectrometer operates in scheduled MRM (Multiple Reaction Monitoring) mode described in Table 3, by monitoring 2 transitions and selecting the optimum voltage of declustering potential, collision energies and collision cell exit potentials for each compound. Sciex Analyst software 1.7 (AB SCIEX instruments, Foster, CA, USA) and Multiquant software 3.0 (AB SCIEX instruments, Foster, CA, USA) were applied for data acquisition and data processing respectively.

Table 3. MRM parameters for mycotoxin detection.

Analyte	Type	Q1 (m/z)	Q3 (m/z)	Retention Time (min)	DP (Volts)	CE (Volts)	CXP (Volts)
DON	$[M+H]^+$	297.2	249.2 / 231.1	1.93	37	16 / 18	16 / 16
DAS	$[M+H]^+$	384.0	307.1 / 105.1	3.00	54	9 / 40	27 / 20
FB1	$[M+H]^+$	722.2	334.3 / 352.4	4.23	75	53 / 50	16 / 16
FB2	$[M+H]^+$	706.1	336.3 / 318.4	5.67	73	48 / 50	14 / 14
FB3	$[M+H]^+$	706.1	336.3 / 318.4	5.04	73	48 / 50	14 / 14
HT-2	$[M+H]^+$	442.2	263.2 / 215.1	3.71	39	25 / 28	14 / 11
T-2	$[M+H]^+$	484.2	215.2 / 185.1	4.42	38	29 / 38	15 / 16
OTA	$[M+H]^+$	404.0	239.0 / 358.1	5.14	42	31 / 19	13 / 15
ZON	$[M+H]^+$	319.2	283.2 / 187.1	5.26	60	17 / 26	15 / 15
AFG1	$[M+H]^+$	329.0	243.1 / 283.1	2.78	73	37 / 35	16 / 16
AFG2	$[M+H]^+$	331.1	245.1 / 257.1	2.64	70	41 / 42	16 / 16

Table 3. *Cont.*

Analyte	Type	Q1 (m/z)	Q3 (m/z)	Retention Time (min)	DP (Volts)	CE (Volts)	CXP (Volts)
AFB1	[M+H]$^+$	313.1	285.1 241.1	3.17	58	32 51	14 18
AFB2	[M+H]$^+$	315.2	287.2 259.1	2.99	43	36 41	12 17
NIV	[M+H]$^+$	313.1	175.1 115.1	1.77	96	21 73	12 8

Q1: first quadrupole; Q3: third quadrupole; DP: declustering potential; CE: collision energy; CXP: collision cell exit potential.

4.6. Method Validation

The proposed method was validated by an in-house quality control procedure. Instrumental linearity, method detection limit, reporting limit, accuracy and precision were estimated.

4.6.1. Instrumental Linearity

A quadratic regression of the calibration data with all levels was used with weighted 1/x. Correlation coefficient (R^2) was higher than 0.995 in all cases. The linearity was evaluated based on using eight-point calibration curves.

4.6.2. Method Detection and Reporting Limit

Method detection limit (MDL) refers to the lowest concentration of the analyte that a method can detect reliably. To determine the MDL, 7 corn matrix blank samples were spiked at a different concentration for each analyte and processed through the entire method along with blank. The standard deviation derived from the spiked sample recoveries was used to calculate the MDL using this equation:

$$\text{MDL} = tS \ (n = 7 \text{ replicates, } t = 3.143)$$

Reporting limit refers to a level at which reliable quantitative results may be obtained. The MDL is used as a guide to determine the RL. The RL is the two times the MDL in this work. The calculated MDL and RL for all mycotoxins are shown in Table 4.

Table 4. MDL Study for Mycotoxins in Corn ($n = 7$).

Toxin	Spike (µg/mL)	SD	MDL	RL
DON	0.08	0.00023	0.00713	0.01426
DAS	0.04	0.00115	0.00360	0.00720
FB1	0.24	0.01588	0.04991	0.09983
FB2	0.24	0.01357	0.04266	0.08533
FB3	0.08	0.00348	0.01094	0.02189
HT-2	0.08	0.00115	0.00362	0.00724
T-2	0.08	0.00122	0.00383	0.00766
OTA	0.04	0.00159	0.00498	0.00997
ZON	0.08	0.00269	0.00845	0.01690
AFG1	0.004	0.00010	0.00032	0.00064
AFG2	0.004	0.00016	0.00049	0.00098
AFB1	0.004	0.00012	0.00037	0.00073
AFB2	0.004	0.00006	0.00018	0.00035
NIV	0.16	0.00456	0.01433	0.02867

SD: Standard Deviation, MDL: Method Detection Limit and RL: Reporting Limit.

4.6.3. Accuracy and Precision

The method validation consisted of five sample sets and each set includes three levels of fortification. All spikes were processed through the entire analytical method. Spike levels, recoveries and standard deviation for the mycotoxins are shown in Table 5. The recoveries were within range of 70%–100% for 13 mycotoxins and 55% for NIV. The relative standard deviation of recoveries was lower than 12% for all mycotoxins.

Table 5. Accuracy and precision of LC-MS/MS method for determination of 14 spiked Corn ($n = 5$).

Mycotoxins	Concentration (µg/g)	Mean Recovery (%)	SD (µg/g)	RSD (%)	Mycotoxins	Concentration (µg/g)	Mean Recovery (%)	SD (µg/g)	RSD (%)
DON	0.06	87.3	2.9	3.3	OTA	0.03	86.9	6.1	7.1
	0.08	86.1	2.1	2.4		0.04	89.0	89.0	4.3
	0.2	86.0	1.9	2.2		0.1	94.7	94.7	7.3
DAS	0.03	93.8	5.8	6.2	ZON	0.06	78.7	2.2	2.9
	0.04	96.1	7.6	7.9		0.08	81.9	5.7	7.0
	0.1	91.7	2.3	2.5		0.02	87.2	5.5	6.3
FB1	0.24	75.8	8.9	10.6	AFG1	0.003	84.7	10.5	12.3
	0.6	74.4	6.7	9.1		0.004	86.3	7.3	8.5
	2.4	80.4	1.7	2.1		0.01	81.9	6.1	7.4
FB2	0.24	71.3	5.3	8.7	AFG2	0.003	99.1	8.7	8.8
	0.6	74.0	2.9	4.0		0.004	98.0	4.9	5.0
	2.4	74.6	3.5	5.1		0.01	92.3	2.6	2.9
FB3	0.08	74.2	6.7	9.0	AFB1	0.003	79.6	5.7	7.2
	0.2	77.6	3.9	5.1		0.004	82.0	3.3	4.1
	0.8	78.2	4.6	5.9		0.01	79.3	3.5	4.4
HT-2	0.06	97.5	2.5	2.6	AFB2	0.003	86.1	6.2	7.2
	0.08	97.0	4.2	4.3		0.004	86.5	6.9	7.9
	0.5	84.1	4.4	5.2		0.01	81.9	4.4	5.3
T-2	0.06	96.1	3.5	3.6	NIV	0.12	55.0	3.7	6.7
	0.08	98.4	3.4	3.5		0.16	56.7	3.6	6.3
	0.2	98.4	4.3	4.4		0.4	55.0	4.1	7.5

Author Contributions: Conceptualization, B.N.; methodology, B.N.; validation, B.N. and N.S.A.; formal analysis, B.N and N.S.A.; investigation, B.N.; supervision, M.K.; writing—original draft, B.N.; writing—review and editing, B.N.; project administration, B.N. and M.K. All authors have read and agreed to the published version of the manuscript.

Funding: This research received no external funding.

Acknowledgments: This research was supported by the California Department of Food and Agriculture.

Conflicts of Interest: The authors declare no conflict of interest.

References

1. Jelinek, C.F.; Pohland, A.E.; Wood, G.E.J. Worldwide occurrence of mycotoxins in foods and feeds-an update. *Assoc. Off. Anal. Chem.* **1989**, *72*, 223–230. [CrossRef]
2. Bennett, J.W. Mycotoxins, mycotoxicoses, mycotoxicology and mycopathologia. *Mycopathlogia* **1987**, *100*, 3–5. [CrossRef]
3. Abdulkadar, A.H.W.; Al-Ali, A.A.; Al-Kildi, A.M.; Al-Jedah, J.H. Mycotoxins in food products available in Qatar. *Food Control* **2004**, *15*, 543–548. [CrossRef]
4. WHO Food Additives Series: 47 Safety Evaluation of Certain Mycotoxins in Food, Prepared by the Fifty-Sixth Meeting of the Joint FAO/WHO Expert Committee on Food Additives (JECFA). Available online: http://www.inchem.org/documents/jecfa/jecmono/v47je01.htm (accessed on 14 March 2014).

5. Miraglia, M.; Marvin, H.J.P.; Kleter, G.A.; Battilani, P.; Brera, C.; Coni, E.; Cubadda, F.; Croci, L.; De Santis, B.; Dekkers, S.; et al. Climate change and food safety: An emerging issue with special focus on Europe. *Food Chem. Toxicol.* **2009**, *47*, 1009–1021. [CrossRef] [PubMed]

6. Becker-Algeri, T.A.; Castagnaro, D.; De Bortoli, K.; de Souza, C.; Drunkler, D.A.; Badiale-Furlong, E. Mycotoxins in bovine milk and dairy products: A review. *J. Food Sci.* **2016**, *81*, 544–552. [CrossRef] [PubMed]

7. Pereira, C.S.; Cunha, S.C.; Fernandez, J.O. Prevalent mycotoxins in animal feed: Occurrence and analytical method. *Toxins* **2019**, *11*, 290. [CrossRef] [PubMed]

8. Milićević, D.R.; Škrinjar, M.; Baltić, T. Real and Perceived Risks for Mycotoxin Contamination in Foods and Feeds: Challenges for Food Safety Control. *Toxins* **2010**, *2*, 572–592. [CrossRef] [PubMed]

9. Juan, C.; Gonzalez, L.; Soriano, J.M.; Molto, J.C.; Manes, J. Accelerated solvent extraction of ochratoxin A from rice samples. *J. Agric. Food Chem.* **2005**, *53*, 9348–9351. [CrossRef]

10. Li, C.; Wu, Y.L.; Yang, T.; Huang-Fu, W.G. Rapid determination of fumonisins B1 and B2 in corn by liquid chromatography-tandem mass spectroscopy with ultrasonic extraction. *J. Chromatogr. Sci.* **2012**, *50*, 57–63. [CrossRef] [PubMed]

11. Bauer, J.; Gareis, M. Ochratoxin A in the food chain. *J. Vet. Med. Ser. B* **1987**, *34*, 613–627. [CrossRef]

12. Tang, Y.Y.; Lin, H.Y.; Chen, Y.C.; Su, W.T.; Wang, S.C.; Chiueh, L.C.; Shin, Y.C. Development of a quantitative multi-mycotoxin method in rice, maze, wheat and peanut using UPLC-MS/MS. *Food Anal. Methods* **2013**, *6*, 727–736. [CrossRef]

13. Malysheva, S.V.; Di Mavungu, J.D.; Boonen, J.; De Spiegeleer, B.; Goryacheva, I.Y.; Vanhaecke, L.; De Saeger, S. Improved positive electrospray ionization of patulin by adduct formation: Usefulness in liquid chromatography-tandem mass spectroscopy multi-mycotoxin analysis. *J. Chromatogr. A* **2012**, *1270*, 334–339. [CrossRef] [PubMed]

14. Trufelli, H.; Palma, P.; Famiglini, G.; Cappiello, A. An overview of matrix effects in liquid chromatography-mass spectrometry. *Mass Spectrom. Rev.* **2011**, *30*, 491–509. [CrossRef] [PubMed]

15. Matuszewski, B.K.; Constanzer, M.L.; Chavez-Eng, C.M. Strategies for the assessment of matrix effect in quantitative bioanalytical methods based on HPLC-MS/MS. *Anal. Chem.* **2003**, *75*, 3019–3030. [CrossRef] [PubMed]

16. Li, D.; Steimling, J.A.; Konschnik, J.D.; Grossman, S.L.; Kahler, T.W. Quantitation of mycotoxins in four food matrices comparing stable isotope dilution assay (SIDA) with matrix-matched calibration methods by LC-MS/MS. *J. AOAC. Int.* **2019**, *102*, 1673–1680. [CrossRef] [PubMed]

17. Turner, N.W.; Subrahmanyam, S.; Piletsky, S.A. Analytical methods for determination of mycotoxins: A review. *Anal. Chimica Acta* **2009**, *632*, 168–180. [CrossRef] [PubMed]

18. Zheng, M.Z.; Richard, J.L.; Binder, J. A review of rapid methods for the analysis of mycotoxins. *Mycopathologia* **2006**, *161*, 261–273. [CrossRef]

19. Soleimany, F.; Jinap, S.; Rahmani, A.; Khatib, A. Simultaneous detection of 12 mycotoxins in cereals using RP-HPLC-PDA-FLD with PHRED and a post-column derivatization system. *Food Addit. Contam.* **2011**, *28*, 494–501. [CrossRef]

20. Ofitserova, M.; Nerkar, S.; Pickering, M.; Torma, L.; Thiex, N. Multiresidue mycotoxin analysis in corn grain by column high-performance liquid chromatography with postcolumn photochemical and chemical derivatization: Single-laboratory validation. *J. AOAC Int.* **2009**, *92*, 15–25. [CrossRef]

21. Monbaliu, S.; Van Poucke, C.; Detavernier, C.; Dumoulin, F.; Van De Velde, M.; Schoeters, E.; Van Dyck, S.; Averkieva, O.; Van Peteghem, C.; De Seger, S. Occurrence of mycotoxins in feed as analyzed by a multi-mycotoxin LC-MS/Ms method. *J. Agric. Food Chem.* **2010**, *58*, 66–71. [CrossRef]

22. Ediage, E.N.; Di Mavungu, J.D.; Monbaliu, S.; Van Peteghem, C.; De Saeger, S. A validated multianalyte LC–MS/MS method for quantification of 25 mycotoxins in cassava flour, peanut cake and maize samples. *J. Agric. Food Chem.* **2011**, *59*, 5173–5180. [CrossRef] [PubMed]

23. Lattanzio, V.M.; Solfrizzo, M.; Powers, S.; Viscounti, A. Simultaneous determination of aflatoxins, ochratoxin A and Fusarium toxins in maize by liquid chromatography tandem mass spectrometry after multitoxin immunoaffinity clean-up. *Rapid Commun. Mass Spectrom.* **2007**, *21*, 3253–3261. [CrossRef]

24. Solfrizzo, M.; Gambacorta, L.; Bibi, R.; Ciriaci, M.; Paoloni, A.; Pecorelli, I. Multimycotoxin analysis by LC-MS/MS in cereal food and feed: Comparison of different approaches for extraction, purification, and calibration. *J. AOAC Int.* **2018**, *101*, 647–657. [CrossRef] [PubMed]

25. Agilent Application Note 5991-8962EN. Analysis of mycotoxins in Food matrices using the Agilent Ultivo triple Quadrupole LC/MS. 2019. Available online: https://www.agilent.com/cs/library/applications/5991-8962EN_Ultivo_AppNote.pdf (accessed on 2 July 2019).
26. Rasmussen, R.R.; Storm, I.M.L.D.; Rasmussen, P.H.; Smedsgaard, J.; Nielsen, K.F. Multi-mycotoxin analysis of maize silage by LC-MS/MS. *Anal. Bioanal. Chem.* **2010**, *397*, 765–776. [CrossRef] [PubMed]
27. Malachova, A.; Stranska, M.; Vaclavikova, M.; Elliott, C.T.; Black, C.; Meneely, J.; Hajslova, J.; Ezekiel, C.N.; Schuhmacher, R.; Krska, R. Advanced LC-MS-based methods to study the co-occurrence and metabolization of multiple mycotoxins in cereals and cereal-based food. *Anal. Bioanal. Chem.* **2018**, *410*, 801–825. [CrossRef]
28. Kah, M.; Brown, C.D. Log D: Lipophilicity for ionisable compounds. *Chemosphere* **2008**, *72*, 1401–1408. [CrossRef]

Review

Advances in Colorimetric Strategies for Mycotoxins Detection: Toward Rapid Industrial Monitoring

Marjan Majdinasab [1], Sondes Ben Aissa [2] and Jean Louis Marty [2,*]

1 Department of Food Science & Technology, School of Agriculture, Shiraz University, Shiraz 71441-65186, Iran; majdinasab@shirazu.ac.ir

2 BAE-LBBM Laboratory, University of Perpignan via Domitia, 52 Avenue Paul Alduy, CEDEX 9, 66860 Perpignan, France; sondes.benaissa@univ-perp.fr

* Correspondence: jlmarty@univ-perp.fr; Tel.: +33-468662254

Abstract: Mycotoxins contamination is a global public health concern. Therefore, highly sensitive and selective techniques are needed for their on-site monitoring. Several approaches are conceivable for mycotoxins analysis, among which colorimetric methods are the most attractive for commercialization purposes thanks to their visual read-out, easy operation, cost-effectiveness, and rapid response. This review covers the latest achievements in the last five years for the development of colorimetric methods specific to mycotoxins analysis, with a particular emphasis on their potential for large-scale applications in food industries. Gathering all types of (bio)receptors, main colorimetric methods are critically discussed, including enzyme-linked assays, lateral flow-assays, microfluidic devices, and homogenous in-solution strategies. This special focus on colorimetry as a versatile transduction method for mycotoxins analysis is comprehensively reviewed for the first time.

Keywords: mycotoxins; colorimetric detection; rapid tests; ELISA; lateral flow assays; microfluidics; nano-materials; food safety; commercialization

Key Contribution: This paper exhaustively reviews the recent trends in the rapid colorimetric (bio)sensing of mycotoxins in the last five years (2015–2020). Latest figures of merit of colorimetric methods are thoroughly discussed, highlighting their great potential for industrial applications.

check for
updates

Citation: Majdinasab, M.; Ben Aissa, S.; Marty, J.L. Advances in Colorimetric Strategies for Mycotoxins Detection: Toward Rapid Industrial Monitoring. *Toxins* **2021**, *13*, 13. https://dx.doi.org/10.3390/toxins1301 0013

Received: 30 November 2020
Accepted: 22 December 2020
Published: 24 December 2020

Publisher's Note: MDPI stays neutral with regard to jurisdictional claims in published maps and institutional affiliations.

1. Introduction

Widespread mycotoxins contamination of food and feed poses a serious menace for human's health and contributes to massive economic losses in the agriculture industry. Mycotoxins are chemically diverse groups of low molecular weight fungal metabolites that are almost unpredictable and unavoidable in crops and have a wide variety of toxic effects [1]. These thermal-stable fungal toxins affect a broad range of agricultural products including cereals, cereal-based foods, dried fruits, wine, milk, coffee beans, cocoa bakery, and meat products [2].

Hitherto, over 300 kinds of mycotoxins have been characterized, but only about a dozen have led the priority list of risk assessment due to their high occurrence in food staples and severe health effects [3]. Representative mycotoxins include aflatoxins (AFs), ochratoxins (OTA), fumonisins (FB), zearalenone (ZEN), patulin (PAT), deoxynivalenol (DON), and trichothecenes. According to the International Agency for Research on Cancer (IARC), some are proved to be strong carcinogenic agents such as aflatoxin B1 (AFB1) while others are under suspicion to have carcinogenic effects [4]. Nonetheless, all of them have shown acute and chronic toxicities [5]. Hence, stringent regulations relating to mycotoxins have been established in many countries to protect the consumer from their harmful effects [6]. The established maximum limits (MLs) differ depending on the mycotoxin and the targeted foodstuff. In particular, the strictest regulations have been set for aflatoxins in the processed food products for infants [2]. In addition to the regulatory framework, consumers have become recently more aware of health and food quality. Therefore, research on the development of high-throughput, real-time, and reliable portable detection methods for food safety augmented [7].

The operation procedure should be simplified continuously for users' convenience, avoiding the need for laboratory-based techniques. Many instrumental methods have been used from the very early discovery of mycotoxins till now, such as thin-layer chromatography (TLC), high-performance liquid chromatography (HPLC), in combination with different detectors (e.g., fluorescence, diode array, UV), liquid chromatography coupled with mass spectrometry (LC-MS), liquid chromatography-tandem mass spectrometry (LC-MS/MS) and gas chromatography-tandem mass spectrometry (GC-MS/MS) for mycotoxin analysis [2]. Owing to their high sensitivity and precise analysis, such techniques present the gold methods to control mycotoxins levels in food samples in compliance with the regulatory framework. Reviews of these methods have been summarized and published elsewhere [8,9]. Despite their analytical merits, chromatographic methods involve tedious multistep processes that are time-consuming and require highly skilled personnel. Moreover, expensive and bulky instruments restrict their use for in-situ mycotoxin analysis. Therefore, more convenient and user-friendly methods were still highly desirable for the rapid monitoring of mycotoxins' traces in food and feed.

Consequently, optical methods have received great attentions in developing rapid detection kits specific to common mycotoxins. Among different sensing strategies, colorimetric detection methods are particularly well-suited for on-site biosensing due to their simple readout and operation. They can serve for qualitative, semi-quantitative or quantitative methods for a rapid screening in field, in silo, or during the agri-food processing. The portability of such miniaturized tools is profitable for industrials to validate their products' conformity in accordance with regulatory limits.

Colorimetric methods can be classified based on the type of color-generating probes (dyes, enzymes, nanomaterials) and the sensing reaction phase (solution-based and solid substrate-based). Enzyme-linked immunosorbent assays (ELISA) are the most popular colorimetric screening tools that reached successfully the commercialization stage for mycotoxins analysis along with some lateral flow immunoassays (LFIA). Thanks to their unique features, detection kits relying on these two techniques are being manufactured by multiple companies worldwide. Despite the current market competitiveness, the colorimetric methods dedicated to the determination of representative mycotoxins in foodstuffs continue to attract industrials for a reliable and cost-effective monitoring. Campbell et al. have recently reviewed the available commercial kits [10], emphasizing that antibody-based schemes conquer the most part of the market owing to their superior specificity for real-world applications.

However, many successful proofs of concept were described in the recent literature using either chemical sensing or other bioreceptors. In particular, aptamers are short single-stranded oligonucleotides (DNA or RNA) which can replace antibodies as recognition element in sensing strategies. They exhibit several advantages such as high stability, low production cost, high affinity and specificity, and high sensitivity. Parallelly, emerging nanomaterials had led to an unprecedented improvement of these alternative sensing strategies in food safety control.

Overall, the rapid development of colorimetric methods has brought many opportunities for rapid mycotoxins detection. Many emerging and novel (bio)assays have been reported as competitive analytical tools with easy operation and fast visible response. However, to date, there are very few reviews that focused on the colorimetric transduction application for mycotoxins analysis in food matrices regardless the bioreceptor nature and the target mycotoxin type. Thus, it is necessary to give a comprehensive summarization. This helps to understand the current trends and assist decision-makers to apply such cost-effective technologies in agri-food industries. The aim of the present review is to place the diverse colorimetric methods (solution-based (bio)assays, ELISA, lateral flow assays, microfluidics) within a critical framework that compares the merits and limitations of each methodology and highlights the progress that has been made in recent five years. Current figures of merit of rapid colorimetric methods with great potential for industrial applications are thoroughly discussed.

2. Common Colorimetric Probes

2.1. Enzymes-Based Probes

Enzymes are robust signal amplification systems in bioassays and biosensors. They can be used in both optical and electrochemical sensing strategies. The principle of the enzyme-based colorimetric assays is to detect target analyte through the enzymatic conversion of a chromogenic substrate into a colored product. The produced color can be detected by the naked eye (qualitative methods) and through spectrophotometry or colorimetric analysis software (semi-quantitative or quantitative methods). The three most common types of enzyme-based colorimetric probes—including enzyme horseradish peroxidase (HRP), G-quadruplex sequences or DNAzymes, and acetylcholinesterase (AChE)—have been applied for colorimetric detection of mycotoxins and are hence discussed here.

HRP, found in the roots of horseradish plant, is the most popular enzymatic marker in bioassays due to ability to be conjugated with antibodies or other recognition elements, while preserving its activity, low-cost, and versatility. HRP can catalyze the reaction of hydrogen peroxide with certain organic, electron-donating substrates to yield highly colored products. An extensive range of electron-donating dye substrates are commercially available for use as HRP detection reagents. Some of them can be employed to form soluble colored products suitable for use in spectrophotometric detection methods, while other substrates form insoluble products that are mainly appropriate for staining techniques. Among them, 3,3′,5,5′-Tetramethylbenzidine or TMB is widely used as a soluble chromogenic substrate for colorimetric detection in enzyme-linked immunosorbent assay (ELISA) and other bioassays. However, some HRP-based bioassays suffer from limited sensitivity due to small amount of enzyme (i.e., HRP) that catalyzes chromogenic substrate. To address this issue, Lin et al. presented a method to combine the analyte-recognition element complex with a large number of enzymes [11]. They developed a liposome-based colorimetric aptasensor for ochratoxin A (OTA) detection in a TMB-H_2O_2 reaction medium. In this context, liposome as a sphere-shaped vesicle with hydrophobic and hydrophilic character was used for encapsulation of HRP. The main component of the detection system was a dumbbell-shaped probe including magnetic beads (MBs), double-stranded DNA (dsDNA), and HRP-encapsulated liposome (Figure 1a). The dsDNA was formed by the hybridization between OTA aptamer and its complementary probes ssDNA-1 and ssDNA-2. ssDNA-2 was conjugated with liposome and used as detection probe. In the presence of OTA, the aptamer combined with OTA to form G-quadruplex, resulting in the release of the ssDNA-2 and the HRP-encapsulated liposome. Each liposome containing a large quantity of HRP was lysed by adding the mixed solution of TMB and H_2O_2. HRP catalyzed H_2O_2-mediated oxidation of TMB and resulted in color change from colorless to blue. The assay was highly sensitive due to the signal amplification caused by the large amount of HRP embedded in liposome. The limit of detection (LOD) was obtained 0.023 ng·mL^{-1}. The assay was simple, low-cost, highly selective and reliable for the analysis of real samples. However, the reaction time for the G-quadruplex formation (40 min) and TMB oxidation (20 min) was too long. The aptasensor was also applied for OTA detection in corn samples.

Haem peroxidases such as HRP use protein scaffolds that activate heme to react with H_2O_2 [12]. There is extensive information on the reaction mechanism and properties of protein-based peroxidases. It was recently revealed that certain nucleic acid sequences have the ability to catalyze reactions similar to those carried out by heme. These nucleic acid sequences are non-canonical Guanine-rich structures with stacked G-tetrads assembled by Hoogsteen hydrogen-bonding. These sequences, named as G-quadruplex (G4), are able to bind hemin (iron (III)-protoporphyrin IX) to form a unique type of G4 DNAzyme or RNAzyme with powerful peroxidase-mimicking activity [13]. In comparison with natural protein peroxidases, G4 DNAzymes/RNAzymes show several advantages such as small size, easy synthesis, more stability, and facile manipulation, which make them good candidates in biosensing [13]. However, they suffer from relatively low catalytic activity compared to protein peroxidases which restricts their further development and application [14]. To overcome this limitation, several strategies have been developed to improve the catalytic efficiency of G4 DNAzymes/RNAzymes. These include (1) addition of polycationic amines such as spermine, spermidine, and putrescine; (2) addition of the nucleotide ATP to DNAzyme reactions; (3) conjugation of hemin with the G4-quadruplex moiety through covalent linkage or with cationic peptides; and (4) flanking adenine or cytosine nucleotides on G-quadruplex activities [12]. Incorporation of aptamers and DNAzymes as functional nucleic acids results in simple detection of target analyte by visual color development.

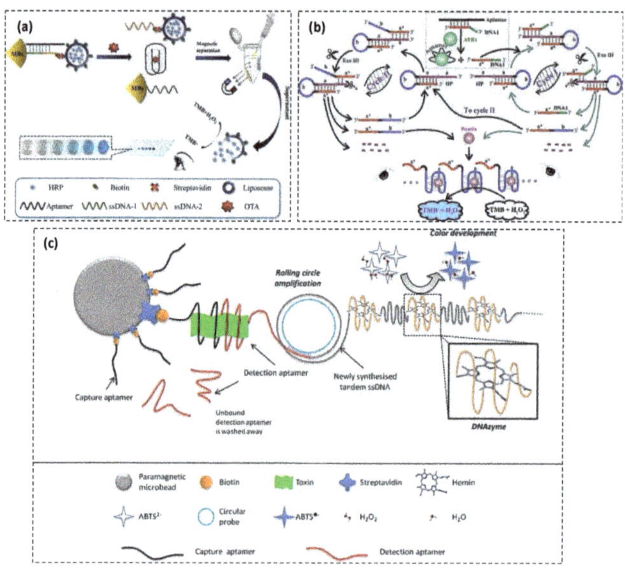

Figure 1. (**a**) Schematic illustration of the colorimetric aptasensor for OTA detection based on HRP-encapsulated liposome; (**b**) developed aptasensor for AFB1 detection using G-quadruplex as the signal reporter. Domain a is complementary to domain a*. Domain b is the caged G-rich sequence. Exo III performs the cyclic cleavage reactions in Cycles I and II; (**c**) aptasensor for OTA detection, based on rolling circle amplification and an auto-catalytic DNAzyme structure. Reproduced with permission from [11,15,16], respectively.

Using G-quadruplex as the signal reporter, a colorimetric aptasensor was developed for AFB1 detection [15]. The aptasensor was fabricated by the combination of an ingenious hairpin DNA probe with exonuclease III (Exo III)-assisted signal amplification. The hairpin DNA probe contained a $3'$-protruding segment (domain a) as the recognition unit, the stem zone (domains a and a*), and a caged G-rich sequence located in the loop region (domain b). The presence of the AFB1 activated the continuous cleavage reactions by Exo III toward a hairpin probe, resulting in the autonomous accumulation of numerous free G-quadruplex sequences, which catalyzed the oxidation of TMB by H_2O_2 to generate a colorimetric signal (Figure 1b). The aptasensor represented many advantages including high sensitivity (LOD of 1 pM), good selectivity, simple operation, wash-free, label-free format, low-cost, naked-eye detection, and applicability to samples with complex matrices. However, the assay time was long (incubation time 40 min). The assay was used for AFB1 detection in peanut samples.

Detection with the aid of magnetic beads-based separation has emerged as a rapid, simple, reliable, and efficient alternative to conventional immobilization methods. In this regard, a colorimetric aptasensor based on apta-magnetic separation assisted with DNAzyme was developed for AFB1 detection [17]. The procedure consisted of one-step separation of AFB1 by biotinylated aptamer conjugated to streptavidin magnetic beads which was followed by the addition of DNAzyme modified aptamer in the presence hemin and TMB/H_2O_2 to produce a colorimetric signal. The aptasensor was able to detect as low as 40 ppb and 22.6 ppb OTA visually and by spectrophotometer, respectively. The developed assay was selective, reliable, inexpensive, and rapid (incubation time 15 min). However, the incubation time of DNAzyme was long (30 min). The aptasensor was able to detect AFB1 in food samples.

Sensitivity of DNA-based biosensors can be significantly increased using a technique known as rolling circle amplification (RCA). RCA is an isothermal enzymatic amplification process of DNA where a short DNA or RNA primer is amplified using circular DNA template. With proper application of this technique, it is possible to synthesize large quantities of any type of nucleic acid strand. In this context, a highly sensitive aptasensor based on RCA and an auto-catalytic DNAzyme structure was designed for OTA detection [16]. In this work, a capture aptamer was linked to paramagnetic beads for specific capturing of OTA while a second aptamer was applied for OTA detection. The detection aptamer contained a DNAzyme producing sequence and an RCA priming sequence for the isothermal DNA amplification triggered by a circular ssDNA. When OTA was captured, the circular DNA was amplified, generating a single-stranded and tandem repeated long homologous copy of its sequence. In the DNA strand, a self-catalytic structure was formed with hemin as the catalytic core causing a blue color in the presence of 2,2′-azino-bis(3-ethylbenzothiazoline-6-sulfonic acid) (ABTS) and H_2O_2 (Figure 1c). A low LOD of 1.09×10^{-12} ng·mL^{-1} was obtained. Although the aptasensor was highly sensitive and selective, it suffered from long incubation time and complicated operation with multiple steps of washing and separation. The aptasensor was used for OTA detection in urine samples.

There are various enzymes—such as cholinesterase, urease, glucose oxidase, etc.—which have been employed in mycotoxin detection methods based on enzymatic inhibition. AChE (obtained from electric eel) is the most commonly used enzyme due to its susceptibility toward mycotoxin [7]. It can be used for the detection of aflatoxin B1 (AFB1) due to the inhibitory effect of AFB1 to AChE enzymatic activity [18]. It has been proven that AChE is inhibited by the AFB1 due to non-covalently binding of toxin at the external site, which is placed on the active site gorge entrance (located at the tryptophan residue) [7]. Based on AChE inhibition, AFB1 was determined by a colorimetric method (Ellman's method) developed on chromatography paper [19]. In this work, genipin cross-linked chitosan was used for AChE immobilization. For the colorimetric detection of AFB1 on microfluidic paper-based analytical device (µPAD), AChE immobilized on cross-linked chitosan was loaded on the edges of the flower-shaped µPAD. Then, AFB1 solution and 5,5-dithiobis-2-nitrobenzoic acid (DTNB, Ell-man's reagent) solution were applied at the center of flower-shaped µPAD. After 3-min incubation, acetylthiocholine iodide (ATCh) solution was also added at the center, and incubated for 5 min. The Ellman's colorimetric assay is based on the reaction of thiocholine (a product of enzymatic hydrolysis of ATCh) with DTNB to form a colored product. In the presence of AFB1, the AChE activity on ATCh substrate is inhibited resulting in failure to form a colored product. Cross-linking of chitosan resulted in a colorimetric signal enhancement. The assay was simple, low-cost, rapid (detection time ≈ 8 min), and fairly selective. However, the sensitivity of the assay was not reported. The assay was used for the detection of AFB1 in spiked corn samples.

AChE is considered very stable but lack of selectivity towards many toxins such as carbamates, organophosphate pesticides, anatoxin-a (a natural neurotoxic), and mycotoxins, which restrict its applicability. To address this issue, many efforts have been made to produce mutants of AChEs to improve the selectivity of enzyme against a specific toxin. Genetic modification of enzyme can also improve its stability and the assay sensitivity [20].

Representative examples of recent developed enzyme-based probes for the colorimetric detection of mycotoxins are reported in Table 1.

Table 1. Representative examples of recent developed enzyme-based probes for the colorimetric detection of mycotoxins.

Strategy	Detection Probe	Target	LOD	Linear Range	Specificity	Sample	Advantages/Disadvantages	Ref.
Colorimetric aptasensor based on HRP-encapsulated liposome which catalyzed TMB oxidation	HRP	OTA	$0.023 \ ng \cdot mL^{-1}$	$0.05\text{--}2.0 \ ng \cdot mL^{-1}$	High	Corn	Simple, low-cost, highly selective, sensitive and reliable/long detection time (60 min)	[11]
Combination of an ingenious hairpin DNA probe with exonuclease III (Exo III)-assisted signal amplification and TMB oxidation	DNAzyme	AFB1	1 pM	1 pM–100 nM	High	Peanut	High sensitivity, good selectivity, simple operation, wash-free, label-free format, low-cost, applicability to samples with complex matrices/long incubation time (40 min)	[15]
Dual aptamer-DNAzyme in combination with apta-magnetic separation	DNAzyme	AFB1	22.6 ppb	0–200 ppb	Good	Corn, rice, groundnut, black pepper, chili	High sensitivity, good selectivity, low-cost, reliable/long incubation time of DNAzyme (30 min)	[17]
Detection aptamer containing a DNAzyme sequence and an RCA priming sequence for the isothermal DNA amplification	DNAzyme	OTA	$1.09 \times 10^{-12} \ ng \cdot mL^{-1}$	$10^{-12}\text{--}10 \ ng \cdot mL^{-1}$	High	Urine	High sensitivity and selectivity, applicability to biological samples with complex matrices/long incubation time and complicated operation with multiple steps of washing and separation	[16]
Combination of DNA aptamer and two split hemin-binding DNAzyme halves and G-quadruplex formation of a split DNAzyme-hemin/aptamer complex with peroxidase mimicking activity in the absence of AFB1	DNAzyme	AFB1	$0.1 \ ng \cdot mL^{-1}$	$0.1\text{--}10^{4} \ ng \cdot mL^{-1}$	High	Corn	High sensitivity and selectivity simple and low-cost/long incubation time (70 min)	[21]
Inhibition of chitosan-immobilized AchE activity by AFB1and Ellman's method	AchE	AFB1	Not reported	Not reported	Good	Corn	Simple, rapid (detection time $\approx$ 8 min), low-cost, portable/failure to report LOD and linear range	[19]

2.2. Nanomaterial-Based Probes

Accelerated by the advances in nanomaterials (NMs), colorimetric methods for the detection of mycotoxins have undergone a rapidly developing stage in the past few years [22]. Their nanometric size (less than 100 nm) and unique physicochemical features, including distinctive optical and catalytic properties, have promoted the extensive use of nanostructured materials in colorimetric methods. Accordingly, researchers handled each nanomaterial differently to adapt it with the desired function in the sensing assay. It is widely reported that NMs are attractive candidates to immobilize bioreceptors, including enzymes, antibodies, and aptamers, thanks to their large size to volume ratio, which provides a high specific active surface. In particular, magnetic nanoparticles (MNPs) from iron-based nanoparticles are widely used in colorimetric bioassays to capture, separate, and enrich target analytes, especially when a low detection limit is required. However, we emphasize in this section the signaling roles of NMs in colorimetric methods dedicated to mycotoxins detection. Glimpsing at the relevant literature, two prominent roles of NMs are depicted. NMs mainly based on metal nanoparticles show color switching tunable properties and are thereby used as direct colorimetric probes. Enzyme-like NMs (or nanozymes) also contribute to the advances in colorimetric assays, particularly through peroxidase and oxidase-like catalysis that generate colored products. Some NMs can be also employed as signal mediators to enhance assay sensitivities in cascade amplification systems.

As optical signal generators, noble metal nanoparticles—including gold NPs, silver NPs, etc.—are majorly used in mycotoxins' (bio)assays due to their unique physicochemical properties. In particular, detection strategies based on changes in the localized surface plasmon resonance (LSPR) signal caused by the aggregation of noble metal NPs have shown suitable sensitivities to detect mycotoxins [23]. In such systems, NPs can be dispersed in colloidal solution via surface anionic repulsion. In the presence of electrolytes containing salt cations or cationic polymers, charges are stabilized, and NPs tend to aggregate. This aggregation alters the LSPR effect, resulting in a red shift of the UV-vis absorption spectrum [24]. Harnessing this property, AuNPs have been extensively tested in the plasmonic sensing of some fungal toxins, owing to their easy synthesis, high extinction coefficients, photostability, and non-toxicity. AuNPs have been considered as ideal signal generating probes because of the visible color change from red to blue through salt-induced nanoparticles assembly, or inversely through their redispersion [23].

Specifically, the advances of nucleic acid manipulation and aptamers selection have powerfully accelerated the progress in plasmonic mycotoxins detection [25]. Nucleic acid strands are more convenient than antibodies for unmodified AuNPs aggregation-based assays, with promising results in the semi-quantitative and quantitative real-sample application [26]. For instance, A label-free optical sensor was reported for the selective detection of AFB1 using a DNA-based aptamer along with unmodified spherical colloidal AuNPs (diameter ~ 13 nm). Recognition of AFB1 was achieved based on the salt-induced AuNPs aggregation. High selectivity was observed against the presence of OTA. Low detection-limit of 0.025 ng·mL^{-1} AFB1 was reported with the linear dynamic determination range of 0.025–100 ng·mL^{-1} [27]. More recently, Phanchai et al. [28] have performed in silico studies to investigate the molecular dynamics (MD) of this detection approach using AuNPs aggregation taking as an example anti-OTA aptamer (Figure 2a). This offered new insights into the mechanism of recognition highlighting the effect of the ionic composition of solvent as well as the kinetics behind the interaction between the three molecular partners—i.e., AuNPs, aptamer, and the mycotoxin. The reported MD simulation revealed an insightful analysis of the interaction mechanisms in the AuNP-based aptasensing platforms that can be projected to any other similar pattern.

Another strategy of colorimetric signal generating relies on nanomaterial-based labels like common in lateral flow immunochromatographic assays (LFIAs). A number of NMs was described as antibodies' nano-labels for the visual rapid detection of mycotoxins [29], such as AuNPs [30], graphene oxide (GO) [31], Prussian blue nanoparticles (PBNPs) [32], etc. In such devices, the color of test lines is usually drawn by the labeled antibodies involved in specific immunoreactions (cf. Section 3.3). Typical mycotoxins' LFIAs use AuNPs as convenient nano-labels. Interestingly, Kong et al. [33] described a semi-quantitative and quantitative AuNPs-based LFIA for the simultaneous detection of 20 types of mycotoxins from five classes—including zearalenones, deoxynivalenols, T-2 toxins, aflatoxins, and fumonisins—in cereal food samples (Figure 2b). The whole detection process took 20 min in total and was used for the reliable detection of mycotoxins in cereal samples. The LOD of three mycotoxins (AFB1, ZEN, and OTA) were far below the European maximum residue limits.

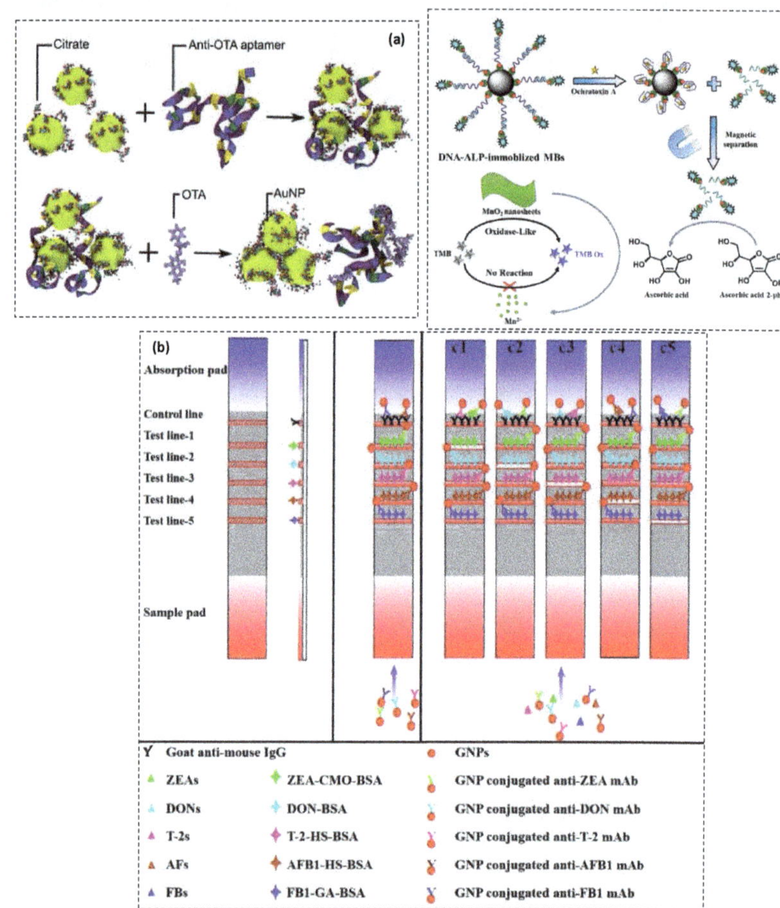

Figure 2. (**a**) Simulation of the molecular interactions involved in the aggregation of citrate-capped AuNPs for the rapid aptasensing of OTA; (**b**) A gold nanoparticle-based semi-quantitative and quantitative LFIA for the simultaneous detection of 20 mycotoxins; (**c**) Mechanism of MnO_2 nanozyme-based cascade colorimetric aptasensor for OTA detection. Reproduced with permission from [28,33,34].

Nanozymes are unique nanomaterials that have been proven to show catalytic activities in a similar way to biological enzymes with greater stability. This particular feature enables the enhancement of enzymatic response or the development of enzyme-free colorimetric methods. Accordingly, some nanozymes owing peroxidase-like and oxidase-like activities have been used as colorimetric probes for mycotoxins analysis [35]. They are widely used in different formats to afford rapid colorimetric observation, sensitive response, and cost-effective analysis.

For example, Tian et al. [34] developed a sensitive OTA aptasensor harnessing the oxidase-mimicking activity of MnO_2 nanosheets to catalyze the TMB oxidation (Figure 2c). In this assay, ascorbic acid generated under ALP action reduces MnO_2 nanosheets to Mn^{2+} ions, and thereby inhibits the catalytic activity of MnO_2 in the presence of TMB. With the increasing amount of OTA, a highly sensitive color change from blue to colorless was obtained. This sensing method enabled competitive LOD (0.07 nM) compared to conventional single enzymatic colorimetric schemes. While in the colorimetric sensing system based on peroxidase-like nanomaterials, the detection of targets is performed through measuring the absorption variation of the TMB-H_2O_2 reaction. According to this scheme, analysis of OTA has been demonstrated using a hybrid recognition matrix composed of Fe_3O_4 doped with AuNPs, amino-modified capture DNA and anti-OTA aptamer deposited on glass beads (GB-aptamer/cDNA-Au@Fe_3O_4) [36]. The peroxidase-like activity of Au@Fe_3O_4 NPs was effectively enhanced due to the synergistic effect between the AuNPs and Fe_3O_4 NPs. Low detection limit of 30 pg mL^{-1} OTA was achieved with a linear current response range of 0.05–200 ng·mL^{-1}. Selectivity has been proven in the presence of OTB, FB1, and AFB1. Sensor performance for the determination of OTA from real samples has been demonstrated with peanut and corn samples.

It is worth noting that other detection strategies can implicate NMs as signal mediators in combination with other main colorimetric probes for sensitivity enhancement. As an example, MnO_2 nanosheets can be used in new colorimetric methods based on AuNPs aggregation schemes since MnO_2 nanoflakes can produce abundance metal ions Mn^{2+} after decomposition [37], or combined with enzymes to react with catalysis products [34].

Representative examples of NMs-based assays from the recent literature are summarized in Table 2.

Table 2. Representative examples of recent developed nanomaterial-based methods for the colorimetric detection of mycotoxins.

Strategy	Detection Probe	Target	LOD	Linear Range	Specificity	Sample	Advantages/Disadvantages	Ref.
Aptamer assay based on AuNPs aggregation by poly diallyldimethyl ammonium chloride polymer (PDDA).	AuNPs	OTA	$0.009\ \mathrm{ng \cdot mL^{-1}}$	0.05–$50\ \mathrm{ng \cdot mL^{-1}}$	High	Chinese liquor sample	Cost-effectiveness, few steps, rapid detection (15 min), good sensitivity/Possible cross-reactivity at high target concentrations.	[38]
AuNP dimer disassembly by the target-induced release of complementary DNA probes.	AuNPs	OTA	$0.05\ \mathrm{nM}$	0.2–$250\ \mathrm{nM}$	High	Red wine	Improved sensitivity, low-cost, short detection time (15 min)/Difficult applicability to colored complex samples	[39]
Double calibration curve of label free aptasensing assay based on salt-induced aggregation of AuNPs.	AuNPs	OTA	$0.03\ \mathrm{ng \cdot mL^{-1}}$	0.03–$316\ \mathrm{ng \cdot mL^{-1}}$	Good	Corn	Widened detection range, enhanced sensitivity, reliability, rapid detection/low selectivity	[40]
Peroxidase-like activity of AuNPs in the presence of H_2O_2 and TMS substrate.	AuNPs	ZEN	$10\ \mathrm{ng \cdot mL^{-1}}$	10–$250\ \mathrm{ng \cdot mL^{-1}}$	High	Corn and corn oil	Simple one-step assay, short detection time/Relatively high detection limit	[41]
Chemical nano-sensor based on cysteamine-modified AuNPs aggregation via electrostatic interaction with hydrolyzed target.	AuNPs	FB1	$0.90\ \mathrm{\mu g \cdot kg^{-1}}$	2–$8\ \mathrm{\mu g \cdot kg^{-1}}$	Low	Corn	Simple one-step assay, Rapid homogenous test (3 min)/Real sample interferences, low sensitivity and selectivity.	[42]
ALP- induced gold nanoparticle aggregation mediated by MnO_2 nanosheets reduction in the presence of generated ascorbic acid.	AuNPs, MnO_2 nanosheets	OTA	$5.0\ \mathrm{nM}$	6.25–$750\ \mathrm{nM}$	High	Grape juice & red wine	Enzymatic amplification, high selectivity/multi reaction steps, possible cross reactivity in real samples	[37]
Colorimetric aflatoxins immunoassay by using mesoporous silica nanoparticles decorated with gold nanoparticles.	AuNPs@m-SiNPs nanocomposite	AFs (AFB1, AFB2, and AFG2)	$0.16\ \mathrm{ng \cdot mL^{-1}}$	1–$75\ \mathrm{ng \cdot mL^{-1}}$ for AFB1	High	Nuts, cornflakes, cornmeal, peanuts, peanut butter and pecan nuts	High sensitivity, versatile real matrix applicability, 30 min incubation time/long synthesis and modification of transducer	[43]
AFB1 hydrolyzed to phenolate anions react with curcumin enol form-Zn red complex to give curcumin enol form-ZnO-Phenol yellow complex.	ZnO NPs	AFB1	$11\ \mathrm{\mu g \cdot kg^{-1}}$	0–$36\ \mathrm{\mu g \cdot kg^{-1}}$	Good	Rice	Simple and rapid detection, bioreceptor-free sensor, HPLC validation/Chemical modification of target	[44]
Cascade aptasensor by double catalytic amplifications using ALP activity combined to the inhibition of the MnO_2 oxidase-mimicking activity.	MnO_2 nanosheets	OTA	$0.07\ \mathrm{nM}$	1.25–$250\ \mathrm{nM}$	Excellent	Grape juice	Amplified colorimetric signal, high sensitivity and selectivity/Many washing and addition steps	[34]
Simultaneous dual target detection via the combination of two the catalysis of TMB under acidic conditions and the release of TP under alkaline conditions.	Fe_3O_4-GO nanocomposite and AuNPs	AFB1	$1.5\ \mathrm{ng \cdot mL^{-1}}$	5–$250\ \mathrm{ng \cdot mL^{-1}}$	High	Peanuts	Multiplexed detection, high sensitivity and selectivity/Tedious probes synthesis, specific pH and temperature conditions, long incubation time (90 min), multiple steps of washing and separation	[45]
		OTA	$0.15\ \mathrm{ng \cdot mL^{-1}}$	0.5–$80\ \mathrm{ng \cdot mL^{-1}}$				
Salt-induced coagulation of iron-modified 2D covalent triazine framework nanosheets (2D Fe-CTFs) that showed strong peroxidase-like activity	2D Fe-CTFs	OTA	NR	0.2–$0.8\ \mathrm{\mu M}$	NR	NR	Promising proof of concept/limited detection range, analytical performances not reported	[46]
PAD sensor array based on silver and gold nanoparticles aggregation synthesized by three different capping agents.	AgNPs and AuNPs	AFB1	$2.7\ \mathrm{ng \cdot mL^{-1}}$	$3.1\ \mathrm{ng \cdot mL^{-1}}$ –$7.8\ \mathrm{\mu g \cdot mL^{-1}}$	Excellent	Mixtures of pistachio, wheat and coffee, milk	Very fast colorimetric response (5 s), multiplexed detection of five mycotoxins, low cost, device portability/Optical nanoprobes fabrication	[47]
		AFG1	$7.3\ \mathrm{ng \cdot mL^{-1}}$	$8.2\ \mathrm{ng \cdot mL^{-1}}$–$8.4\ \mathrm{\mu g \cdot mL^{-1}}$				
		AFM1	$2.1\ \mathrm{ng \cdot mL^{-1}}$	$2.5\ \mathrm{ng \cdot mL^{-1}}$–$8.2\ \mathrm{\mu g \cdot mL^{-1}}$				
		OTA	$3.3\ \mathrm{ng \cdot mL^{-1}}$	$4.0\ \mathrm{ng \cdot mL^{-1}}$–$3.8\ \mathrm{\mu g \cdot mL^{-1}}$				
		ZEN	$7.0\ \mathrm{ng \cdot mL^{-1}}$	$8.0\ \mathrm{ng \cdot mL^{-1}}$–$7.9\ \mathrm{\mu g \cdot mL^{-1}}$				

3. Colorimetric Strategies for Mycotoxins Detection

Colorimetric assays in mycotoxins detection have attracted much attention due to their simple sensing mode where colorimetric signal can be detected by the naked eye and without the need to complicated and expensive instruments. They are developed in two modes including solution-based and flat substrate-based assays. Solution-based assays involve free colloidal reagents in the same homogenous phase of targets. Such mycotoxin detection strategies can be performed using organic dyes, colored enzymatic products, or nanomaterial probes. On the other hand, three kinds of colorimetric flat substrate-based assays are more common in mycotoxin detection. They include enzyme-linked immunosorbent assay (ELISA), lateral flow assays (LFAs), and microfluidic-based assays.

3.1. Solution-Based Assays

In-solution assays rely on the colloidal interaction between different biomolecules without their immobilization on a substrate. The colorimetric signal is usually generated after cascade additions of reagents in a total volume of some hundreds of microliters where the colorimetric probe, the bioreceptor, and the target can meet. Numerous reports of homogenous solution-based assays for mycotoxins detection have been developed owing to their rapid operation and facile design. Most of these patterns are based on target-induced/disabled nanoparticles aggregation, enzymes or enzyme-like catalytic activities, or chemical dyes in label-based or label-free formats.

As described earlier, noble nanoparticles are characterized by an intrinsic size- and distance-dependent optical signal. Particularly, AuNPs showed a great success in the design of solution-based colorimetric assays using different aggregation approaches.

As an example, a simple colorimetric assay has been described by Chotchuang et al. for the detection of fumonisin B1 using dispersed cysteamine-functionalized gold nanoparticles (Cys-AuNPs) [42]. The target mycotoxin was first hydrolyzed (HFB1) to then induce NPs aggregation via hydrogen bonding. At an optimal pH of 9, color change from wine-red to blue-gray and absorption spectra from 520 nm to 645 nm can be either observed visually or measured by a UV-vis spectrophotometer. This 3 min sensing approach achieved satisfactory results between 2–8 μg kg^{-1} FB1 concentration range and a detection limit of 0.90 μg kg^{-1}. Although this method was successfully applied to corn samples, its specificity could be decreased in the presence of interfering molecules that are able to aggregate AuNPs in the absence of target. Therefore, the use of specific bioreceptors such as aptamers is more common in AuNPs-based detection of mycotoxins.

Nucleic acid strands are known to protect AuNPs against salt-induced aggregation because of strong van der Waals interactions between DNA bases and gold [48]. This electrostatic affinity induces aptamer's adsorption which stabilizes the dispersed nanoparticles. Upon mycotoxin recognition, aptamers desorb from the surface of AuNPs to preferentially complex with the target. Subsequently, stable gold aggregates are formed under the action of electrolytes or cationic polymers leading to the solution color changing. According to this aptasensing strategy, label-free AuNPs-based aptasensors were frequently reported for the rapid detection of mycotoxins [49], including ochratoxin A [38,40,50], aflatoxin B1 [27,51], and zearalenone [52].

Interestingly, Liu and collaborators have found that aromatic targets sch as ochratoxin A can also adsorb on the surface of AuNPs after aptamer folding and further inhibit salt-induced aggregation [40]. This limitation renders the assay unreliable at high target concentrations. To expand the detection range, they described a double calibration curve method in which both aggregation mechanisms are combined using two experimental conditions (Figure 3a). Using this system, they claimed that the OTA concentration range could be widened from $10^{-10.5}$–10^{-8} to $10^{-10.5}$–$10^{-6.5}$ g·mL^{-1}.

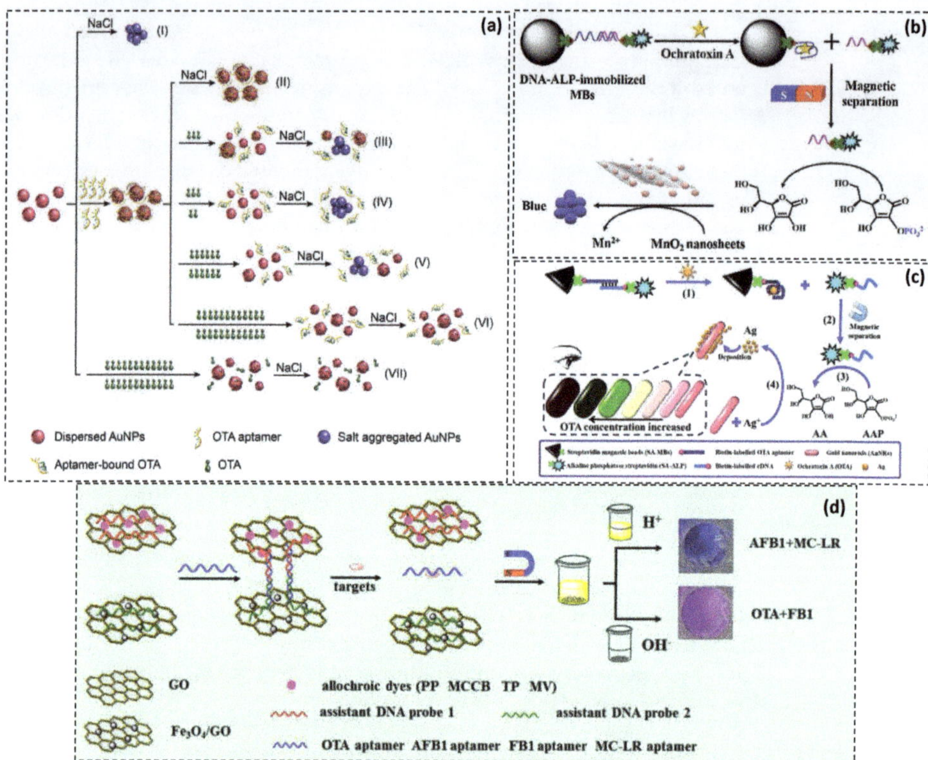

Figure 3. (**a**) Gold nanoparticle-aptamer-based LSPR sensing of ochratoxin a at a widened detection range by double calibration curve method; (**b**) Colorimetric aptasensor for the ochratoxin A (OTA) assay based on the structure-switching of OTA aptamer coupling with alkaline phosphatase (ALP)-MnO$_2$ cascade catalytic reaction; (**c**) Multicolor colorimetric detection of OTA via structure-switching aptamer and enzyme-induced metallization of gold nanorods; (**d**) pH-Resolved simultaneous detection of four targets based on magnetic separation of two GO platforms with allochroic dyes. Reproduced with permission from [37,40,53,54].

Although the achieved limits of detection are demonstrated to be in compliance with regulatory levels, such colorimetric assays present relatively high LOD values compared to other optical or electrochemical approaches. The lack of sensitivity was explained by the number of NPs required to generate a significant color change. Aiming to overcome this constraint, Xiao et al. [39] described a colorimetric aptasensor based on the disassembly of aggregates of oriented AuNP dimers by target molecules. This AuNPs dimer-based sensor has shown better stability, sensitivity (LOD = 0.02 µg·L^{-1}) and OTA detection dynamic range (0.08–100.8 µg·L^{-1}). Furthermore, it was noted that the disassembly of AuNPs dimers was faster than that of large aggregates reducing thus the analysis time [41].

In another option, the colorimetric signal of substrate-free assays can be amplified by catalytic reactions using either enzymes or nanozymes. Harnessing the inherent peroxidase-like activity of AuNPs, Sun et al. [41] developed a rapid apta-assay specific to zearalenone. In this assay, ZEN aptamer inhibits the catalytic activity of AuNPs in the presence of H$_2$O$_2$ and TMB. The solution remains red until the target binding to the aptamer, which restores the peroxidase-mimicking of nanozymes to oxidize the colorless TMB into blue oxTMB. Quantitative analysis was reported in the ZEN concentration range of 10–250 ng·mL^{-1}, and the limit of detection is 10 ng·mL^{-1}. The assay was applied to test ZEN in corn and corn oil samples, but high sensitivity was still challenging.

As an alternative, a combination of enzymatic action and gold nanoparticles aggregation was suggested by He et al. [37]. This colorimetric method was developed to detect OTA indirectly via the activity of alkaline phosphatase (ALP) (Figure 3b). Briefly, aptamer-modified magnetic beads (MBs) were conjugated to DNA-linked ALP by hybridization. After OTA recognition, magnetic separation allowed to collect the released quantity of enzyme. The ALP can then hydrolyze ascorbic acid 2-phosphate (AAP) to ascorbic acid (AA), which mediates the reduction of MnO_2 nanosheets to Mn^{2+}. These metal cations allow thereby the aggregation of AuNPs and lead to vivid color changes in the sensing system [37]. The dynamic range extends from 6.25 to 750 nM and an improved LOD of 2 ng·mL^{-1} was recorded. This colorimetric method was applied to grape juice and red wine matrix with satisfactory recoveries.

A comparable approach was also described by the same group while replacing MnO_2 nanosheets by gold nanorods (AuNRs) and silver ions [53]. After magnetic separation, generated AA acted as reducing agents that transform Ag^+ to metal silver forming an Ag shell on the surface of AuNRs (Figure 3c). This caused a blue-shift of the longitudinal AuNRs' LSPR and a rainbow-like multicolor change.

Multicolor detection of OTA was also reported by AuNRs etching (diameter ~ 14 nm) mediated by G-quadruplex (AG4-OTA)-hemin DNAzyme and exonuclease I [55]. The product of peroxidase-like activity in acidic solution TMB^{2+} can etch the AuNRs by oxidizing Au(0) into Au(I). Variation of the optical characteristics of AuNRs arising from the change in interparticle distance and the number of hydrogen bonds has been reported as the key sensing strategy. A linear response range of 10–200 nM OTA was found with a LOD of 30 nM by visual observation and a lower LOD of 10 nM by spectrophotometry. The selectivity towards OTA was tested with the interfering mycotoxins AFB1, ZEN, and OTB. The method was successfully applied to the determination of OTA in spiked beer samples.

Besides enzymes and nanomaterials, commercially available organic dyes have also been used to conceive solution-based colorimetric methods. Interestingly, Wang's research group developed some multiplexed assays for the real-time detection of different mycotoxins based on allochroic dyes [45,54]. For instance, Hao et al. proposed a pH-resolved colorimetric aptasensing method for the simultaneous detection of four targets, including three mycotoxins, ochratoxin A, aflatoxins B1, fumonisin B1, and a marine toxin, microcystin-LR [54]. This assay involves four allochroic dyes—namely, phenolphthalein (PP), malachite green carbinol base (MGCB), thymolphthalein (TP), and methyl violet (MV)—as multiple signal indicators with colors of different wavelengths. Two DNA-GO platforms were prepared; the first was modified with Fe_3O_4 for magnetic separation while the second adsorbed the hydrophobic dyes (Figure 3d). Both platforms were linked by partial hybridization to a target-specific aptamer. Upon target recognition, aptamer structure switching disabled hybridization and dissociated GO platforms. The subsequent magnetic separation followed by centrifugation allowed the spectroscopic analysis of supernatant in acidic solution and precipitate in alkaline solution. The absorption of supernatant solutions was directly proportional to AFB1 and MC-LR concentrations because of the MGCB and MV release at pH 3. Whereas the absorption of precipitates containing PP and TP adsorbed dyes was inversely proportional to OTA and FB1 mycotoxins in an alkaline pH of 12. This approach enabled the simultaneous detection of OTA and AFB1 in peanut samples with satisfactory recoveries (97.8–104.3%).

More recently, a derived nanocomposite-based strategy was described by Zhu et al. [45] employing TP dye signaling in acidic conditions for AFB1 detection and AuNPs as nanozymes to detect OTA via TMB catalysis in the alkaline precipitate. Competitive limits of detection as low as 1.5 ng·mL^{-1} and 0.15 ng·mL^{-1} were thus obtained for AFB1 and OTA, respectively.

3.2. Enzyme-Linked Immunosorbent Assay (ELISA)

According to the literatures, ELISA is the most popular and most frequently used technique for mycotoxin analysis especially aflatoxins [56]. Among different types of ELISA, the direct competitive ELISA is commonly used in mycotoxin detection. In recent years, only a few studies have been focused on mycotoxin detection using ELISA method. However, there are a large number of commercial ELISA kits produced by different companies worldwide.

Traditional ELISA uses antibody as recognition element and HRP-catalyzed TMB to generate color as a signal reporter. Although ELISA has been recognized as an excellent and accurate method for mycotoxin analysis, but the procedure is somewhat time-consuming (incubation time of approximately 1–2 h), uneconomical, unsuitable for field testing due to the need for specialist plate readers, and unreliable due to the similarity of the structure of mycotoxins, which causes false positive results [57]. Therefore, many efforts have been made to improve the shortcomings mentioned. One attempt is to improve the colorimetric signal. Conventional colorimetric signal using HRP and TMB is not suitable for naked-eye detection in deprived areas with limited resources because a plate reader is required to distinguish the tonality of analytes with similar concentrations. Recently, colorimetric ELISA has gained considerable attention due to its simple readout without specialist devices. Acid–base indicators are ideal signal reporters for naked-eye distinction because most of them provide a significant contrasting color at their titration end points under a narrow pH range. Several enzymes including alkaline phosphatase, urease and penicillinase have been used in ELISA to change the pH through catalyzing the related specific substrate to produce hydrogen or hydroxide ions [58]. In this regard, Xinog et al. developed a direct competitive colorimetric ELISA using glucose oxidase (GOx) as an alternative to HRP for glucose oxidation into gluconic acid and H_2O_2 (Figure 4a) [58]. The pH indicator bromocresol purple (BCP), which was highly sensitive to pH variation, was applied as signal output. BCP indicator showed a vivid color change from yellow to grayish purple in the presence of 100 pg·mL^{-1} AFB1. Therefore, the cutoff limit was determined to be 100 pg·mL^{-1} by the naked eye. The developed GOx-based colorimetric ELISA exhibited high sensitivity and excellent selectivity with IC50 value at 66.27 pg·mL^{-1}, which was approximately 10-fold lower than that of traditional HRP-based ELISA. However, long incubation time and multi-step washing were still the major limitations of the ELISA method. The proposed assay was applied for AFB1 determination in corn samples with acceptable accuracy and precision.

Among colorimetric ELISA methods, plasmonic ELISA is another attempt with simple readout format suitable for on-site detection. Gold nanoparticles (AuNPs) are good candidate as colorimetric indicator in plasmonic ELISA due to high molar extinction coefficient and localized surface plasmon resonance (LSPR) characteristics. The LSPR of AuNPs is related to their size, shape, composition, and agglomerate mood [59]. The LSPR variation of AuNPs generates a significant color change that is easily observable by the naked eyes. Based on differences in producing LSPR mechanism, plasmonic ELISA is classified into four types that employ the aggregation, etching, controlled growth kinetics, and AuNPs metallization. Among them, enzyme-induced silver metallization on the AuNPs surface can produce a remarkable LSPR, and provide a multicolor change in the solution [60]. Several enzymes GOx, alkaline phosphatase and β-galactosidase have been used to catalyze their substrates and produce reducing agents such as H_2O_2, ascorbic acid, and p-aminophenol which can reduce the silver ions on the AuNPs surface. In this regard, Pei et al. developed a colorimetric plasmonic ELISA for OTA detection based on the urease-induced metallization of gold nanoflowers (AuNFs) [60]. OTA-labeled urease was employed as competing antigen to hydrolyze urea into ammonia. In the presence of ammonia, silver ions were reduced by the formyl group from glucose to produce a silver shell around AuNFs resulted in the solution color change from blue to brownish red (Figure 4b). The plasmonic ELISA exhibited high sensitivity with a cutoff limit of 40 pg·mL^{-1} and LOD at 8.205 pg·mL^{-1} (19-folds lower than those of HRP-based ELISA). The proposed procedure provided a highly selective and sensitive, simple, robust, and high-throughput screening method for the quantitative determination of OTA in food and feed samples. However, it suffered from long incubation time.

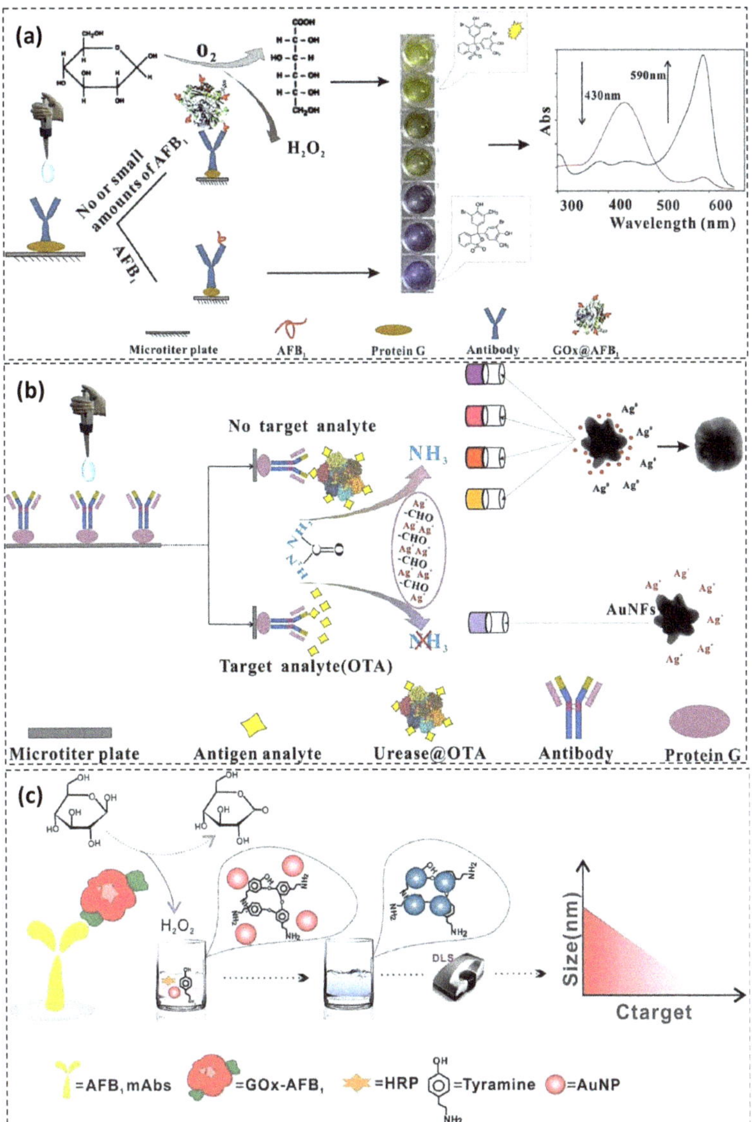

Figure 4. (**a**) Colorimetric GOx-based ELISA using acid-base indicator bromocresol purple (BCP) for AFB1 detection; (**b**) Plasmonic ELISA based on the urease-induced metallization of gold nanoflowers for OTA detection; (**c**) DLS-ELISA method associated with H_2O_2-mediated tyramine signal amplification system for AFB1 detection. Reproduced from [58,60,61], respectively, with permission.

In another plasmonic ELISA, aggregation-induced color change of AuNPs, as a main strategy to regulate the plasmonic signal, was employed for the ultrasensitive detection of AFB1 using dynamic light scattering (DLS) signal instead of absorbance (Figure 4c) [61]. In the developed DLS-ELISA, GOx-AFB1 was used as competing antigen because the GOx can effectively convert glucose to H_2O_2. Then, the produced H_2O_2 converted into hydroxyl radical in the presence of HRP to induce AuNPs aggregation. Indeed, H_2O_2-mediated TYR was used as signal amplification system. The DLS-ELISA exhibited a LOD as low as 0.12 pg·mL^{-1} which was about 153- and 385-folds lower than those of conventional plasmonic and colorimetric ELISA, respectively. The ultrahigh sensitivity is attributed to the high sensitivity of light-scattering intensity to particle size changes. The DLS-ELISA was employed for AFB1 detection in corn samples with good reliability and precision.

Mukherjee et al. compared aptamer-based enzyme linked apta-sorbent assay (ELASA) with antibody-based ELISA and its potential to replace antibodies in usual immunoassay formats either as capture probe or detection probe without affecting the sensitivity [62]. The ELASA was based on the principle of target capture by aptamer where, OTA specific aptamer was used for toxin detection. Then, anti-OTA IgG primary antibody and anti-rabbit secondary antibody labeled with alkaline phosphatase (ALP) where added as detection agents. The colorimetric signal was produced under addition of para-nitrophenyl phosphate (p-NPP) as substrate. The LOD was obtained 0.84 pg·mL^{-1}. The developed ELASA exhibited a similar sensitivity to the conventional antibody-based ELISA with a LOD of 1.13 pg mL^{-1}. However, the OTA aptamer showed about 40% cross-reactivity with aflatoxins. By selecting aptamer with a low percentage of cross-reactivity, ELASA can be a good alternative to the conventional ELISA. The proposed ELASA was used for OTA detection in groundnut and coffee bean.

Another innovation in improving the ELISA characteristics is to replace nanomaterial-based enzyme mimics (nanozymes) as artificial enzymes with natural enzymes. Nanozymes exhibit excellent properties such as easy synthesis, high stability, low cost, and design flexibility. Different kinds of nanomaterials—including noble metal nanoparticles (e.g., AuNP and AgNPs), graphene oxide, magnetic iron oxide, etc.—have been used in sensing methods. Xu et al. proposed a nanozyme-linked immunosorbent assay using metal–organic frameworks (MOFs) for AFB1 detection [63]. MOF with peroxidase-like activity was replaced with HRP for antibody labeling and catalyzing TMB to generate colorimetric signal. The MOF-ELISA system increased the accuracy of detection and inhibited false positive problems in the detection method, indicating that MOFs exhibited better catalytic activity and more stability than HRP. The LOD was obtained 0.009 ng·mL^{-1} which was 20-folds lower than those of HRP-based ELISA. The proposed ELISA was employed for AFB1 detection in peanut milk and soymilk.

Representative examples of recent developed ELISA methods for the detection of mycotoxins are reported in Table 3.

Table 3. Representative examples of recent developed ELISA methods for the detection of mycotoxins.

Strategy	Target	LOD	Linear Range	Specificity	Sample	Advantages/Disadvantages	Ref.
Colorimetric ELISA based on glucose oxidase-regulated the color of bromocresol purple acid-base indicator	AFB1	Cutoff limit: 100 pg·mL^{-1}	25–200 pg·mL^{-1}	Excellent	Corn	High sensitivity and selectivity, good repeatability/long incubation time, multi-step washing	[58]
Plasmonic ELISA based on the urease-induced metallization of gold nanoflowers	OTA	8.205 pg·mL^{-1}	5.0–640 pg·mL^{-1}	Excellent	Rice, corn, wheat, white wine	High sensitivity and selectivity, robust, and high-throughput, good repeatability/long incubation time, multi-step washing	[60]
DLS-ELISA based on AuNPs aggregation via hydroxyl radicals produced by HRP activity on H$_2$O$_2$	AFB1	0.12 pg·mL^{-1}	Not reported	Excellent	Corn	High sensitivity and selectivity, reliable, good repeatability, high accuracy/long incubation time, multi-step washing	[61]
Apta-ELISA based on target capture by OTA specific aptamer and color development by anti-rabbit secondary antibody labeled with ALP	OTA	0.84 pg·mL^{-1}	1 pg·mL^{-1}–1 μg·mL^{-1}	Good	Groundnut, coffee bean	High sensitivity, low-cost/long incubation time, multi-step washing, cross-reactivity, matrix interference	[62]
A nanozyme-linked immunosorbent assay based on metal–organic frameworks (MOFs) instead of enzyme to catalyze chromogenic TMB	AFB1	0.009 ng·mL^{-1}	0.01–20 ng·mL^{-1}	Excellent	Peanut milk, soymilk	High sensitivity and selectivity, high catalytic activity and stability of MOF, good recovery rate and accuracy, avoiding false positive and false negative results/long incubation time, multi-step washing	[63]
Multiplexed ELISA based on immobilization of protein-analyte conjugates in separate wells	OTA AFB1 ZEN	4.0 ng·mL^{-1} 0.1 ng·mL^{-1} 0.3 ng·mL^{-1}	4.0–120 ng·mL^{-1} 0.1–1.0 ng·mL^{-1} 0.3–50.0 ng·mL^{-1}	Not reported	Poultry, corn	High sensitivity, reduction of simultaneous detection time of three mycotoxins/long incubation time, multi-step washing	[64]
Multiplex nanoarray based on ELISA technique via nano-spotting of mycotoxin-protein conjugates into single wells of a microplate	ZEN T-2 toxin FB1	IC50: 197.4 0.7 216.7 μg·kg^{-1} in wheat	Not reported	High	Wheat, corn	High sensitivity and selectivity, reduction of simultaneous detection time of three mycotoxins, throughput, easily adaptable by end users/long incubation time, multi-step washing	[65]
Plasmonic ELISA based on the HRO-assisted etching of gold nanorods	AFB1	Cutoff limit: 12.5 pg·mL^{-1}	3.1–150 pg·mL^{-1}	Excellent	Corn	High sensitivity and selectivity, high accuracy and precision, portable, equipment-free/long incubation time, multi-step washing	[66]

3.3. Lateral Flow Assays

Lateral-flow assays (LFAs), also known as immunochromatographic assays (ICAs), are among the most widely used and popular methods in detecting various analytes such as microorganisms, pesticides, heavy metals, diseases biomarkers, and mycotoxins. In recent years, researchers have paid more attention to screening mycotoxins by LFA. LFA is based on the movement of fluid sample across the membrane by capillary force and binding reaction between antibody-antigen or nucleic acid-target analyte [67]. The standard LFA strip is comprised of four parts including a sample pad (the area where the sample is dropped); a conjugate pad (the area where biorecognition element conjugated with label is immobilized); a reaction nitrocellulose membrane (the area containing test line and control line for target binding to antibody or nucleic acid probe); and absorbent pad (as a wick to reserve additional fluid flow) [68,69]. Sandwich mode (for large analytes) and competition mode (for small analytes) are the two most widely used detection formats. Competitive mode is suitable for mycotoxins with low molecular weight and single epitope.

The optimization of the experimental conditions is crucial to develop a LFA with excellent performance and high sensitivity. High sensitivity, low immunoreagent consumption, and ideal color intensity are major parameters for the construction of LFAs. Utilization of an appropriate label is important for a sensitive analysis. Different colored labels such as colored latex beads, AuNPs, magnetic particles (MPs), carbon nanostructure, and enzymes have been used for developing LFA. In addition to sensitivity, label should not change the features of biorecognition element, and it must create stable conjugation with recognition element. AuNPs have been frequently used colorimetric labels in developing LFA strip due to having all the mentioned features [68–70]. Di Nardo et al. developed a novel LFA using dual color AuNPs and a single Test line for simultaneous determination of AFB1 and type-B fumonisins (FMBs) [71]. In this assay, red (spherical, mean diameter $\approx$ 30 nm) and blue (desert rose-like, mean diameter $\approx$ 75 nm) AuNPs were conjugated to anti-aflatoxin and anti-fumonisins antibodies, respectively. The single test line was formed by spraying the mixture of two antigens including AFB1-BSA and FMB-BSA. According to the competitive format, mycotoxin-free samples provided a purple test line due to the combination of the red and blue AuNPs. Contaminated samples with AFB1 or FMBs resulted in the blue and red color Test line, respectively. The simultaneous presence of both mycotoxins provided the usual disappearance of the Test line. (Semi-) quantitative analysis was obtained using a simple smartphone and RGB colorimetric analysis. The use of a single strip to multiplex analysis provided a simple, rapid, low-cost and reagent-saving assay. The developed strips with LOD at 0.5 and 20 ng·mL^{-1} for AFB1 and FMB1, respectively, were employed to determine these two mycotoxins in wheat and pasta samples.

Conventional AuNPs-based LFA suffers from a major challenge in measuring target concentration in complex food matrices with dark color due to its poor resistance to the background matrix and color interference. To address this issue, Hao et al. developed a novel LFA using bifunctional magneto-gold nanohybrid (MGNH) label as a hetero-structured nanomaterial for the simultaneous magnetic separation and colorimetric detection of OTA in grape juice [72]. In this assay, MGNH-labeled monoclonal antibodies (mAb) were used for the MGNH-mAb-OTA complex formation and subsequently rapid separation of the complex from sample using an external magnetic field. Then, MGNH-mAb-OTA complex was resuspended in buffer and applied on LFA strip for colorimetric detection (Figure 5a). Grape juice with purple color and high concentrations of sugar, pigment, and tannins was used as complex matrix to evaluate the designed method. The novel LFA was highly sensitive with LOD at 0.094 ng·mL^{-1}. The assay showed high accuracy, reproducibility, practicability, and short detection time (10 min of magnetic separation and 5 min of immunoreaction).

Most of the multiplex LFAs for mycotoxins analysis have been designed for detection of only two or three kinds of mycotoxins [73], while sometimes more than this occurs in some foods such as cereals. On the other hand, quantitative analysis is a main issue in LFA technology which is often carried out by desktop readers or handheld readers. These devices are slightly inferior in terms of popularity, portability, and timely data sharing compared to smartphone-based analysis [74]. Therefore, these existing limitations must be overcome to receive a practical LFA for multiplex and on-site detection. For this purpose, Liu et al. developed two kinds of multiplexed LFA strips using AuNPs and time-resolved fluorescence microspheres (TRFMs) as label for the detection of AFB1, zearalenone (ZEN), deoxynivalenol (DON), T-2 toxin (T-2), and fumonisin B1 (FMB1) in cereals (Figure 5b) [30]. Five test lines were sprayed on a single test strip for each mode of detection. Quantitative results were obtained using a smartphone dual detection mode device. The visual LODs of AuNPs-LFA were 10, 2.5, 1, 10, and 0.5 ng·mL^{-1} for AFB1, ZEN, DON, T-2 and FMB1, respectively. In the TRFMs-LFA format, LODs were 2.5, 0.5, 0.5, 2.5, and 0.5 ng·mL^{-1}, respectively for the mentioned mycotoxins. Quantitative LODs (qLODs) for these myco-toxins were obtained 0.59, 0.24, 0.32, 0.90, and 0.27 ng·mL^{-1} (in AuNPs-LFA), and 0.42, 0.10, 0.05, 0.75, and 0.04 ng·mL^{-1} (in TRFMs-LFA). TRFMs-LFA was more sensitive than AuNPs-LFA due to large surface area and stokes shift of TRFMs. On the other hand, AuNPs was low-cost, more popular, stable and easy to synthesize. The assay was reliable, quantitative and highly sensitive for on-site detection of multiple mycotoxins. However, a main problem of a multiplex LFA is the cross reactivity between Ag-Ab pairs, so that the developed LFA was able to detect 20 mycotoxins from five classes.

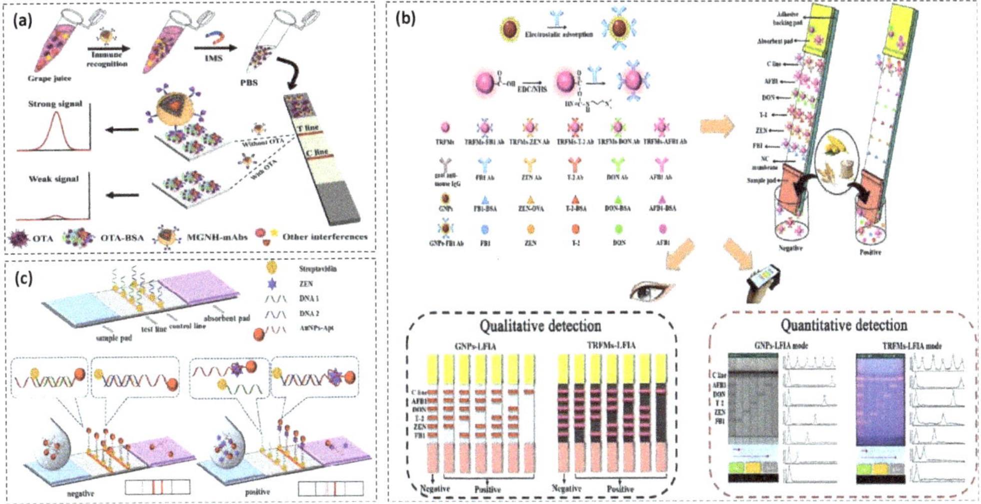

Figure 5. (**a**) Schematic illustration of the detection principle of the MGNH-based LFA strip; (**b**) smartphone-based AuNPs and TRFMs-LFAs for multiplex mycotoxins detection; (**c**) aptamer-based LFA for ZEN detection in the presence and absence of ZEN analyte. Reproduced from [30,72,75], respectively, with permission.

In addition to AuNPs, colloidal carbon can be used as a colored label in LFAs. It is comparatively inexpensive and can be synthesized in a large scale. Furthermore, it shows high chemical stability and recognizable color to develop LFA with high sensitivity. Many colloidal carbon-based-LFA have been developed for detection of different analytes. For example, Yu et al. proposed a LFA using Graphene oxide (GO) and carboxylated GO as labels for AFB1 detection [31]. GO can be easily conjugated with biomolecules without any additional activation due to having a large variety of oxygen-containing chemical groups. Moreover, it shows excellent hydrophilicity and high stability at room temperature. In this study, mA against AFB1 was conjugated with GO. The vLOD and cut-off values for AFB1 were 0.3 and 1 ng·mL^{-1}, respectively. It was exhibited that GO and carboxylated GO can be used as viable black labels to develop a low-cost LFA compared to the AuNPs labels. The method was successfully applied for AFB1 detection in peanut oil, maize, and rice.

In recent years, many efforts have been made to replace aptamer with antibody in LFA technology due to potential advantages of aptamers (as mentioned earlier). However, aptamer-based assays mostly require laboratory infrastructure, which limits their application. Incorporating the advantages of aptamers and LFA technology is a basic step to complete the user-friendliness of aptamers. In this context, Wu et al. developed a competitive aptamer-based LFA for rapid and sensitive detection of ZEN [75]. The assay was based on competition between the DNA 1 on the test line and ZEN in the for binding to AuNPs-labeled aptamer. In the absence of ZEN, AuNPs-labeled aptamer hybridized with DNA 1 on the test line and DNA 2 on the control line, resulting in two colored lines on the strip while, in the presence of high concentration of ZEN, the test line was colorless (Figure 5c). The proposed aptamer-based LFA with high specificity and sensitivity (vLOD and qLOD of 20 and 5 ng·mL^{-1}, respectively) and short detection time (5 min) was applied for ZEN detection in spiked corn samples.

Representative examples of recent developed LFA test strips for the detection of mycotoxins are reported in Table 4.

Table 4. Representative examples of recent developed LFA strip tests for the detection of mycotoxins.

Strategy	Target	LOD	Linear Range	Specificity	Sample	Advantages/Disadvantages	Ref.
Dual color AuNPs and a single test line for simultaneous detection of two mycotoxins	AFB1 FBs	$0.5\ \text{ng·mL}^{-1}$ $20\ \text{ng·mL}^{-1}$	Not reported	Not reported	Wheat, pasta	High sensitivity, simple, low-cost, rapid (detection time of 10 min), semi-quantitative, reliable/no evaluation of selectivity and stability	[71]
Magneto-gold nanohybrid based LFA for simultaneous separation and target detection	OTA	$0.094\ \text{ng·mL}^{-1}$	$0.098–12.5\ \text{ng·mL}^{-1}$	High	Grape juice	High sensitivity and selectivity, simple, low-cost, rapid (detection time of 15 min), semi-quantitative, high precision in complex matrices, high reproducibility/no evaluation of stability	[72]
AuNPs and TRFMs-based lateral flow immunoassays for multiplex detection mycotoxins along with a smartphone-based quantitative dual detection mode device	AFB1 ZEN Deoxynivalenol T-2 toxin FB1	AuNPs-LFA: 0.59 0.24 0.32 0.90 $0.27\ \text{ng·mL}^{-1}$ TRFMs-LFA: 0.42 0.10 0.05 0.75 $0.04\ \text{ng·mL}^{-1}$	Not reported	Low	Maize, wheat, bran	High sensitivity, simple, low-cost, rapid, reliable, quantitative, portable/no evaluation of stability, high cross reactivity with other mycotoxin in a single class	[30]
Application of GO and carboxylated GO instead of AuNPs as label	AFB1	$0.3\ \text{ng·mL}^{-1}$	Not reported	Not reported	Peanut oil, maize, rice	High sensitivity, high stability (4 months), simple, low-cost, rapid, reliable, rapid (detection time of 15 min), acceptable reproducibility/qualitative, no evaluation of specificity	[31]
Aptamer-based competitive LFA	ZEN	vLOD: $20\ \text{ng·mL}^{-1}$ qLOD: $5\ \text{ng·mL}^{-1}$	$5–200\ \text{ng·mL}^{-1}$	High	Corn	High sensitivity and selectivity, high stability (2 months at room temperature), simple, low-cost, short detection time (5 min), portable/no evaluation of reproducibility	[75]
AuNPs-aptamer conjugate as recognition element and label; immobilization of biotinylated DNA probe 1 and probe 2 on test and control lines, respectively	OTA	$1\ \text{ng·mL}^{-1}$	Not reported	Excellent	*Astragalus membranaceus*	High sensitivity and selectivity, cost-effective, robust, high stability (6 months at room temperature), simple, short detection time (15 min), portable/no evaluation of reproducibility, qualitative	[76]
Design a smart analysis platform for multiplex LFIA based on AuNPs as label and five test lines	AFB1 ZEN Deoxynivalenol T-2 toxin FB1	4 40 200 10 $20\ \mu\text{g·kg}^{-1}$	Not reported	Low	Wheat	High sensitivity, cost-effective, robust, simple, short detection time (15 min), portable, quantitative/no evaluation of reproducibility and stability, high cross reactivity with other mycotoxin in a single class	[77]
Enhanced signal sensitivity in LFIA using multi-branched gold nanoflowers	AFB1	$0.32\ \text{pg·mL}^{-1}$ in rice	$0.5–25\ \text{pg·mL}^{-1}$	Low	Rice	Excellent sensitivity, cost-effective, simple, short detection time (15 min), portable, quantitative/no evaluation of reproducibility and stability, high cross reactivity with AFG2	[78]
Multiplex LFIA based on AuNPs as label	AFB1 ZEN OTA	0.1–0.13 0.42–0.46 0.19–0.24 $\mu\text{g·kg}^{-1}$	Not reported	Not reported	Corn, rice, peanut	High sensitivity, cost-effective, simple, short detection time (15 min), portable, quantitative/no evaluation of reproducibility, selectivity and stability	[79]

Table 4. *Cont.*

Strategy	Target	LOD	Linear Range	Specificity	Sample	Advantages/Disadvantages	Ref.
AuNPs-based LFIA with silver staining for signal amplification	FB1 Deoxynivalenol	cut-off values: 2.0 40 ng·mL^{-1}	Not reported	Not reported	Maize	High sensitivity, low-cost, simple, short detection time, portable/no evaluation of reproducibility, selectivity and stability	[80]
LFIA based on multifunctional photothermal contrast Fe$_3$O$_4$@Au supraparticle (Fe$_3$O$_4$@Au SP)	OTA	0.12 pg·mL^{-1}	1 pg·mL^{-1}–1 µg·mL^{-1}	Good	Corn, peanut, soybean	High sensitivity and selectivity, low-cost, simple, reliable/no evaluation of reproducibility and stability	[81]
A LFIA using Prussian blue nanoparticle (PBNP) as a peroxidase mimicking label for TMP catalysis	OTA	10 pg·mL^{-1}	10 pg·mL^{-1}–1 µg·mL^{-1}	High	Human serum	High sensitivity and selectivity, low-cost, simple, reliable/no evaluation of reproducibility and stability	[32]

3.4. Microfluidics-Based Assays

In addition to LFA, microfluidic technology has attracted much attention in recent years to detect a variety of analytes. According to the definition provided by Whitesides from Harvard University, "microfluidic is the science and technology of systems that process or manipulate small amounts (10^{-9} to 10^{-18} L) of fluids, using channels with dimensions of tens to hundreds of micrometers". This technique shows a great potential to control the concentrations of molecules in space and time [82]. High surface-to-volume ratios, small consumption of reagents, prevalence of viscous and capillary forces and laminar flows are the major features of microfluidic-based systems [83]. Based on such properties, microfluidic can be integrated with biosensor technology in order to develop analytical devices with high sensitivity, reproducibility, portability, low-cost, short detection time, and high throughput [84]. Early microfluidic systems were fabricated of silicon and glass. Because of high cost of silicon and fragility of glass, polymer-based devices were then offered in the late 1990s which were cheaper than glass and silicon and provided an extensive range of chemical materials expanded from polydimethylsiloxane (PDMS) to thermoplastics [85,86].

Recently, microfluidic-based assays have attracted a large amount of interest in the detection of mycotoxins. The incorporation of microfluidic system and immunoassay is considered as one of the most popular platforms for detecting mycotoxins with high sensitivity and short detection time. For example, Machado et al. developed a PDMS-based microfluidic immunoassay with four chambers for simultaneous detection of OTA, AFB1, and DON [83]. The competitive immunoassay was developed by the immobilization of BSA-mycotoxin conjugates onto separate chambers (Figure 6a). The first inlet was considered for sample loading. In the presence of the given mycotoxin, the free toxin competed with the toxin-BSA immobilized on the PDMS surface for the specific binding to IgG-HRP conjugate. Therefore, a high concentration of a target free toxin resulted in a low density of IgG-HRP captured by the immobilized BSA-toxin. After addition of TMB, a colorimetric signal was observed which was inversely proportional to the mycotoxin concentration. Smartphone was used to obtain semi-quantitative results. The proposed assay exhibited LODs at <40, 0.1–0.2 and <10 ng·mL^{-1} for OTA, AFB1, and ZEN, respectively. Furthermore, the immunoassay was applied for the simultaneous detection of these three mycotoxins in corn samples after a simple sample preparation method. The multiplexed analysis with a relatively low cost and simple operation can be performed in less than 10 min. However, these methods can be simplified by reducing the number of user-intervention steps such as pipetting.

In another microfluidic device, AuNPs were used as colored labels for indicating various concentrations of alternariol monomethyl ether (AME), one of the most frequently occurred Alternaria mycotoxins [87]. Microfluidic chip was fabricated using Norland Optical Adhesive 81 and glass substrate (Figure 6b). AuNPs a conjugated with AME specific mAb and magnetic nanoparticles (MNPs)-BSA-AME conjugates were used as capture probe and competitive antigen, respectively. In the presence of AME, it firstly bound to the AuNPs-mAbs in conjugate pad and micro-mixing channel. Therefore, large numbers of free AuNPs-mAbs-AME conjugates were kept in supernatant after magnetic separation. Then, the supernatant was transferred into immunogold amplification solution containing ascorbic acids as reducing agent and hexadecyltrimethylammonium bromide as a surfactant to stabilize the amplified AuNPs-mAbs. In this solution, the free AuNPs-mAbs-AME conjugates were used as gold seeds for the signal amplification. UV spectroscopy and smartphone imaging APP were used for monitoring of the AuNPs color change. The assay was able to analyze six samples in parallel within 15 min. The fabricated microfluidic immunoassay exhibits LODs at 12.5 pg·mL^{-1} and 200 pg·mL^{-1}, by UV spectroscopy and smartphone imaging, respectively. It was successfully applied for AME detection in spiked fruit samples. The device can be used for sensitive, rapid, low-cost, and on-site detection of mycotoxins.

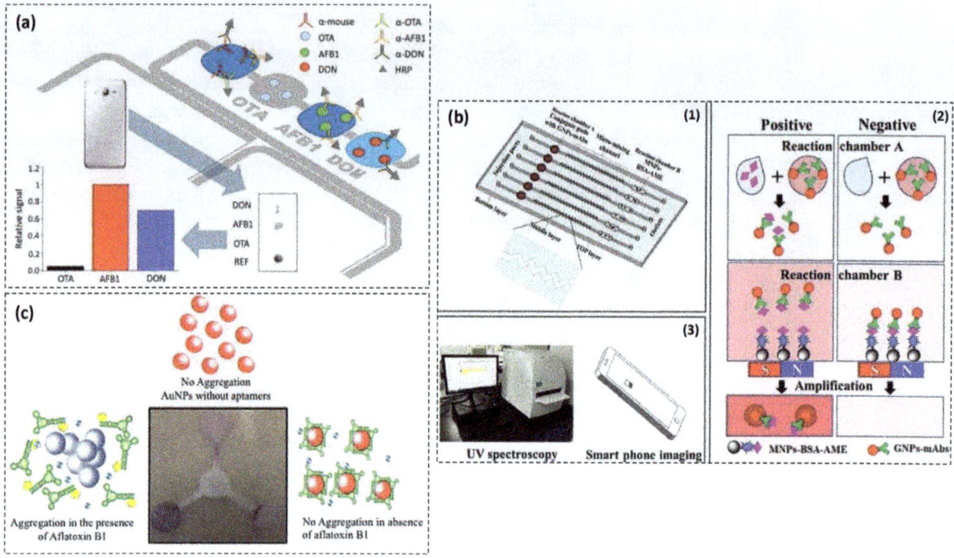

Figure 6. (**a**) Schematic representation of the microfluidic immunoassay with smartphone data acquisition for multiplexed mycotoxin detection; (**b**) microfluidic immunoassay for AME detection: (**1**) The 3D structural diagram of the fabricated microfluidic chip, (**2**) The schematic of the AME detection using the amplified microfluidic immunoassay, and (**3**) The colorimetric detection by UV spectroscopy and smartphone imaging APP; (**c**) Development of a µPAD using aptamer and for AFB1 detection based on salt-induced aggregation of AuNPs in the presence of analyte. Reproduced from [83,87,88], respectively, with permission.

Paper is an ideal substrate to construct microfluidic devices. It is a good alternative to glass and polymer. Paper-based microfluidic systems were introduced by Whitesides group in 2007 as lab-on-chip (LOC) devices. The paper-based microfluidic devices are cheaper, easier to fabricate, easier to use, easier to dispose, compatible to chemicals/biochemicals used in bio-medical applications and environmentally friendly. Paper segmentation to hydrophobic and hydrophilic regions by hydrophobic materials can provide hydrophilic channels for fluid flow via capillary action and without the need for a pump. However, despite all these advantages, paper-based microfluidic devices are only suitable for semi-quantitative rather than quantitative analysis [85,86,89].

Although there are a number of well-developed systems for immunoassay in microfluidic (lab-on-chip) format, the use of aptamers in similar devices is on very beginning. Kasoju et al. developed a microfluidic paper-based analytical device (µPAD) for AFB1 detection using aptamer as recognition element [88]. The hydrophobic barriers were developed on the Whatman filter paper using photolithography. Two control zones (negative and positive) and one analyte zone were designed on the paper. Detection was performed based on salt-induced aggregation of AuNPs in the presence of analyte. The aptamer-AuNPs conjugate was adsorbed onto the paper through physical adsorption and AFB1 was allowed to flow over the µPAD. In the presence of AFB1, the aptamer combined with AFB1 and bare AuNPs was aggregated in the presence of NaCl (Figure 6c). The developed assay showed a LOD of 10 nM in spiked samples. The developed µPAD was suitable for rapid (detection time < 1 min), simple, label-free, and on-site detection of mycotoxins.

Representative examples of recent developed microfluidic-based assays for the detection of mycotoxins are reported in Table 5.

Table 5. Representative examples of recent developed microfluidic-based assays for the detection of mycotoxins.

Strategy	Target	LOD	Linear Range	Specificity	Sample	Advantages/Disadvantages	Ref.
PDMS-based microfluidic immunoassay using antibody labeled with HRP	OTA AFB1 ZEN	<40 0.1–0.2 <10 ng·mL^{-1}	–	High	Corn	Good sensitivity, high selectivity, low-cost, short detection time (10 min), portable/no evaluation of reproducibility and stability, lots of user-intervention steps, semi-quantitative	[83]
Microfluidic immunoassay based on combination target binding to AuNPs-mAb and immunogold amplification of AuNPs-mAbs-AME	Altenariol monomethyl ether	12.5 pg·mL^{-1} 200 pg·mL^{-1}	12.5–200 pg·mL^{-1} 200–1000 pg·mL^{-1}	High	Apple, cherry, orange	High sensitivity and selectivity, low-cost, short detection time (15 min), portable, simultaneous analysis of six samples, quantitative/no evaluation of reproducibility and stability	[87]
Development of a μPAD based on salt-induced aggregation of AuNPs in the presence of analyte	AFB1	10 nM	1 pM–1 μM	High	Water	High sensitivity and selectivity, cost-effective, short detection time (> 1 min), portable/no evaluation of reproducibility and stability, no evaluation of food matrix, qualitative	[88]
Colorimetric competitive immunoassay into a paper microfluidic device using AuNPs as signal indicator	Deoxynivalenol	0.644 ng·mL^{-1}	0.01–20 ng·mL^{-1}	High	Wheat, corn	High sensitivity and selectivity, rapid (detection time of 12 min), low-cost, portable, and reliable/qualitative, no evaluation of reproducibility and stability	[90]

4. Recent Advances toward Practical Applications

Currently there are several reference methods for mycotoxin analysis such as HPLC and LC-MS/MS. However, their usage is dependent on a variety of factors including expensive laboratory equipment, academic laboratories with skilled personnel, and time-consuming analysis [10]. In these cases, biosensors and rapid detection kits can overcome such limitations. The primary goal in developing a biosensor or bioassay is making an analytical device that can rapidly and accurately quantify target analytes in the field at a low cost. A biosensor/bioassay with such features can have the potential to be practical and commercial. As described above, the colorimetry sensing technology is springing up showing excellent sensitivity as a powerful tool for mycotoxins detection. However, sample pretreatment is fundamental for the extraction of mycotoxins from complex food and feed matrices. This preliminary step is necessary due to many interfering compounds—such as proteins, lipids, sugars, and salts in complex food/feed samples—leading to matrix effects, signal interference, instruments contamination, and even false-positive results [91].

4.1. Mycotoxin Extraction Methods

Food and feed sample pretreatment mainly requires the selective isolation and enrichment of target analytes from the complex matrix. After the size reduction of solid samples, extraction of mycotoxins can be performed. Most of the mycotoxins are soluble in organic solvents—including methanol, acetone, acetonitrile, ethyl acetate, and dichloromethane—but are hardly soluble in water except fumonisins and patulin [92]. Solvent extraction is the most common method for the extraction and purification of mycotoxins in colorimetric assays. In a representative extraction procedure, liquid samples such as milk, juice, and wine are directly subjected to liquid-liquid extraction for the initial isolation of mycotoxins. However, solid-liquid extraction is used for the extraction of mycotoxins from solid samples such as grains and cereals. Generally, a mixture of organic solvent and acidic buffer or water is extensively used to extract mycotoxins. In this mixture, the water improves the organic solvent's penetration in the food/feed matrix, while the acidic solvent can decompose the strong bonds between the analyte and other food components (e.g., protein and sugars), resulting in increasing the extraction efficiency [93].

Because of the destructive effect of organic solvents on enzyme or antibody bioreceptors (e.g., denaturing properties), many efforts have been made to minimize these effects or replace new extraction methods. In the first case, diluting the sample after solvent extraction can reduce the solvent's damaging effect. In the latter case, supercritical CO_2 extraction and microwave-assisted extraction can be promising methods for replacement with conventional solvent extraction.

After mycotoxin extraction, filtration and centrifugation are used to remove any suspended particles. In most cases, no further purification steps are required, and the sample is used for detection. However, in biosensors/bioassays with high detection limits, clean-up procedures such as solid-phase extraction (SPE), dispersive solid-phase extraction (DSPE), solid-phase micro-extraction (SPME), and immunoaffinity column (IAC) can be performed in order to improve the detection sensitivity and specificity [9]. These advanced extraction methods were recently reviewed by Agriopoulou et al. [94].

4.2. Inventory of Commercially Available Kits for Mycotoxins Detection

Commercial test kits for mycotoxin detection are utilized as an appropriate alternative for more user-friendly, inexpensive, robust, and rapid analysis. There are a large number of commercial detection kits available for mycotoxin analysis in the current market, as summarized in the Table 6. They commonly include ELISA kits, membrane-based immunoassays such as lateral flow immunoassays (LFIAs), fluorescence polarization immunoassays (FPIAs), and immunoaffinity column coupled with fluorometric assay. The majority of these test kits are based on an immunoassay format which relies on the specific interaction of antigen and antibody. Moreover, colorimetric detection kits are most preferred because of the ability to see results with the naked eye. Among them, LFIAs with acceptable sensitivity, good accuracy, portability, short detection time, ease of use, and no need for specialized personnel have become strong competitors on the market for mycotoxin analysis. Several companies worldwide produce LFIA test strip to analyze different kinds of mycotoxins. Charm Sciences Inc., Pribolab, EnviroLogix, Romer labs, Vicam, CUSABIO, etc. are among LFIA strips producer for aflatoxins, DON, ZEN, T2, and OTA. ROSA (rapid one step assay) lateral flow strips developed by Charm Sciences Inc. are the leading mycotoxin test worldwide.

Commercial LFIAs test strips usually use AuNPs as colored label. The method can provide qualitative and/or semi-quantitative results within minutes (e.g., Afla-V and AflaCheck strip tests by Vicam, 5 and 3 min, respectively). For semi-quantitative analysis, portable readers have been developed for on-site detection. For example, PerkinElmer's QuickSTAR Horizon strip reader provides quantitative results for mycotoxins including AFB1, AFB2, AFG1, and AFG2; at detection levels of 2 to 300 ppb within 6 min. Charm EZ-M Reader is another type of portable strip reader which can show results within 3–5 min. The color-coded strips allow the EZ-M reader to automatically recognize which mycotoxin group you are testing and will adjust the reader temperature and time accordingly. Some companies such as R-Biopharm provided a mobile app on a smartphone to analyze color signal instead of a reader, specifically for aflatoxins, T2/HT2, ZEN, and FMN.

Commercially available test kits have been developed for determination of individual mycotoxins or for multiple mycotoxins in one group (e.g., aflatoxins B1, B2, G1, and G2). Since we are usually faced with contamination of food and feed to more than one mycotoxin, the current trend in LFIA technology is to develop strips with multiple test lines for the simultaneous detection of multiple mycotoxins.

After LFIA test strips, ELISA-based kits have allocated a major portion of market amongst other mycotoxin detection methods. Many companies—such as Sigma, Elabscience, Eurofins, Romer Labs, ELISA Technologies, Cusabio, Astori Tecnica, etc.—offer ELISA kits to detect the most common types of mycotoxins. Some of these kits can detect several types of mycotoxins in one group (e.g., aflatoxins B1, B2, G1, and G2).

Table 6. Main companies providing commercial colorimetric immuno-kits for mycotoxins analysis.

Company	Kit	Type of Detection	Mycotoxins						Time (min)	Multiplexing	Ref.
			AFs	OTA	FUM	ZEN	DON	T2/HT2			
Astori Tecnica	ELISA	QNT, Semi-QNT for OTA	✓	✓	✓	✓	✓	✓	20–30	No	[95]
	LFIA	QLT, QNT	AFM1	(−)	(−)	(−)	(−)	(−)	10		
Charm Sciences Inc.	LFIA	QNT	✓	✓	✓	✓	✓	✓	3–5	No	[96]
CUSABIO	ELISA	QNT	AFB1	✓	(−)	✓	✓	T2	20	No	[97]
	LFIA	QNT	AFB1	✓	(−)	✓	✓	T2	3–5		
Elabscience	ELISA	QNT	✓	✓	✓	✓	✓	✓	75	No	[98]
	LFIA	QNT	✓	✓	✓	✓	✓	✓	8		
EnviroLogix	LFIA	QLT, QNT	✓	✓	✓	✓	✓	✓	2–4	Yes (AF, ZEN, DON, FUM)	[99]
Eurofins	ELISA	QNT	✓	✓	✓	✓	✓	✓	15–75	No	[100]
	LFIA	QNT	✓	(−)	✓	✓	✓	(−)	5		
Helica	ELISA	QNT	✓	✓	✓	✓	✓	(−)	NR	No	[101]
Neogen	ELISA	QNT	✓	✓	✓	✓	✓	✓	10–20	No	[102]
	LFIA	QLT, QNT	✓	✓	✓	✓	✓	✓	3–8		
R-Biopharm	ELISA	QNT	✓	✓	✓	✓	✓	✓	45–150	No	[103]
	LFIA	Semi-QNT, QNT	✓	(−)	✓	✓	✓	✓	5		
Romer Labs	ELISA	QNT	✓	✓	✓	✓	✓	✓	15	Yes (up to 6 mycotoxins)	[104]
	LFIA	QLT, QNT	✓	✓	✓	✓	✓	(−)	3	No	
Vicam	LFIA	Semi-QNT, QNT	✓	✓	✓	✓	✓	✓	5	No	[105]
Beacon Analytical Systems	ELISA	QNT	✓	✓	✓	✓	✓	✓	15–75	No	[106]
Creative Diagnostics	ELISA	QNT	✓	✓	✓	(−)	✓	✓	15–120	No	[107]
	LFIA	QLT, Semi-QNT, QNT	✓	✓	✓	✓	✓	✓	3–10		
PerkinElmer	ELISA	QNT	✓	✓	✓	✓	✓	✓	15–60	No	[108]
	LFIA	QNT	✓	✓	✓	✓	✓	✓	4–6		
Unisensor	LFIA	QNT	✓	(−)	(−)	(−)	(−)	(−)	10	No	[109]
Pribolab	ELISA	QNT	✓	✓	✓	✓	✓	✓	NR	No	[110]
	LFIA	QLT, QNT	✓	✓	✓	✓	✓	(−)	10–12		
Randox	ELISA	QNT	✓	(−)	(−)	(−)	(−)	(−)	NR	No	[111]
	Biochip Arrays	QNT	✓	✓	✓	✓	✓	✓	120	Yes (up to 10 mycotoxins)	
Novakits	ELISA	Semi-QNT, QNT	✓	✓	✓	✓	✓	✓	15–70	No	[112]
	LFIA	QNT	✓	✓	✓	✓	✓	✓	5	Yes (7 mycotoxins)	
Sigma	ELISA	QNT	✓	✓	✓	✓	✓	(−)	NR	No	[113]
Bio-Check	ELISA	QNT	✓	✓	✓	✓	✓	(−)	NR	No	[114]
	LFIA	QNT	✓	✓	✓	(−)	✓	(−)	3–5	No	

QNT = Quantitative; Semi-QNT = Semi-quantitative; QLT = Qualitative; NR = Not reported; ✓ = Available detection kit for the specific mycotoxin, (−) = Not available.

The HRP-TMB system is the most common method for colorimetric signal generation in ELISA-based kits. Commercial ELISA kits are sensitive, selective, high throughput, with minimum sample preparation steps. Moreover, the detection time in commercial kits has been shortened so that, most of them are able to detect target mycotoxin within 1–2 h. Romer labs has produced ELISA kits for aflatoxins, OTA, DON, T2, ZEN, and FMN with incubation periods of 15 min. In some kits, cross-reactivity of antibodies leads to overestimation of results while matrix effect plays a key role in providing false-positive results. To avoid such effects, most kits define the limited matrices to which the ELISA kit can be used [115]. However, this feature can be considered a limitation for such ELISA kits. On the other hand, some companies such as Eurofins have developed sensitive ELISA method for a wide range of matrices.

As with the LFIA kits, the current trend in ELISA technique is to develop commercial multiplex assays. Multiplex ELISA can be developed by immobilizing different toxins in different wells of a single microplate in competitive format.

4.3. Interesting Examples from Literature with Great Potential for Industrial Applications

As it can be observed from literatures, there are many new studies on mycotoxin detection with the potential for industrial application. However, several parameters should be considered before a bioassay or biosensor can reach the commercialization stage. These parameters are briefly described in the following.

LOD and sensitivity: Sensitivity is a main factor in mycotoxin detection due to the presence of mycotoxins in small amounts in food and their toxic hazards to the consumer in very low concentrations. Regarding this issue, most of developed biosensors/bioassays show high sensitivity which makes them suitable for point of care testing (POCT) application. For quantitative assays, the LOD is defined as the mean value of the blank (matrix blank) reading in analyte concentration, plus three times the standard deviations [115]. In the case of sensitivity, false negative and false positive results and sample matrix which can affect the assay sensitivity should be considered.

Specificity and cross reactivity: Since antibodies are the most used recognition element in the development of mycotoxin bioassays, the use of monoclonal antibodies with high specificity and low cross-reactivity is of great importance in the development of commercial kits. Cross-reactivity of antibodies can lead to overestimation of results and inaccurate overall risk assessment for consumers. Generally, individual mycotoxin assays show higher specificity compared to multiplex assays. Because detection of multiple mycotoxins with a single test an important feature in developing commercial devices, the use of antibody with acceptable cross-reactivities to detect groups of related mycotoxins can be affordable.

Accuracy and precision: accuracy and precision are important parameters for practical application of a developed assay. Accuracy is proximity of the measurements to a specific value obtained by a reliable method. A method for expressing the accuracy of the developed analytical approach is by establishing a correlation between results of the developed and the reference methods. In the case of mycotoxin detection, HPLC or LC-MS can be considered as a reference method. Repeatability and reproducibility are mostly applied as indicators of method precision. Lot to lot reproducibility and shelf-life stability can influence accuracy and precision of the method.

Other parameters that should be noticed in developing a practical method for mycotoxin include portability for on-site applications, user-friendliness, and low-cost detection.

According to the above-mentioned factors, some of the researches reported in this review have the potential to be industrialized. Most of enzymatic detection methods based on DNAzyme, reported in Table 1, show high sensitivity and selectivity and good precision which make them suitable for mycotoxin detection. However, long incubation time, multi-step washing, and complex operation are considered as main limitations toward their practical applications. Although reported AchE-based assay is simple, rapid, and low-cost, it cannot be considered as a very selective method due to the presence of other analytes with AchE inhibitory action such as pesticides and so on.

Reported ELISA-based methods (Table 2) with high sensitivity and selectivity can be ideal for industrialization after a few modifications related to the detection time. In terms of meeting current trends, multiplex ELISA demonstrated by Urusov et al. [64] for simultaneous detection of AFB1, OTA, and ZEN with a high detectable signal and high sensitivity is an ideal platform for practical multiplex analysis. The proposed assay was successfully validated for food samples with complex matrices. Multiplex nanoarray based on ELISA technique developed by McNamee et al. [65] was able to detect ZEA, T-2 toxin, and FMB1 in a simple way with high sensitivity and accuracy. The established protocol offered a higher throughput of samples and potential feasibility for easy to use and multiplex detection compared to the other developed ELISA methods.

Developed ELISA protocols based on nanomaterials as enzyme substitute can be also considered for practical applications due to greater stability than conventional ELISA. In this regard, the nanozyme-linked immunosorbent assay based on MOFs (Table 2) for AFB1 detection is a good example [63]. The developed method showed high sensitivity and selectivity, high accuracy and excellent stability without false positive and false negative results. Plasmonic ELISAs using enzyme and nanomaterials can be suitable for naked-eye detection and on-site application with no need for ELISA reader devices. Quantitative results can be obtained with smartphone-based signal readout systems. In this case, plasmonic ELISA methods demonstrated by Xiong et al. [66] and Pei et al. [60] can be mentioned.

LFIAs are strong competitors on the market for mycotoxin detection due to their unique features such as high sensitivity and selectivity, low-cost, short detection time, portability, and user-friendliness [29]. Therefore, research and development in the field of these popular detection kits is very important. Many efforts have been made for the development of quantitative and multiplex LFIA test strips. In this regard, LFIA based on AuNPs and TRFMs for multiplex detection of AFB1, ZEN, FMB1, DON, and T-2 toxin along with a smartphone-based quantitative dual detection mode device [30] is a good example of a multiplex and quantitative analysis (Table 3). Developed strips showed high sensitivity and reliability. Application of smartphone provided a low-cost and portable quantitative method. Stability of strips during storage is a main parameter in practical application which should be checked in this study.

Aptamer-based LFAs can be suitable alternatives for antibody-based LFA in the future due to lower cost and higher stability of aptamers compared to antibodies. For example, aptamer-based LFA developed by Wu et al. [75] exhibited high sensitivity and high stability (2 months at room temperature), and short detection time (5 min) for the determination of ZEN. Another example is aptamer-based LFA for OTA detection with high sensitivity and excellent selectivity [76]. It was stable for 6 months at room temperature.

Only a few studies are available in the field of microfluidic-based assays for mycotoxin detection. However, this method could have a good future for mycotoxins detection. Among developed microfluidic methods in mycotoxin analysis, paper-based systems show greater potential for commercialization due to simplicity, low-cost and portability. They do not need additional equipment such as pump to generate flow. On the other hand, quantification of the results can be made with a smartphone. As reported in Table 4, the proposed µPAD by Kasoju et al. [88] for AFB1 detection showed high sensitivity, low-cost, short detection time (>1 min), and portability. Moreover, the proposed µPAD for DON detection has a great potential for industrialization [90].

5. Conclusions and Future Directions

Development of a suitable biosensor or bioassay for mycotoxin analysis as alternative approaches to conventional sophisticated techniques such as chromatography-based methods is of great importance in the field of biosensing research. Among different sensing strategies, colorimetric approaches are very popular, simple, and convenient and present great value for on-site detection. Colorimetric methods are categorized based on type of colored label and the medium in which the reaction develops (solution-based and solid substrate-based). ELISA and LFA are considered as the most common solution-based and solid-substrate-based colorimetric detection methods due to their unique features. In addition to a wide variety of researches in these areas, multiple companies worldwide are developing and producing detection kits relying on these two techniques. The products from different suppliers can be different in terms of sensitivity, selectivity, the type of matrix used, detection time, and number of detectable mycotoxins.

Notwithstanding the great success and advance in this field, hand-held digital biosensors and smartphone-based quantitative detection are desired for the future market, aiming for an intuitive user experience. Furthermore, because of the very competitive market for mycotoxin test kits, the future market welcomes products capable of performing multiplex analysis and high portability.

It is also possible that new recognition elements, such as aptamer and molecularly imprinted polymer (MIP), will replace the antibody commercially available tests for a more convenient future technology. Despite their merits, aptasensing assays remain relatively immature for industrial monitoring. This is probably due to the short history of reproducible aptasensors in real sample analysis and the lack of broad dissemination of results on the market saturated with immune kits. Nevertheless, owing to the advantages and prospects offered by the mass production of specific aptamers, its contribution to the colorimetric sensors market in food safety is expected to evolve rapidly and shape the future market.

Funding: This research received no external funding.

Acknowledgments: This study was supported by Shiraz University and University of Perpignan.

Conflicts of Interest: The authors declare no conflict of interest.

References

1. Lee, H.J.; Ryu, D. Worldwide Occurrence of Mycotoxins in Cereals and Cereal-Derived Food Products: Public Health Perspectives of Their Co-occurrence. *J. Agric. Food Chem.* **2017**, *65*, 7034–7051. [CrossRef] [PubMed]
2. Alshannaq, A.; Yu, J.-H. Occurrence, Toxicity, and Analysis of Major Mycotoxins in Food. *Int. J. Environ. Res. Public Health* **2017**, *14*, 632. [CrossRef] [PubMed]
3. Kuiper-Goodman, T. Risk assessment and risk management of mycotoxins in food. In *Mycotoxins in Food*; Magan, N., Olsen, E.M., Eds.; Woodhead Publishing: Cambridge, UK, 2004; pp. 3–31. ISBN 978-1-85573-733-4.
4. Ostry, V.; Malir, F.; Toman, J.; Grosse, Y. Mycotoxins as human carcinogens—the IARC Monographs classification. *Mycotoxin Res.* **2017**, *33*, 65–73. [CrossRef] [PubMed]
5. Edite Bezerra da Rocha, M.; Freire, F.D.C.O.; Erlan Feitosa Maia, F.; Izabel Florindo Guedes, M.; Rondina, D. Mycotoxins and their effects on human and animal health. *Food Control* **2014**, *36*, 159–165. [CrossRef]
6. Van Egmond, H.P.; Schothorst, R.C.; Jonker, M.A. Regulations relating to mycotoxins in food: Perspectives in a global and European context. *Anal. Bioanal. Chem.* **2007**, *389*, 147–157. [CrossRef] [PubMed]
7. Rovina, K.; Nurul Shaeera, S.; Merrylin Vonnie, J.; Xin Yi, S. Recent Biosensors Technologies for Detection of Mycotoxin in Food Products. In *Mycotoxins and Food Safety*; IntechOpen: London, UK, 2020.
8. Turner, N.W.; Subrahmanyam, S.; Piletsky, S.A. Analytical methods for determination of mycotoxins: A review. *Anal. Chim. Acta* **2009**, *632*, 168–180. [CrossRef] [PubMed]
9. Yang, Y.; Li, G.; Wu, D.; Liu, J.; Li, X.; Luo, P.; Hu, N.; Wang, H.; Wu, Y. Recent advances on toxicity and determination methods of mycotoxins in foodstuffs. *Trends Food Sci. Technol.* **2020**, *96*, 233–252. [CrossRef]
10. Nolan, P.; Auer, S.; Spehar, A.; Elliott, C.T.; Campbell, K. Current trends in rapid tests for mycotoxins. *Food Addit. Contam. Part A Chem. Anal. Control. Expo. Risk Assess.* **2019**, *36*, 800–814. [CrossRef] [PubMed]
11. Lin, C.; Zheng, H.; Sun, M.; Guo, Y.; Luo, F.; Guo, L.; Qiu, B.; Lin, Z.; Chen, G. Highly sensitive colorimetric aptasensor for ochratoxin A detection based on enzyme-encapsulated liposome. *Anal. Chim. Acta* **2018**, *1002*, 90–96. [CrossRef] [PubMed]
12. Adeoye, R.I.; Osalaye, D.S.; Ralebitso-Senior, T.K.; Boddis, A.; Reid, A.J.; Fatokun, A.A.; Powell, A.K.; Malomo, S.O.; Olorunniji, F.J. Catalytic activities of multimeric G-quadruplex dnazymes. *Catalysts* **2019**, *9*, 613. [CrossRef]
13. Li, W.; Li, Y.; Liu, Z.; Lin, B.; Yi, H.; Xu, F.; Nie, Z.; Yao, S. Insight into G-quadruplex-hemin DNAzyme/RNAzyme: Adjacent adenine as the intramolecular species for remarkable enhancement of enzymatic activity. *Nucleic Acids Res.* **2016**, *44*, 7373–7384. [CrossRef] [PubMed]
14. Kosman, J.; Zukowski, K.; Juskowiak, B. Comparison of Characteristics and DNAzyme Activity of G4–Hemin Conjugates Obtained via Two Hemin Attachment Methods. *Molecules* **2018**, *23*, 1400. [CrossRef] [PubMed]
15. Wu, J.; Zeng, L.; Li, N.; Liu, C.; Chen, J. A wash-free and label-free colorimetric biosensor for naked-eye detection of aflatoxin B1 using G-quadruplex as the signal reporter. *Food Chem.* **2019**, *298*, 125034. [CrossRef] [PubMed]

16. Santovito, E.; Greco, D.; D'Ascanio, V.; Sanzani, S.M.; Avantaggiato, G. Development of a DNA-based biosensor for the fast and sensitive detection of ochratoxin A in urine. *Anal. Chim. Acta* **2020**, *1133*, 20–29. [CrossRef] [PubMed]

17. Setlem, K.; Mondal, B.; Shylaja, R.; Parida, M. Dual Aptamer-DNAzyme based colorimetric assay for the detection of AFB1 from food and environmental samples. *Anal. Biochem.* **2020**, *608*, 113874. [CrossRef]

18. Chrouda, A.; Zinoubi, K.; Soltane, R.; Alzahrani, N.; Osman, G.; Al-Ghamdi, Y.O.; Qari, S.; Al mahri, A.; Algethami, F.K.; Majdoub, H.; et al. An acetylcholinesterase inhibition-based biosensor for aflatoxin B1 detection using sodium alginate as an immobilization matrix. *Toxins* **2020**, *12*, 173. [CrossRef] [PubMed]

19. Mirón-Mérida, V.A.; Wu, M.; Gong, Y.Y.; Guo, Y.; Holmes, M.; Ettelaie, R.; Goycoolea, F.M. Genipin cross-linked chitosan for signal enhancement in the colorimetric detection of aflatoxin B1 on 3MM chromatography paper. *Sens. Bio-Sens. Res.* **2020**, *29*, 100339. [CrossRef]

20. Bazin, I.; Tria, S.A.; Hayat, A.; Marty, J.L. New biorecognition molecules in biosensors for the detection of toxins. *Biosens. Bioelectron.* **2017**, *87*, 285–298. [CrossRef]

21. Seok, Y.; Byun, J.Y.; Shim, W.B.; Kim, M.G. A structure-switchable aptasensor for aflatoxin B1 detection based on assembly of an aptamer/split DNAzyme. *Anal. Chim. Acta* **2015**, *886*, 182–187. [CrossRef]

22. Li, Y.; Wang, Z.; Sun, L.; Liu, L.; Xu, C.; Kuang, H. Nanoparticle-based sensors for food contaminants. *TrAC Trends Anal. Chem.* **2019**, *113*, 74–83. [CrossRef]

23. Ghosh, S.K.; Pal, T. Interparticle Coupling Effect on the Surface Plasmon Resonance of Gold Nanoparticles: From Theory to Applications. *Chem. Rev.* **2007**, *107*, 4797–4862. [CrossRef] [PubMed]

24. Aldewachi, H.; Chalati, T.; Woodroofe, M.N.; Bricklebank, N.; Sharrack, B.; Gardiner, P. Gold nanoparticle-based colorimetric biosensors. *Nanoscale* **2018**, *10*, 18–33. [CrossRef] [PubMed]

25. Zhang, N.; Liu, B.; Cui, X.; Li, Y.; Tang, J.; Wang, H.; Zhang, D.; Li, Z. Recent advances in aptasensors for mycotoxin detection: On the surface and in the colloid. *Talanta* **2021**, *223*, 121729. [CrossRef] [PubMed]

26. Evtugyn, G.; Hianik, T. Aptamer-based biosensors for mycotoxin detection. In *Nanomycotoxicology: Treating Mycotoxins in the Nano Way*; Rai, M., Abd-Elsalam, K.A., Eds.; Academic Press: Cambridge, MA, USA, 2019; pp. 35–70. ISBN 9780128179987.

27. Luan, Y.; Chen, Z.; Xie, G.; Chen, J.; Lu, A.; Li, C.; Fu, H.; Ma, Z.; Wang, J. Rapid Visual Detection of Aflatoxin B1 by Label-Free Aptasensor Using Unmodified Gold Nanoparticles. *J. Nanosci. Nanotechnol.* **2015**, *15*, 1357–1361. [CrossRef] [PubMed]

28. Phanchai, W.; Srikulwong, U.; Chompoosor, A.; Sakonsinsiri, C.; Puangmali, T. Insight into the Molecular Mechanisms of AuNP-Based Aptasensor for Colorimetric Detection: A Molecular Dynamics Approach. *Langmuir* **2018**, *34*, 6161–6169. [CrossRef] [PubMed]

29. Tripathi, P.; Upadhyay, N.; Nara, S. Recent advancements in lateral flow immunoassays: A journey for toxin detection in food. *Crit. Rev. Food Sci. Nutr.* **2018**, *58*, 1715–1734. [CrossRef]

30. Liu, Z.; Hua, Q.; Wang, J.; Liang, Z.; Li, J.; Wu, J.; Shen, X.; Lei, H.; Li, X. A smartphone-based dual detection mode device integrated with two lateral flow immunoassays for multiplex mycotoxins in cereals. *Biosens. Bioelectron.* **2020**, *158*, 112178. [CrossRef]

31. Yu, L.; Li, P.; Ding, X.; Zhang, Q. Graphene oxide and carboxylated graphene oxide: Viable two-dimensional nanolabels for lateral flow immunoassays. *Talanta* **2017**, *165*, 167–175. [CrossRef]

32. Tian, M.; Xie, W.; Zhang, T.; Liu, Y.; Lu, Z.; Li, C.M.; Liu, Y. A sensitive lateral flow immunochromatographic strip with prussian blue nanoparticles mediated signal generation and cascade amplification. *Sens. Actuators B Chem.* **2020**, *309*, 127728. [CrossRef]

33. Kong, D.; Liu, L.; Song, S.; Suryoprabowo, S.; Li, A.; Kuang, H.; Wang, L.; Xu, C. A gold nanoparticle-based semi-quantitative and quantitative ultrasensitive paper sensor for the detection of twenty mycotoxins. *Nanoscale* **2016**, *8*, 5245–5253. [CrossRef]

34. Tian, F.; Zhou, J.; Jiao, B.; He, Y. A nanozyme-based cascade colorimetric aptasensor for amplified detection of ochratoxin A. *Nanoscale* **2019**, *11*, 9547–9555. [CrossRef] [PubMed]

35. Zhang, X.; Wu, D.; Zhou, X.; Yu, Y.; Liu, J.; Hu, N.; Wang, H.; Li, G.; Wu, Y. Recent progress in the construction of nanozyme-based biosensors and their applications to food safety assay. *TrAC Trends Anal. Chem.* **2019**, *121*, 115668. [CrossRef]

36. Wang, C.; Qian, J.; Wang, K.; Yang, X.; Liu, Q.; Hao, N.; Wang, C.; Dong, X.; Huang, X. Colorimetric aptasensing of ochratoxin A using Au@Fe$_3$O$_4$ nanoparticles as signal indicator and magnetic separator. *Biosens. Bioelectron.* **2016**, *77*, 1183–1191. [CrossRef] [PubMed]

37. He, Y.; Tian, F.; Zhou, J.; Zhao, Q.; Fu, R.; Jiao, B. Colorimetric aptasensor for ochratoxin A detection based on enzyme-induced gold nanoparticle aggregation. *J. Hazard. Mater.* **2020**, *388*, 121758. [CrossRef] [PubMed]

38. Luan, Y.; Chen, J.; Li, C.; Xie, G.; Fu, H.; Ma, Z.; Lu, A. Highly sensitive colorimetric detection of ochratoxin a by a label-free aptamer and gold nanoparticles. *Toxins* **2015**, *7*, 5377–5385. [CrossRef]

39. Xiao, R.; Wang, D.; Lin, Z.; Qiu, B.; Liu, M.; Guo, L.; Chen, G. Disassembly of gold nanoparticle dimers for colorimetric detection of ochratoxin A. *Anal. Methods* **2015**, *7*, 842–845. [CrossRef]

40. Liu, B.; Huang, R.; Yu, Y.; Su, R.; Qi, W.; He, Z. Gold Nanoparticle-Aptamer-Based LSPR Sensing of Ochratoxin A at a Widened Detection Range by Double Calibration Curve Method. *Front. Chem.* **2018**, *6*. [CrossRef] [PubMed]

41. Sun, S.; Zhao, R.; Feng, S.; Xie, Y. Colorimetric zearalenone assay based on the use of an aptamer and of gold nanoparticles with peroxidase-like activity. *Microchim. Acta* **2018**, *185*, 535. [CrossRef]

42. Chotchuang, T.; Cheewasedtham, W.; Jayeoye, T.J.; Rujiralai, T. Colorimetric determination of fumonisin B1 based on the aggregation of cysteamine-functionalized gold nanoparticles induced by a product of its hydrolysis. *Microchim. Acta* **2019**, *186*, 655. [CrossRef]

43. Althagafi, I.I.; Ahmed, S.A.; El-Said, W.A. Colorimetric aflatoxins immunoassay by using silica nanoparticles decorated with gold nanoparticles. *Spectrochim. Acta Part A Mol. Biomol. Spectrosc.* **2021**, *246*, 118999. [CrossRef]

44. Khansili, N.; Rattu, G.; Kumar, A.; Krishna, P.M. Development of colorimetric sensor with zinc oxide nanoparticles for rapid detection of aflatoxin B1 in rice. *Mater. Today Proc.* **2020**, *21*, 1846–1855. [CrossRef]

45. Zhu, W.; Li, L.; Zhou, Z.; Yang, X.; Hao, N.; Guo, Y.; Wang, K. A colorimetric biosensor for simultaneous ochratoxin A and aflatoxins B1 detection in agricultural products. *Food Chem.* **2020**, *319*, 126544. [CrossRef] [PubMed]

46. Su, L.; Zhang, Z.; Xiong, Y. Water dispersed two-dimensional ultrathin Fe(III)-modified covalent triazine framework nanosheets: Peroxidase like activity and colorimetric biosensing applications. *Nanoscale* **2018**, *10*, 20120–20125. [CrossRef] [PubMed]

47. Sheini, A. Colorimetric aggregation assay based on array of gold and silver nanoparticles for simultaneous analysis of aflatoxins, ochratoxin and zearalenone by using chemometric analysis and paper based analytical devices. *Microchim. Acta* **2020**, *187*, 167. [CrossRef] [PubMed]

48. Demers, L.M.; Östblom, M.; Zhang, H.; Jang, N.-H.; Liedberg, B.; Mirkin, C.A. Thermal Desorption Behavior and Binding Properties of DNA Bases and Nucleosides on Gold. *J. Am. Chem. Soc.* **2002**, *124*, 11248–11249. [CrossRef]

49. Rhouati, A.; Bulbul, G.; Latif, U.; Hayat, A.; Li, Z.-H.; Marty, J.; Rhouati, A.; Bulbul, G.; Latif, U.; Hayat, A.; et al. Nano-Aptasensing in Mycotoxin Analysis: Recent Updates and Progress. *Toxins* **2017**, *9*, 349. [CrossRef]

50. YIN, X.; WANG, S.; LIU, X.; HE, C.; TANG, Y.; LI, Q.; LIU, J.; SU, H.; TAN, T.; DONG, Y. Aptamer-based Colorimetric Biosensing of Ochratoxin A in Fortified White Grape Wine Sample Using Unmodified Gold Nanoparticles. *Anal. Sci.* **2017**, *33*, 659–664. [CrossRef]

51. Hosseini, M. Aptamer-based Colorimetric and Chemiluminescence Detection of Aflatoxin B1 in Foods Samples. *Acta Chim. Slov.* **2015**, *62*, 721–728. [CrossRef]

52. Zhang, Y.; Lu, T.; Wang, Y.; Diao, C.; Zhou, Y.; Zhao, L.; Chen, H. Selection of a DNA Aptamer against Zearalenone and Docking Analysis for Highly Sensitive Rapid Visual Detection with Label-Free Aptasensor. *J. Agric. Food Chem.* **2018**, *66*, 12102–12110. [CrossRef]

53. Tian, F.; Zhou, J.; Fu, R.; Cui, Y.; Zhao, Q.; Jiao, B.; He, Y. Multicolor colorimetric detection of ochratoxin A via structure-switching aptamer and enzyme-induced metallization of gold nanorods. *Food Chem.* **2020**, *320*, 126607. [CrossRef]

54. Hao, N.; Lu, J.; Zhou, Z.; Hua, R.; Wang, K. A pH-Resolved Colorimetric Biosensor for Simultaneous Multiple Target Detection. *ACS Sens.* **2018**, *3*, 2159–2165. [CrossRef] [PubMed]

55. Yu, X.; Lin, Y.; Wang, X.; Xu, L.; Wang, Z.; Fu, F.F. Exonuclease-assisted multicolor aptasensor for visual detection of ochratoxin A based on G-quadruplex-hemin DNAzyme-mediated etching of gold nanorod. *Microchim. Acta* **2018**, *185*. [CrossRef] [PubMed]

56. Omar, S.S.; Haddad, M.A.; Parisi, S. Validation of HPLC and Enzyme-Linked Immunosorbent Assay (ELISA) techniques for detection and quantification of aflatoxins in different food samples. *Foods* **2020**, *9*, 661. [CrossRef] [PubMed]

57. Singh, J.; Mehta, A. Rapid and sensitive detection of mycotoxins by advanced and emerging analytical methods: A review. *Food Sci. Nutr.* **2020**, *8*, 2183–2204. [CrossRef]

58. Xiong, Y.; Pei, K.; Wu, Y.; Xiong, Y. Colorimetric ELISA based on glucose oxidase-regulated the color of acid–base indicator for sensitive detection of aflatoxin B1 in corn samples. *Food Control* **2017**, *78*, 317–323. [CrossRef]

59. Daniel, M.C.; Astruc, D. Gold Nanoparticles: Assembly, Supramolecular Chemistry, Quantum-Size-Related Properties, and Applications Toward Biology, Catalysis, and Nanotechnology. *Chem. Rev.* **2004**, *104*, 293–346. [CrossRef] [PubMed]

60. Pei, K.; Xiong, Y.; Xu, B.; Wu, K.; Li, X.; Jiang, H.; Xiong, Y. Colorimetric ELISA for ochratoxin A detection based on the urease-induced metallization of gold nanoflowers. *Sens. Actuators B Chem.* **2018**, *262*, 102–109. [CrossRef]

61. Zhan, S.; Hu, J.; Li, Y.; Huang, X.; Xiong, Y. Direct competitive ELISA enhanced by dynamic light scattering for the ultrasensitive detection of aflatoxin B1 in corn samples. *Food Chem.* **2020**, 128327. [CrossRef]

62. Mukherjee, M.; Nandhini, C.; Bhatt, P. Colorimetric and chemiluminescence based enzyme linked apta-sorbent assay (ELASA) for ochratoxin A detection. *Spectrochim. Acta Part A Mol. Biomol. Spectrosc.* **2021**, *244*, 118875. [CrossRef] [PubMed]

63. Xu, Z.; Long, L.L.; Chen, Y.Q.; Chen, M.L.; Cheng, Y.H. A nanozyme-linked immunosorbent assay based on metal–organic frameworks (MOFs) for sensitive detection of aflatoxin B1. *Food Chem.* **2021**, *338*, 128039. [CrossRef]

64. Urusov, A.; Zherdev, A.; Petrakova, A.; Sadykhov, E.; Koroleva, O.; Dzantiev, B. Rapid Multiple Immunoenzyme Assay of Mycotoxins. *Toxins* **2015**, *7*, 238–254. [CrossRef] [PubMed]

65. McNamee, S.E.; Bravin, F.; Rosar, G.; Elliott, C.T.; Campbell, K. Development of a nanoarray capable of the rapid and simultaneous detection of zearalenone, T2-toxin and fumonisin. *Talanta* **2017**, *164*, 368–376. [CrossRef] [PubMed]

66. Xiong, Y.; Pei, K.; Wu, Y.; Duan, H.; Lai, W.; Xiong, Y. Plasmonic ELISA based on enzyme-assisted etching of Au nanorods for the highly sensitive detection of aflatoxin B1 in corn samples. *Sens. Actuators B Chem.* **2018**, *267*, 320–327. [CrossRef]

67. Majdinasab, M.; Zareian, M.; Zhang, Q.; Li, P. Development of a new format of competitive immunochromatographic assay using secondary antibody–europium nanoparticle conjugates for ultrasensitive and quantitative determination of ochratoxin A. *Food Chem.* **2019**, *275*, 721–729. [CrossRef] [PubMed]

68. Bahadır, E.B.; Sezgintürk, M.K. Lateral flow assays: Principles, designs and labels. *TrAC Trends Anal. Chem.* **2016**, *82*, 286–306. [CrossRef]

69. Singh, J.; Sharma, S.; Nara, S. Evaluation of gold nanoparticle based lateral flow assays for diagnosis of enterobacteriaceae members in food and water. *Food Chem.* **2015**, *170*, 470–483. [CrossRef] [PubMed]
70. Sajid, M.; Kawde, A.-N.; Daud, M. Designs, formats and applications of lateral flow assay: A literature review. *J. Saudi Chem. Soc.* **2015**, *19*, 689–705. [CrossRef]
71. Di Nardo, F.; Alladio, E.; Baggiani, C.; Cavalera, S.; Giovannoli, C.; Spano, G.; Anfossi, L. Colour-encoded lateral flow immunoassay for the simultaneous detection of aflatoxin B1 and type-B fumonisins in a single Test line. *Talanta* **2019**, *192*, 288–294 [CrossRef] [PubMed]
72. Hao, L.; Chen, J.; Chen, X.; Ma, T.; Cai, X.; Duan, H.; Leng, Y.; Huang, X.; Xiong, Y. A novel magneto-gold nanohybrid-enhanced lateral flow immunoassay for ultrasensitive and rapid detection of ochratoxin A in grape juice. *Food Chem.* **2021**, *336*, 127710. [CrossRef]
73. Shao, Y.; Duan, H.; Zhou, S.; Ma, T.; Guo, L.; Huang, X.; Xiong, Y. Biotin–Streptavidin System-Mediated Ratiometric Multiplex Immunochromatographic Assay for Simultaneous and Accurate Quantification of Three Mycotoxins. *J. Agric. Food Chem.* **2019**, *67*, 9022–9031. [CrossRef]
74. Li, X.; Wang, J.; Yi, C.; Jiang, L.; Wu, J.; Chen, X.; Shen, X.; Sun, Y.; Lei, H. A smartphone-based quantitative detection device integrated with latex microsphere immunochromatography for on-site detection of zearalenone in cereals and feed. *Sens. Actuators B Chem.* **2019**, *290*, 170–179. [CrossRef]
75. Wu, S.; Liu, L.; Duan, N.; Li, Q.; Zhou, Y.; Wang, Z. Aptamer-Based Lateral Flow Test Strip for Rapid Detection of Zearalenone in Corn Samples. *J. Agric. Food Chem.* **2018**, *66*, 1949–1954. [CrossRef] [PubMed]
76. Zhou, W.; Kong, W.; Dou, X.; Zhao, M.; Ouyang, Z.; Yang, M. An aptamer based lateral flow strip for on-site rapid detection of ochratoxin A in Astragalus membranaceus. *J. Chromatogr. B* **2016**, *1022*, 102–108. [CrossRef] [PubMed]
77. Xing, C.; Dong, X.; Xu, T.; Yuan, J.; Yan, W.; Sui, X.; Zhao, X. Analysis of multiple mycotoxins-contaminated wheat by a smart analysis platform. *Anal. Biochem.* **2020**, *610*, 113928. [CrossRef] [PubMed]
78. Ji, Y.; Ren, M.; Li, Y.; Huang, Z.; Shu, M.; Yang, H.; Xiong, Y.; Xu, Y. Detection of aflatoxin B1 with immunochromatographic test strips: Enhanced signal sensitivity using gold nanoflowers. *Talanta* **2015**, *142*, 206–212. [CrossRef] [PubMed]
79. Chen, Y.; Chen, Q.; Han, M.; Zhou, J.; Gong, L.; Niu, Y.; Zhang, Y.; He, L.; Zhang, L. Development and optimization of a multiplex lateral flow immunoassay for the simultaneous determination of three mycotoxins in corn, rice and peanut. *Food Chem.* **2016**, *213*, 478–484. [CrossRef] [PubMed]
80. Yu, Q.; Li, H.; Li, C.; Zhang, S.; Shen, J.; Wang, Z. Gold nanoparticles-based lateral flow immunoassay with silver staining for simultaneous detection of fumonisin B1 and deoxynivalenol. *Food Control* **2015**, *54*, 347–352. [CrossRef]
81. Zhang, T.; Lei, L.; Tian, M.; Ren, J.; Lu, Z.; Liu, Y.; Liu, Y. Multifunctional Fe$_3$O$_4$@Au supraparticle as a promising thermal contrast for an ultrasensitive lateral flow immunoassay. *Talanta* **2021**, *222*, 121478. [CrossRef]
82. Wang, J.; Ren, Y.; Zhang, B. Application of Microfluidics in Biosensors. In *Advances in Microfluidic Technologies for Energy and Environmental Applications*; IntechOpen: London, UK, 2020.
83. Machado, J.M.D.; Soares, R.R.G.; Chu, V.; Conde, J.P. Multiplexed capillary microfluidic immunoassay with smartphone data acquisition for parallel mycotoxin detection. *Biosens. Bioelectron.* **2018**, *99*, 40–46. [CrossRef]
84. Liu, W.; Zhu, Y. Development and application of analytical detection techniques for droplet-based microfluidics-A review. *Anal. Chim. Acta* **2020**, *1113*, 66–84. [CrossRef]
85. Alsaeed, B.; Mansour, F.R. Distance-based paper microfluidics; principle, technical aspects and applications. *Microchem. J.* **2020**, *155*, 104664. [CrossRef]
86. LIANG, S.-J.; MAO, J.-K.; GONG, C.; YU, D.-D.; ZHOU, J.-G. Research and Application Progress of Paper-based Microfluidic Sample Preconcentration. *Chinese J. Anal. Chem.* **2019**, *47*, 1878–1886. [CrossRef]
87. Man, Y.; Li, A.; Li, B.; Liu, J.; Pan, L. A microfluidic colorimetric immunoassay for sensitive detection of altenariol monomethyl ether by UV spectroscopy and smart phone imaging. *Anal. Chim. Acta* **2019**, *1092*, 75–84. [CrossRef] [PubMed]
88. Kasoju, A.; Shrikrishna, N.S.; Shahdeo, D.; Khan, A.A.; Alanazi, A.M.; Gandhi, S. Microfluidic paper device for rapid detection of aflatoxin B1 using an aptamer based colorimetric assay. *RSC Adv.* **2020**, *10*, 11843–11850. [CrossRef]
89. Jayawardane, B.M.; Wei, S.; McKelvie, I.D.; Kolev, S.D. Microfluidic Paper-Based Analytical Device for the Determination of Nitrite and Nitrate. *Anal. Chem.* **2014**, *86*, 7274–7279. [CrossRef] [PubMed]
90. Jiang, Q.; Wu, J.; Yao, K.; Yin, Y.; Gong, M.M.; Yang, C.; Lin, F. Paper-Based Microfluidic Device (DON-Chip) for Rapid and Low-Cost Deoxynivalenol Quantification in Food, Feed, and Feed Ingredients. *ACS Sens.* **2019**, *4*, 3072–3079. [CrossRef]
91. Zhang, Y.; Li, G.; Wu, D.; Li, X.; Yu, Y.; Luo, P.; Chen, J.; Dai, C.; Wu, Y. Recent advances in emerging nanomaterials based food sample pretreatment methods for food safety screening. *TrAC Trends Anal. Chem.* **2019**, *121*, 115669. [CrossRef]
92. Liu, M.; Wang, J.; Yang, Q.; Hu, N.; Zhang, W.; Zhu, W.; Wang, R.; Suo, Y.; Wang, J. Patulin removal from apple juice using a novel cysteine-functionalized metal-organic framework adsorbent. *Food Chem.* **2019**, *270*, 1–9. [CrossRef]
93. Rahmani, A.; Jinap, S.; Soleimany, F. Qualitative and Quantitative Analysis of Mycotoxins. *Compr. Rev. Food Sci. Food Saf.* **2009**, *8*, 202–251. [CrossRef]
94. Agriopoulou, S.; Stamatelopoulou, E.; Varzakas, T. Advances in Analysis and Detection of Major Mycotoxins in Foods. *Foods* **2020**, *9*, 518. [CrossRef]
95. TECNICA ASTORI Aflatoxins/Mycotoxins ELISA Kits. Available online: https://www.astorilab.com/index.php/en/products-catalogue/11-aflatoxins-mycotoxins-elisa-kits (accessed on 15 November 2020).

96. Charm Sciences INC Mycotoxin Tests. Available online: https://www.charm.com/products/test-and-kits/mycotoxin-tests/ (accessed on 15 November 2020).
97. Cusabio Mycotoxin Rapid Test. Available online: https://www.cusabio.com/mycotoxin.html (accessed on 15 November 2020).
98. Elabscience Mycotoxin Testing. Available online: https://www.elabscience.com/Products-mycotoxins-215.html (accessed on 15 November 2020).
99. EnviroLogix Mycotoxin Test Kits. Available online: https://www.envirologix.com/mycotoxin-testing/mycotoxin-test-kits/ (accessed on 15 November 2020).
100. Eurofins Mycotoxins. Available online: https://tecna.eurofins-technologies.com/home/products/mycotoxins/ (accessed on 15 November 2020).
101. HELICA ELISA Kits, Research Products. Available online: https://gentaur.com/search/Supplier/Helica (accessed on 15 November 2020).
102. NEOGEN Mycotoxin Testing. Available online: https://www.neogen.com/en-gb/categories/mycotoxins/?q=12&s=MostPopular (accessed on 15 November 2020).
103. R-Biopharm Tests for the detection of mycotoxins in food. Available online: https://food.r-biopharm.com/analytes/mycotoxins/ (accessed on 16 November 2020).
104. Romer Labs Mycotoxin Test Kits | Fast & Reliable Mycotoxin Detection. Available online: https://www.romerlabs.com/en/products/test-kits/mycotoxin-test-kits/ (accessed on 16 November 2020).
105. Vicam Mycotoxin Testing Solutions. Available online: http://vicam.com/products (accessed on 16 November 2020).
106. Beacon Anaytical Systems Mycotoxins Plate Kits. Available online: https://www.beaconkits.com/mycotoxins (accessed on 16 November 2020).
107. Creative Diagnostics Mycotoxins Analysis. Available online: https://www.creative-diagnostics.com/food-analysis/tag-mycotoxins-2.htm (accessed on 16 November 2020).
108. PerkinElmer Mycotoxins Analysis in Food. Available online: https://www.perkinelmer.com/uk/category/mycotoxins-in-food (accessed on 16 November 2020).
109. Unisensor Mycotoxins Sensor Kits. Available online: https://unisensor.be/en (accessed on 16 November 2020).
110. Pribolab—Solutions in Food Safety and Testing-mycotoxin Testing. Available online: http://www.pribolab.com/ (accessed on 16 November 2020).
111. Randox Feed & Cereals—Randox Food. Available online: https://www.randoxfood.com/feed-and-cereals-analysis/#1521620960005-d87fa0aa-85bd (accessed on 16 November 2020).
112. Novakits Mycotoxines. Available online: http://www.novakits.com/20-mycotoxines (accessed on 16 November 2020).
113. Sigma-Aldrich Mycotoxin Testing. Available online: https://www.sigmaaldrich.com/life-science/cell-biology/antibodies/antibody-products.html?TablePage=114571894 (accessed on 16 November 2020).
114. Bio-Check Mycotoxin Testing Kits. Mycotoxin Lab and on-site Detection Tests. Available online: https://www.biocheck.uk/mycotoxins (accessed on 16 November 2020).
115. Li, W.; Powers, S.; Dai, S.Y. Using commercial immunoassay kits for mycotoxins: 'joys and sorrows'? *World Mycotoxin J.* **2014**, *7*, 417–430. [CrossRef]

Article

Sensitive Aflatoxin B1 Detection Using Nanoparticle-Based Competitive Magnetic Immunodetection

Jan Pietschmann [1], Holger Spiegel [1], Hans-Joachim Krause [2], Stefan Schillberg [1] and Florian Schröper [1,*]

1   Fraunhofer Institute for Molecular Biology and Applied Ecology IME, Forckenbeckstraße 6, 52074 Aachen, Germany; jan.pietschmann@ime.fraunhofer.de (J.P.); holger.spiegel@ime.fraunhofer.de (H.S.); stefan.schillberg@ime.fraunhofer.de (S.S.)
2   Institute of Biological Information Processing, Bioelectronics IBI-3, Forschungszentrum Jülich, 52428 Jülich, Germany; h.-j.krause@fz-juelich.de
*   Correspondence: florian.schroeper@ime.fraunhofer.de

Received: 21 April 2020; Accepted: 20 May 2020; Published: 20 May 2020

Abstract: Food and crop contaminations with mycotoxins are a severe health risk for consumers and cause high economic losses worldwide. Currently, different chromatographic- and immuno-based methods are used to detect mycotoxins within different sample matrices. There is a need for novel, highly sensitive detection technologies that avoid time-consuming procedures and expensive laboratory equipment but still provide sufficient sensitivity to achieve the mandated detection limit for mycotoxin content. Here we describe a novel, highly sensitive, and portable aflatoxin B1 detection approach using competitive magnetic immunodetection (cMID). As a reference method, a competitive ELISA optimized by checkerboard titration was established. For the novel cMID procedure, immunofiltration columns, coated with aflatoxin B1-BSA conjugate were used for competitive enrichment of biotinylated aflatoxin B1-specific antibodies. Subsequently, magnetic particles functionalized with streptavidin can be applied to magnetically label retained antibodies. By means of frequency mixing technology, particles were detected and quantified corresponding to the aflatoxin content in the sample. After the optimization of assay conditions, we successfully demonstrated the new competitive magnetic detection approach with a comparable detection limit of 1.1 ng aflatoxin B1 per mL sample to the cELISA reference method. Our results indicate that the cMID is a promising method reducing the risks of processing contaminated commodities.

Keywords: frequency mixing technology; immunofiltration; magnetic beads; mycotoxin

Key Contribution: Novel nanoparticle-based aflatoxin B1 detection system using magnetic frequency mixing technology for the possibility of highly sensitive on-field testing.

---

1. Introduction

According to the Food and Agriculture Organization of the United Nations and a recent study by Eskola et al. in 2019, approximately 25% of food-crops worldwide are contaminated with mycotoxins, a group of secondary metabolites produced by molds [1,2]. Particularly, the toxins of *Aspergillus*, *Fusarium*, and *Penicillium* species are the most detrimental ones because these so-called aflatoxins, ochratoxins, trichothecenes (especially deoxynivalenol), and zearalenone have various adverse effects on the health of humans and animals [3–8]. From all mycotoxins, aflatoxin B1 (AFB1), mainly produced by *Aspergillus* species, has the strongest adverse effects on health, as it might lead to liver cancer [9–12]. Several factors can promote the enrichment of mycotoxins within food products. Key drivers are late harvests of crops as well as elevated humidity and temperature during storage. Within the EU, but also in the USA, Brazil, and other countries, regulatory requirements regarding the maximum tolerable levels of mycotoxins in various foods have been established. Regulatory limits in the EU for the IARC group 1 carcinogen aflatoxin B1 are set to 2–8 ng/g (ppb; ng·mL^{-1}) in grain, corn, nuts, and fruits, as published in (EG) Nr. 1881/2006. Especially due to those harsh effects and corresponding low regulatory limits, highly sensitive and reliable detection technologies are of primary importance.

Currently, there are basically three analytical technologies used for mycotoxin testing. On the one hand, laboratory-based liquid chromatography (LC) coupled to mass spectrometry (MS) is the most common method, as well as enzyme-linked immunosorbent assay (ELISA), are routinely used for highly sensitive mycotoxin testing. On the other hand, lateral flow assays (LFA) are used for fast but less sensitive on-field testing [13–19]. Although LC-MS/MS methods have the great advantage of high sensitivity and simultaneous detection of currently more than 500 mycotoxins within a single run, expensive equipment, highly qualified staff, and a possible complex sample cleanup inhibit the applicability for fast on-site testing [20]. In comparison to LC-based methods, ELISA techniques are cheaper and, according to Renauld and colleagues (2019), faster in assay procedure, but also require a lab with the corresponding equipment for sample preparation and analysis [21].

In the field of ELISA, the most commonly used format for aflatoxin detection is the competitive assay because of the small molecular structure of the antigen that prevents the simultaneous binding of two antibodies, which would be the prerequisite for a sandwich ELISA. The basic working principle is the coating of mycotoxin-conjugate onto a microtiter plate followed by the addition and incubation of a mycotoxin-specific antibody in the presence of a liquid extract of sample material. In this step, mycotoxins in the sample compete with coated mycotoxins for antibody-binding. Subsequently, antibodies saturated with soluble mycotoxins are removed during a washing step and cannot contribute to a signal, achieved by either direct readout or indirect readout by means of a labeled secondary antibody. In this format, a low signal corresponds to a high concentration of mycotoxin within the sample, and vice versa. Most of the currently commercially available ELISA-based assay formats have a limit of detection (LOD) ranging from 1 µg·mL^{-1} (equals ng·mL^{-1} or ppb) up to 50 µg·mL^{-1}, for example, Ridascreen Aflatoxin B1 30/15 test kit (r-biopharm, Darmstadt, Germany) or AgraQuant Aflatoxin B1 ELISA test (Romer Labs, Butzbach, Germany). Nevertheless, due to the requirement for laboratory equipment, their usability for on-site testing is very limited. In such cases, commercially available LFA have the highest potential for fast and cost-effective on-site mycotoxin testing. However, most of these assay systems provide only qualitative results with LODs ranging from 4 ppb to 5 ppb, as achieved with RIDA®Quick Aflatoxin test kit(r-biopharm, Darmstadt, Germany) or AFB1 (Aflatoxin B1) lateral flow assay kit (Elabscience, Texas, USA), respectively. Primarily due to matrix interference effects, these assays can result in false-positive detection [22,23].

A novel, portable magnetic immunodetection approach has been described in previous studies [24–28]. The detection of human and plant pathogens, as well as various proteins by sensing and quantifying superparamagnetic particles (MP) with the help of a portable magnetic reader (Figure 1A), have been successfully demonstrated. This device can be operated using a conventional external power adapter or a portable battery, allowing an on-site readout without electrical infrastructure. For this immunomagnetic detection approach, MPs were functionalized with monoclonal antibodies directed against target molecules, retained in a sandwich-based manner within an immunofiltration column, and can be detected by means of frequency mixing magnetic detection (FMMD) technology [29].

In this technique, magnetic particles are subjected to two sinusoidal magnetic fields of different frequencies generated by two excitation coils, which are schematically shown within the measurement head in Figure 1B. Here, MPs are exposed to a low- and high-frequency magnetic field, so-called driving frequency, generated by the outermost coil, and excitation frequency, generated by the middle-positioned coil. The low frequency with 61 Hz (f_2) has an amplitude of a few millitesla, resulting in alternating positive and negative magnetic saturation of superparamagnetic particles oscillating with a frequency of 2f_2 of 122 Hz [29]. The high-frequency magnetic excitation field

(f_1) with 49 kHz probes the magnetization state of the superparamagnetic particles and yields an iron oxide dose-dependent signal when the low-frequency driving field is close to zero. Finally, the resulting mixing frequency signal of $f_1 + 2f_2$ can be demodulated and detected by the innermost coil, composed of two adjacent sections, so-called detection coil (upper one) and reference coil (lower one). Those sections differ only in the winding-orientation of coils. With this clock- and counterclockwise orientation yielding induced voltages of opposite sign, the directly induced excitation field can be canceled out. By placing the sample carrying the MPs in the detection head, the resulting signal is amplified, measured, and directly visualized at the touchscreen of the handheld, portable magnetic reader (Figure 1). Based on a calibration curve, the detected signal can be attributed to the amount of analyte within the sample.

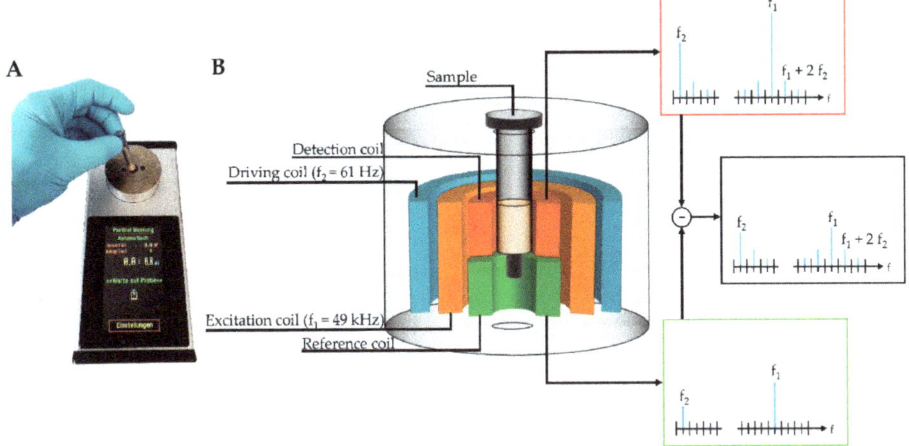

Figure 1. (**A**) Handheld, portable frequency mixing-based readout device and (**B**) schematic cross-section of detection head composed of driving coil providing the low driving frequency (f_2), the excitation coil providing the high excitation frequency (f_1) and detection unit based on a detection coil detecting the resulting mixing frequency signal of magnetic particles (MPs; $f_1 + 2f_2$) together with the directly induced signal and the reference coil detecting only the directly induced signal. The finally resulting measuring signal does not contain the directly induced excitation due to the opposite winding direction of detection and reference coil.

Motivated by the above-described drawbacks of currently used analytical methods for sensitive on-field testing, the aim of this study was to develop a novel, highly sensitive, and portable assay based on competitive magnetic immunodetection (cMID) and FMMD. The sensitivity should be comparable to a laboratory-based ELISA. Hence, initially, a competitive ELISA (cELISA) was established, serving as a reference method. Assay parameters, such as the used coating and antibody concentrations, were optimized to reach a sufficient sensitivity for the detection of aflatoxin B1. Afterward, the cMID assay was established using the same optimization strategy in combination with further evaluation of the required amount of nanoparticles. The basic principle of cMID is the use of biotinylated antibodies, which can be enriched within the coated immunofiltration column by a competitive binding reaction depending on the amount of pre-captured mycotoxin. By flushing magnetic particles functionalized with streptavidin through the column by gravity flow, particles can bind to retained antibodies and subsequently be detected using FMMD. Figure 2 visualizes the basic cMID principle (Figure 2A) and the competitive binding reaction within the column with the corresponding measuring signal (Figure 2B).

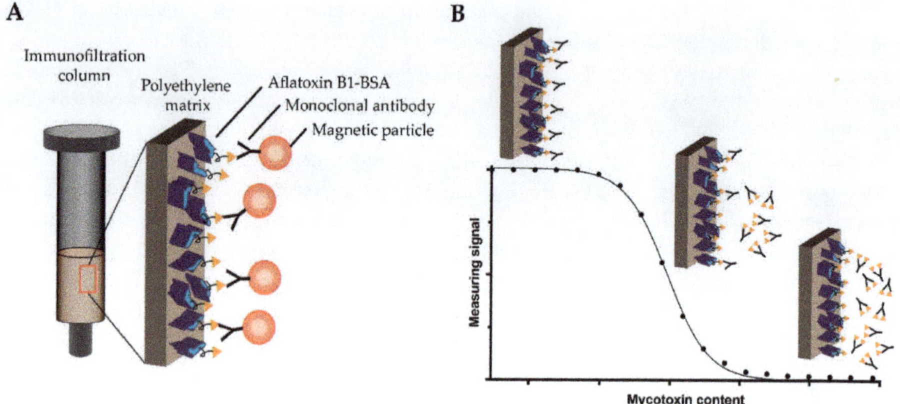

Figure 2. Schematic overview of the competitive magnetic immunodetection principle. (**A**) Immunofiltration column coated with aflatoxin B1-BSA mycotoxin conjugate with bound biotinylated monoclonal antibodies targeting aflatoxin B1. Magnetic particles functionalized with streptavidin bind to antibodies and can be detected by FMMD. (**B**) After pre-incubation of biotinylated, monoclonal antibodies with serially diluted free aflatoxin B1, the sample is flushed through an aflatoxin B1-BSA coated immunofiltration column. Non-saturated antibodies bind to the coated antigen and are retained within the matrix. The higher the mycotoxin content within the sample, the more antibodies are saturated and are flushed through the column. Afterward, streptavidin-functionalized magnetic particles are applied onto the column, bind to retained antibodies and can be detected using FMMD.

2. Results

2.1. Competitive ELISA Conditions

As a reference method to novel cMID, a cELISA was established using optimized conditions defined by checkerboard titration. Here, aflatoxin B1-BSA conjugate was coated, followed by the application of aflatoxin B1-specific monoclonal antibodies pre-incubated with soluble aflatoxin B1 with concentrations ranging from 0.006 ng·mL^{-1} to 5000 ng·mL^{-1} sample buffer. Absorbance was measured at 405 nm by indirect readout using a secondary antibody conjugated to horseradish peroxidase (HRPO) targeting mouse antibodies and application of respective substrate.

2.1.1. ELISA Checkerboard Titration Test

In order to determine optimal aflatoxin B1-BSA (AFB1-BSA) coating concentrations and the appropriate amount of aflatoxin B1-specific monoclonal antibody to obtain the highest possible sensitivity, a checkerboard titration test was performed. As shown in Figure A1, AFB1-BSA coating concentrations ranging from 0.1 µg·mL^{-1} up to 5 µg·mL^{-1} in combination with aflatoxin B1-specific monoclonal antibody (mAb) AFB1_002 concentrations ranging from 19.5 ng·mL^{-1} up to 1250 ng·mL^{-1} were tested in a 96-well plate. Increasing signals were achieved with antibody concentrations above 19.5 ng·mL^{-1} and coating concentrations up to 0.6 µg·mL^{-1}. Coating of AFB1-BSA with higher amounts than 0.6 µg·mL^{-1} resulted in saturated readout signals at all antibody concentrations that did not further increase. Considering sensitivity, the lowest possible amount of antibody should be used with coating concentrations resulting in highest possible signals. As a consequence, combinations of 75 ng·mL^{-1} or 150 ng·mL^{-1} antibody with 0.2 µg·mL^{-1} and 0.4 µg·mL^{-1} coating of AFB1, respectively, seem to be best suited for further experiments.

2.1.2. cELISA Calibration Curve Experiments

To identify the optimal assay parameters, calibration experiments were performed with the previously most promising combinations of AFB1-BSA coating and aflatoxin B1-specific monoclonal antibody AFB1_002 concentrations (Figure 3). Free aflatoxin B1 was used as a competitor, dilutions ranging from 0.006 ng·mL^{-1} to 50,000 ng·mL^{-1}. The combination of 0.2 µg·mL^{-1} coating and 150 ng·mL^{-1} antibody resulted in the highest sensitivity in combination with stable data behavior. Corresponding IC$_{50}$ and Limit of Detection (LOD) values of all four combinatorial experiments are shown in Table 1. Although antibody concentrations below 75 ng·mL^{-1} should theoretically lead to a higher sensitivity, this was actually not observed due to divergent measuring values and thus a non-reliable assay procedure under these conditions (compare 0.4 µg·mL^{-1} coating and 75 ng·mL^{-1} mAb). Furthermore, reducing the amount of mAb resulted in a twofold increase of readout time from approximately 10 min to more than 20 min (data not shown).

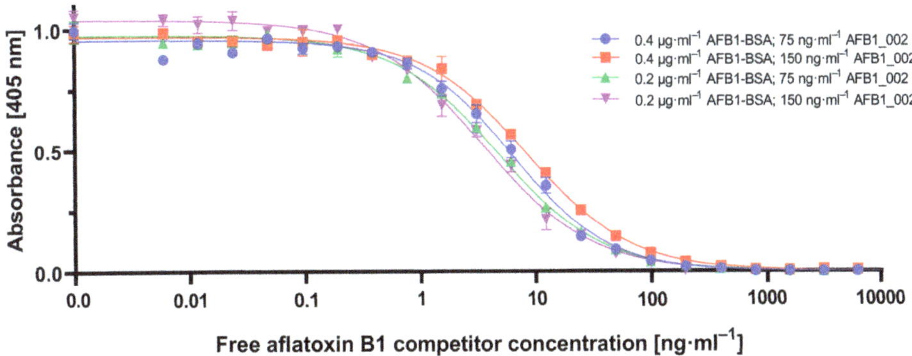

Figure 3. Competitive ELISA-based calibration curves of different pairs of aflatoxin B1-BSA coating and aflatoxin B1-specific monoclonal antibody AFB1_002 concentrations pairs. As a competitor, free aflatoxin B1 in dilutions ranging from 0.006 ng·mL^{-1} up to 50,000 ng·mL^{-1} in sample buffer was used. The indirect readout was done at 405 nm after the application of mouse-specific secondary antibody conjugated with horseradish peroxidase and respective substrate. Each data point represents the mean ± SD ($n = 3$).

Table 1. Sensitivity values of different cELISA-based calibration measurements with different pairs of aflatoxin B1-BSA coating and aflatoxin B1-specific monoclonal antibodies.

ELISA Condition	IC$_{50}$	Limit of Detection
0.2 µg·mL^{-1} Coating 75 ng·mL^{-1} Antibody	4.7 ng·mL^{-1}	0.69 ng·mL^{-1}
0.2 µg·mL^{-1} Coating 150 ng·mL^{-1} Antibody	3.5 ng·mL^{-1}	0.28 ng·mL^{-1}
0.4 µg·mL^{-1} Coating 75 ng·mL^{-1} Antibody	6.3 ng·mL^{-1}	0.75 ng·mL^{-1}
0.4 µg·mL^{-1} Coating 150 ng·mL^{-1} Antibody	8.4 ng·mL^{-1}	0.88 ng·mL^{-1}

Considering the reliability of the developed cELISA, a further repetition of the best-paired concentration setting, as shown in Table 1, was done. Results yielded in an averaged IC$_{50}$ value of 3.79 ng·mL^{-1} and an averaged LOD of 0.39 ng·mL^{-1}, as shown in Figure 4. Those sensitivity values obtained by our ELISA setup were used as a reference for competitive magnetic immunodetection development.

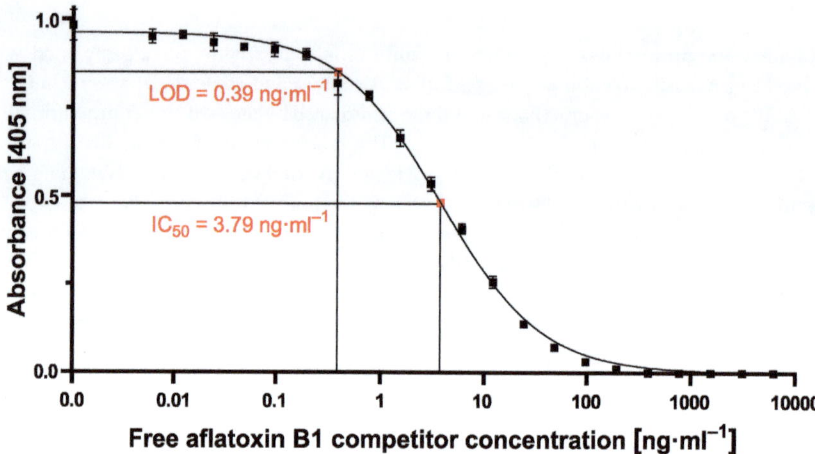

Figure 4. Averaged calibration curve with 0.2 µg·mL^{-1} AFB1-BSA conjugate coating and 150 ng·mL^{-1} mAb with aflatoxin B1 competitor concentrations ranging from 0.006 ng·mL^{-1} to 500,000 ng·mL^{-1}. Measuring signal was achieved by indirect readout at 405 nm with mouse-specific secondary antibody conjugated to horseradish peroxidase and respective substrate LOD: limit of detection; IC$_{50}$: half maximal inhibitory concentration. Each data point represents the mean ± SD (*n* = 12).

2.2. Competitive Magnetic Immunodetection

2.2.1. Development of Competitive Magnetic Immunodetection

In analogy to the development of the cELISA, a similar strategy was used for establishing suitable cMID conditions regarding AFB1-BSA coating and suitable antibody concentrations. For this purpose, equilibrated polyethylene filters were coated with AFB1-BSA conjugate concentrations ranging from 0.5 µg·mL^{-1} up to 10 µg·mL^{-1} with one column per condition. Remaining binding sites were blocked with a BSA solution. Then aflatoxin B1-specific biotinylated monoclonal antibody AFB1_002 in concentrations ranging from 0.3 µg·mL^{-1} up to 10 µg·mL^{-1} were applied onto these columns after pre-incubation with 180 µg·mL^{-1} of 700 nm superparamagnetic streptavidin-functionalized particles (Figure A2). After rinsing the columns with PBS by gravity flow, the columns were inserted into the handheld magnetic readout device and superparamagnetic particles were excited by frequency mixing technology resulting in a response signal in the millivolt (mV) range after amplification.

Especially in the range of low antibody concentrations between 0.3 µg·mL^{-1} and up to 1.3 µg·mL^{-1}, a saturation of measurement signal was reached at higher coating concentrations. In contrast, using higher antibody concentrations, no clear saturation of signal was observed even at higher coating concentrations. Furthermore, using antibody concentrations of 5 µg·mL^{-1} or 10 µg·mL^{-1} resulted in high and erratic signal variability. As explained in Section 2.1.1, the lowest possible antibody concentration should be used for obtaining the highest possible sensitivity. On the other hand, reducing the antibody concentrations leads to a reduction of possible binding sites for magnetic particles and, as a consequence, to a reduction in measuring signal, which may limit the dynamic range as well as the overall robustness of the assay. To obtain the highest possible measuring signal in combination with the highest possible sensitivity, 2 µg·mL^{-1} coating in combination with 2.5 µg·mL^{-1} biotinylated antibody were used.

As shown in Figure A2, the measuring signal increased with higher antibody concentrations. Therefore, it can be assumed that at the used concentrations of 180 µg·mL^{-1} there is still an excess of magnetic particles when using 2.5 µg·mL^{-1} biotinylated mAb. To test this hypothesis, different amounts of 700 nm magnetic particles ranging from 2.5 µg·mL^{-1} to 180 µg·mL^{-1} were flushed through immunofiltration columns coated with 2 µg·mL^{-1} AFB1-BSA conjugate after applying 2.5 µg·mL^{-1}

biotinylated antibody to the matrix (Figure 5). The expected saturation of measuring signal was observed when using more than 80 µg·mL^{-1} magnetic particles. Adding 80 µg·mL^{-1} resulted in 41.2 mV ± 5.5 mV, in comparison to higher bead concentrations with averaged signals of 47.4 mV ± 4.1 mV. For further experiments, 80 µg·mL^{-1} magnetic particle suspension was used.

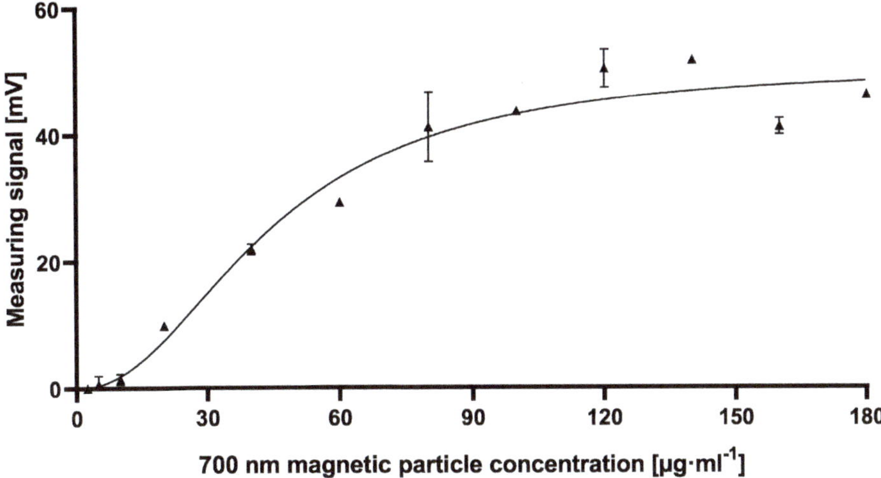

Figure 5. Dose-dependent measuring signal of 700 nm streptavidin-functionalized magnetic particles after applying 2.5 µg·mL^{-1} biotinylated AFB1_002 monoclonal antibody onto 2 µg·mL^{-1} AFB1-BSA coated immunofiltration columns. Each data point represents mean ± SD (*n* = 2).

2.2.2. cMID Calibration Curve Experiments

Once the optimal ratio of assay reagents of 2 µg·mL^{-1} of the coating antigen and 2.5 µg·mL^{-1} of mycotoxin-specific antibody in combination with 80 µg·mL^{-1} magnetic particles was determined based on the above-shown experiments, cMID calibration experiments for the detection of aflatoxin B1 were performed, see Figure 6A. For this, samples with serially diluted free AFB1 in the range of 0.006 ng·mL^{-1} to 50,000 ng·mL^{-1} were prepared. After adding biotinylated mAb to various dilutions of the analyte, a pre-incubation of one hour was done for a complete capturing of mycotoxins. Subsequently, the reaction mixture was applied onto AFB1-BSA coated and blocked columns. Afterward, 700 nm diameter magnetic particles functionalized with streptavidin at the above-determined optimum concentration of 80 µg·mL^{-1} were added. As shown in Figure 6A, the data show the expected reciprocal correlation of mycotoxin concentration and signal but with high variability of signals for dilutions in the low- ng·mL^{-1} range, which prevents a reliable determination of sensitivity parameters. We speculate that this observation could be caused by sterical hindrance within the polyethylene matrix due to the usage of big-sized particles, as can be seen in the standard deviation and data fluctuation with bead concentrations of 80 µg·mL^{-1} or higher (Figure 5). Based on these findings, the assay was repeated with the same parameters except the size of beads used. For this, tenfold smaller magnetic particles were used (70 nm) in order to avoid steric hindrance. As shown in Figure 6B, the use of smaller particles results in an approximately tenfold increased detection signal of roughly 600 mV with reduced variability. With this adapted assay procedure, IC$_{50}$ values of 5.4 ng·mL^{-1} and a LOD of 1.1 ng·mL^{-1} were determined, which is in the same sensitivity range as the cELISA (Figure 4).

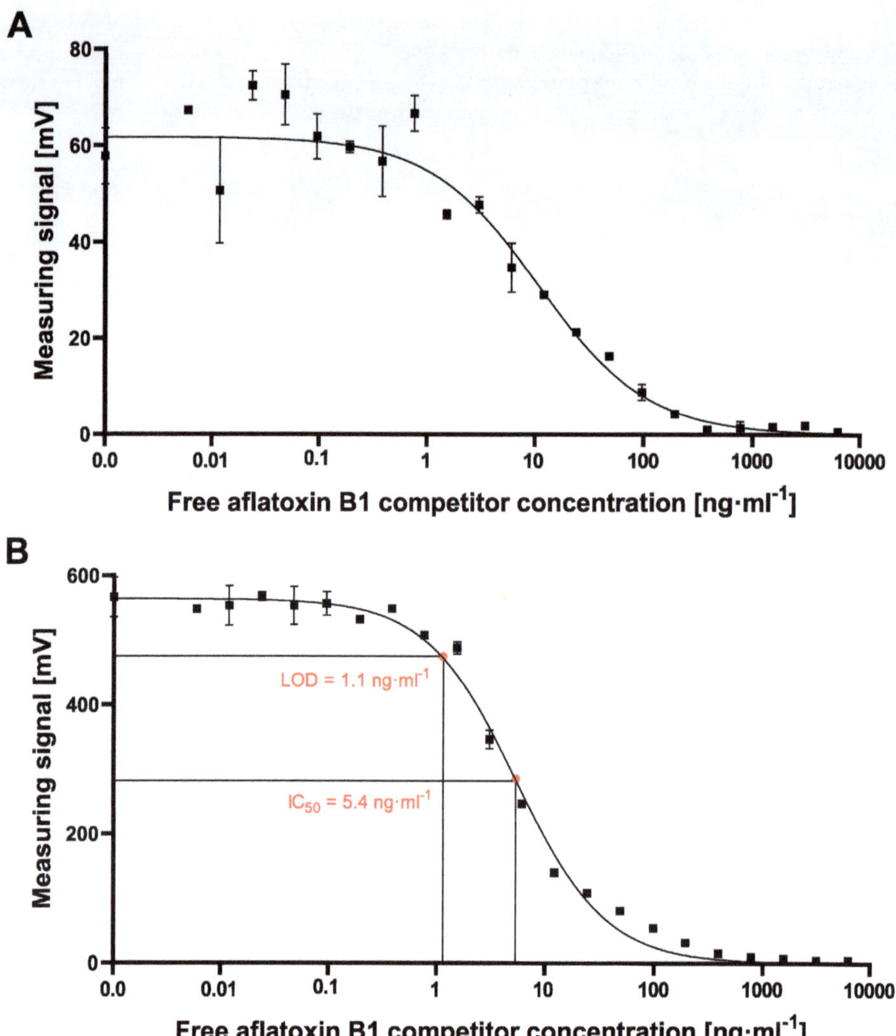

Figure 6. cMID calibration curves with 2 μg·mL^{-1} AFB1-BSA conjugate coating and 2.5 μg·mL^{-1} biotinylated mAb in combination with (**A**) 700 nm streptavidin-functionalized magnetic particles, and (**B**) 70 nm streptavidin-functionalized magnetic particles with aflatoxin B1 competitor concentrations ranging from 0.006 ng·mL^{-1} to 500,000 ng·mL^{-1}. LOD, limit of detection; IC$_{50}$, half maximal inhibitory concentration. Each data point represents the mean ± SD (n = 2).

3. Discussion

In this study, we demonstrated for the first time, a competitive magnetic immunodetection assay for efficient detection and quantification of aflatoxin B1 with comparable sensitivity to described laboratory-based cELISA. Especially due to the handheld reader device which can be operated using a portable battery and without additional laboratory equipment, the method described here is suitable for sensitive on-site aflatoxin B1 testing. Commonly used analytical methods for detection of mycotoxins either exhibit high sensitivity in a laboratory-based setting or can be used on-site with reduced sensitivity [21].

To establish the novel cMID method, several concentration settings of a cELISA as reference method were tested and optimized using checkerboard titration (Figure A1). Here, the most suitable combination of antigen coating concentration and the amount of used antibody was determined. As explained in Section 2.1.1, in all tested antibody concentrations, for a coating of 0.4 $\mu g \cdot mL^{-1}$ and onwards, a saturation in measuring signal could be detected (Figure A1). A clear difference to this can be seen within the checkerboard optimization experiment for cMID (Figure A2). Here, a saturation-like performance can either be seen when using antibody concentrations up to 2.5 $\mu g \cdot mL^{-1}$ or, if further increasing antibody concentrations, saturation cannot be reached. This could be due to the more than 40-fold higher protein binding surface of immunofiltration columns in comparison to the binding surface of an ELISA microtiter plate well. Especially due to the highly porous membrane of the immunofiltration column, a higher protein binding capacity can be reached. High binding 96-well microtiter plates (Greiner Bio-One, article number 655061) have a maximum binding capacity of 600 $ng \cdot cm^2$, whereas ABICAP immunofiltration columns can bind up to 24,000 $ng \cdot cm^2$, which in fact explains the absence of signal saturation at the used conditions. A further difference when using checkerboard titration for assay optimization of cELISA or cMID can be seen in fluctuating readout signals, whereas in ELISA, concentration-dependent signals were obtained with almost no variability between consecutive measuring points. In contrast, signals generated in MID checkerboard titration varied, especially at high antibody concentrations. This could be caused by an overdose of magnetic particles, which might result in sterical hindrances and a partly erratic aggregation of particles within the assay matrix. Based on this hypothesis, a magnetic-particle dose adaptation was made by reducing the applied concentration towards robustly detectable saturated measuring signals (Figure 5).

However, the resulting cMID assay did not lead to the expected results when using 700 nm magnetic beads. Switching to 70 nm magnetic particles instead resulted in a tenfold increase of measuring signal combined with reduced variability and provides a basis for a highly sensitive and reliable assay (Figure 6). The results presented here correlate with data published by Achtsnicht et al. in 2019, where the authors used magnetic frequency mixing for the detection of cholera toxin subunit B (CTB) in a sandwich-based manner [27]. In their experiments, using magnetic particles with hydrodynamic diameters of 75 nm and 1010 nm, they observed an eight-fold higher response signal with smaller particles compared to the larger beads when detecting a concentration of 750 $ng \cdot mL^{-1}$ CTB. The most simple explanation for this striking difference would be the higher absolute number of small particles compared to larger particles when used at identical concentrations (80 $\mu g \cdot mL^{-1}$) since the molar concentration is proportional to the molecular weight of the particles. By applying a higher ratio of small particles onto the column, a higher amount of small MPs could bind to antigens coated on the column surface. Another explanation could be the occurrence of steric hindrance within the ABICAP immunofiltration matrix when using large particles. Especially if there is a low concentration of free mycotoxin within the sample, a high amount of biotinylated antibody will be retained within the matrix in a competitive assay approach, resulting in a high antibody density on the matrix surface. When large MPs are applied, steric hindrance can occur at the surface. Either multiple, closely located biotinylated antibodies could be linked to one MP, or multiple antibodies could be blocked due to the size of large MPs. Here, especially the thousand-fold increased volume of 700 nm diameter MPs in comparison to 70 nm MPs in diameter might have the most important role. By the greatly increased size and volume, several antibodies are covered by one single MP, so that they are no longer accessible for binding of further MPs. By using smaller particles, shielding effects and sterical hindrances at the surface are reduced, resulting in a higher magnetic particle density within the column, and finally, a higher readout signal. A similar effect observed by Achtsnicht et al. (2019) when using small 75 nm instead of large 1010 nm particles in their MID experiments was also explained by the blocking of remaining binding sites [27]. The authors also found similar high standard deviations when large particles remain in a high concentration within the column (Figure 6A). With such a high SD, sensitive detection of either CTB or aflatoxin B1 is almost impossible. In their study, a LOD of 3.1 $ng \cdot mL^{-1}$ was found when using the large particles, but a more than 15-fold lower LOD of 0.2 $ng \cdot mL^{-1}$ could be reached using the small-sized MPs. In our study, especially due to the high standard deviations at high measuring signals when exciting the 700 nm MPs for aflatoxin B1 detection, a reliable calculation of LOD or IC_{50} values was not possible. However, by using 70 nm particles, high and stable readout signals in combination with a perfectly matching nonlinear-fit ($R^2 = 0.9938$) enabled the reliable calculation of LOD and IC_{50} values of 1.1 ng·mL^{-1} and 5.4 ng·mL^{-1}, respectively, which are in the same sensitivity range as our comparative lab-based cELISA

Although the best performance of the cMID assay was achieved with 70 nm small magnetic particles, the general applicability of 700 nm large particles in cMID assays should not be completely neglected since, in contrast to small particles, larger particles can be separated in a gradient magnetic field. However, the possibility of magnetic separation and thereby enrichment of magnetic particles might play a crucial role in sample preparation where cleanup and concentrating of an analyte is required to achieve sufficient assay sensitivity, as it was used for example by Lee et al. (2013) or Xuan and colleagues (2019) [30,31]. In the case of mycotoxin detection, magnetic separation-based sample cleanup could be beneficial, especially when working with partially soluble or insoluble, or roughly homogenized, e.g., grain samples. After magnetic beads captured the mycotoxin molecules in the sample, a magnetic separation step can be applied in which magnetic particles are retained in the magnetic field, while sample debris can be discarded. Additionally, separated mycotoxin-loaded magnetic particles could be resuspended in a much smaller volume, resulting in an enrichment of analyte. As a consequence, this could lead to increased sensitivity when samples are finally analyzed by cMID.

The 70 nm particles used in this study cannot be efficiently separated in a magnetic gradient field, since their magnetic attraction is low, which was shown by Achtsnicht and colleagues (2018) [32]. They correlated the magnetophoretic velocity of superparamagnetic particles when applying a magnetic field and found an increased velocity with increasing particle size [32]. Typically only bigger particles above 700 nm are used for magnetic separation experiments [30,31]. Further testing of different beads or other strategies as combining big and small-sized particles will be addressed in further studies to enable an optimized pairing of separation and cMID.

In conclusion, we successfully demonstrated the development and implementation of a competitive magnetic immunodetection assay for the detection and quantification of aflatoxin B1 with a LOD of 1.1 ng·mL^{-1}. Based on our findings, it can be concluded that the competitive magnetic immunodetection is a powerful tool for portable, easy-to-use on-site monitoring of mycotoxin contamination in various matrices to reduce the risk of processing contaminated food and agricultural products. In future work, we will focus on the further optimization of the cMID procedure using the 70 nm MPs. Especially, a significant reduction of the assay time from currently approximately 4.5 h to less than one hour should be addressed, similar to that described by Rettcher et al. (2015) [26]. There a sandwich-based MID assay lasting less than 30 min was achieved by a more than 50% reduction of initially needed assay time [26]. By using MPs, pre-conjugated with specific anti-mycotoxin antibodies, ready-to-use coated and blocked immunofiltration columns, and testing successive reductions of incubation times of sample-pre-incubation and competitive binding reaction within the matrix, an additional increase of applicability for on-site testing should be possible. The applicability of the described cMID approach will be further studied regarding the detection and quantification of other mycotoxins, as well as a combination of those toxins within a food matrix of one sample. Especially the multiplex detection of various mycotoxins within one sample will be addressed as shown by the multiplex detection of several antibodies in stacked sample matrices by Achtsnicht et al. (2019) [33]. Here the authors detected two different target molecules, namely antibodies, within one sample solution by stacking 3D printed immunofiltration columns coated with different capture antibodies in a sandwich-based MID approach. By adapting this procedure, multiple individually coated matrices could be used for the specific retention of corresponding anti-mycotoxin antibodies in a cMID assay, and with this, a multiplex detection in a food matrix could be obtained.

4. Materials and Methods

4.1. Material and Chemicals

Dimethyl sulfoxide, Tween-20, Aflatoxin B1, Aflatoxin B1-BSA, EZ-Link™ NHS-PEG4 Biotinylation Kit, ABTS buffer, as well as ABTS tablets were purchased from Merck KGaA, Darmstadt, Germany. NaCl, KCl, $Na_2HPO_4 \times 12\ H_2O$, KH_2PO_4, $Na_2(CO_3)$, $NaHCO_3$, Milk powder, and Albumin Fraction V (biotin-free) were acquired from Carl Roth, Karlsruhe, Germany.

The used coupling buffer was prepared by dissolving 15 mM Na_2CO_3 and 35 mM $NaHCO_3$ in MilliQ-water, and pH was set to 9.6 with glacial acetic acid. Phosphate buffered saline (PBS) was prepared by dissolving 137 mM NaCl, 2.7 mM KCl, 8.1 mM $Na_2HPO_4 \times 12\ H_2O$, and 1.5 mM KH_2PO_4 in MilliQ-water and setting pH to 7.4 with hydrochloric acid. As a washing buffer, PBS-T was produced by adding 0.05% (v/v) Tween-20 to PBS. ELISA blocking buffer (EBP) was prepared by adding 5% (w/v) milk powder to PBS. For magnetic immunodetection experiments, blocking solution consists of 1% (w/v) albumin fraction V (biotin free) in PBS, and is called MID-BP. All other chemicals, except Tween-20, were acquired from Roth.

Immunofiltration columns (ABICAP HP columns) were purchased from Senova Gesellschaft für Biowissenschaft und Technik mbH, Weimar, Germany. High binding 96-well microtiter plates (article number 655061) were purchased from Greiner Bio-One GmbH, Frickenhausen, Germany. Anti-mycotoxin monoclonal antibody AFB1_002 was purchased from fzmb GmbH, Bad Langensalza, Germany. Secondary antibody goat anti-mouse IgG-HRPO (article number 115-035-008) was purchased from Jackson ImmunoResearch Europe Ltd. UK. 700 nm streptavidin-functionalized magnetic particles (nanomag®-CLD/synomag®-CLD; article number 05-19-502 S09718) as well as 70 nm streptavidin-functionalized magnetic particles (synomag®-D, article number 104-19-701) were purchased from micromod Partikeltechnologie GmbH, Rostock, Germany.

4.2. Optimization of ELISA

The most suitable aflatoxin B1-BSA coating concentrations for a competitive ELISA protocol in combination with appropriate anti-aflatoxin B1 monoclonal antibody concentrations were determined by checkerboard titration. Throughout the following protocol, all incubation steps were performed at room temperature for one hour in the dark. For coating varying concentrations of AFB1-BSA, the antigen was diluted in a coupling buffer, and 100 μL per well was added to a 96-well highbinding microtiter plate (Greiner Bio-One) and incubated as mentioned above. After washing each well thrice with PBS-T, all wells were blocked using 200 μL of EBP and incubated. Then, the plate was washed again, and concentrations of monoclonal antibody ranging from 19.5 ng·mL^{-1} up to 1250 ng·mL^{-1}, diluted in PBS, were added. After another washing step, as described above, 100 μL of secondary antibody goat anti-mouse IgG-HRPO, diluted 1:10,000 in PBS was added to each well and incubated. Prior to readout with 100 μL of 1 mg·mL^{-1} ATBS substrate in ABTS buffer, the plate was washed again. Absorption was measured at 405 nm after 10 min incubation in the dark.

4.3. cELISA Procedure

For competitive ELISA procedure, AFB1-BSA conjugate was diluted in coupling buffer and plated with 100 μL per well onto a high binding 96-well microtiter plate. As mentioned in Section 4.2, all incubation steps were performed at room temperature for one hour in the dark. After the washing step with PBS-T, each well was blocked with EBP and incubated. Meanwhile, pre-incubation of free aflatoxin B1 with the monoclonal anti-aflatoxin B1 antibody was prepared. For this, 75 μL of a serial dilution of aflatoxin B1 in PBS was prepared, and then 75 μL of antibody solution diluted in PBS was added and incubated. After washing the blocked assay plate three times with PBS-T, 100 μL of pre-incubated samples were transferred to each well, respectively. After incubation, the plate was washed three times with PBS-T. Subsequently, 100 μL of conjugated goat anti-mouse IgG-HRPO secondary antibody, diluted 1:10,000 in PBS was added to each well and incubated. After washing three times with PBS-T,

100 µL of 1 mg·mL^{-1} ABTS substrate in ABTS buffer was added, and absorbance was measured at 405 nm after 15 min incubation in the dark.

4.4. Preparation of Immunofiltration Columns

The equilibration of immunofiltration columns was done, as described by Rettcher et al. (2015) [26]. In brief, after degassing in 96% (v/v) ethanol in a desiccator at –0.8 bar pressure, columns were washed sequentially with 750 µL 50% (v/v) ethanol-water, 750 µL water and twice 750 µL coupling buffer. Afterward, for coupling of aflatoxin B1-BSA conjugate to the matrix, the conjugate was applied to the column in gravity flow diluted in 500 µL coupling buffer and incubated for one hour at room temperature in dark surrounding. For checkerboard titration, a coating concentration, as shown in Figure A2, was used. For the bead-response curve, a coating concentration of 2 µg·mL^{-1} was applied, as well as for cMID assays. Subsequent washing of the columns was performed twice with 750 µL PBS. Remaining binding sites were blocked by applying twice 750 µL of MID-BP. After the first 750 µL flushed through the column by gravity flow, an incubation time of 5 min was set. After the second time, columns were incubated for further 30 min after they were washed again twice with PBS.

After equilibration or blocking, columns can be stored in the coupling buffer or PBS, respectively, at 4 °C for at least 14 days. In this study, a maximum storage time of one day was used.

4.5. Biotinylation of Anti-Aflatoxin B1 Monoclonal Antibody

For biotinylation of antibodies, the EZ-Link™ NHS-PEG4 Biotinylation Kit was used as described by the manufacturer's instruction.

4.6. Optimization of cMID

After coating and blocking of immunofiltration columns, 500 µL samples of biotinylated antibody, diluted in PBS to various final concentrations between 0.3 µg·mL^{-1} and 10 µg·mL^{-1}, were applied and incubated for one hour at room temperature. Afterward, a washing step was performed by rinsing two times 750 µL PBS through the column. Subsequently, for checkerboard titration, 500 µL of 180 µg·mL^{-1} 700 nm magnetic beads suspension in PBS (pH 7.4) was added and flushed through by gravity flow. For bead response analysis, various concentrations of 700 nm magnetic beads were applied. Another washing step, as described above, was performed. For readout, columns were inserted into the portable magnetic reader and the measuring signal in mV was detected, as previously described in Rettcher et al. (2015) [26].

4.7. cMID Calibration Curve Analysis

For cMID calibration curve experiments, a pre-incubation of free mycotoxin and biotinylated antibody was performed. For this, serially diluted aflatoxin B1 samples in PBS, with concentrations ranging from 0.006 ng·mL^{-1} to 100,000 ng·mL^{-1}, were mixed 1:1 with 225 µL biotinylated antibody, also diluted in PBS. After incubation of one hour at room temperature in a dark surrounding, 500 µL of each sample was applied on coated and blocked columns and also incubated as mentioned above. After washing each column twice with 750 µL PBS, 80 µg·mL^{-1} of 70 nm or 700 nm magnetic particles were applied and incubated, as mentioned before. After washing twice, the readout was done as described above.

4.8. Data Analysis

For competitive ELISA as well as for competitive magnetic immunodetection and data analysis, a Hill Slope fit was done with GraphPad Prism 8.0.0. The following formulas were used to determine the LOD on the signal and on the concentration scale:

$$\text{Signal}_{\text{Limit of Detection}} = \text{Average}_{\text{Sample without Competitor}} - 3\text{x SD}_{\text{Sample without Competitor}} \quad (1)$$

$$\text{Concentration}_{\text{Limit of Detection}} = \left(\sqrt[\text{Hill Slope}]{\frac{\text{Maximum Signal} - \text{Lowest Signal}}{\text{Signal}_{\text{Limit of Detection}} - \text{Lowest Signal}} - 1} \right) \times \text{IC}_{50} \text{ ng·mL}^{-1} \quad (2)$$

Author Contributions: Conceptualization, F.S. and J.P.; methodology, J.P.; validation, J.P.; formal analysis, J.P.; resources, F.S., H.S. S.S. and H.-J.K.; investigation, J.P.; data curation, J.P.; writing—original draft preparation, J.P.; writing—review and editing, F.S., H.S., H.-J.K. and S.S.; visualization, J.P.; supervision, F.S. and H.S.; project administration, F.S., H.S. and S.S.; funding acquisition, F.S. All authors have read and agreed to the published version of the manuscript.

Funding: This research was funded by the land North Rhine-Westphalia and the European Regional Development Fund (ERDF; German: EFRE) under grant number EFRE-0801298, reference number LS-2-1-014 and The APC was funded by Fraunhofer IME.

Acknowledgments: The authors would like to thank Stefan Rasche for his helpful advices and support given in discussions.

Conflicts of Interest: The authors declare no conflict of interest. The funders had no role in the design of the study; in the collection, analyses, or interpretation of data; in the writing of the manuscript, or in the decision to publish the results.

Appendix A

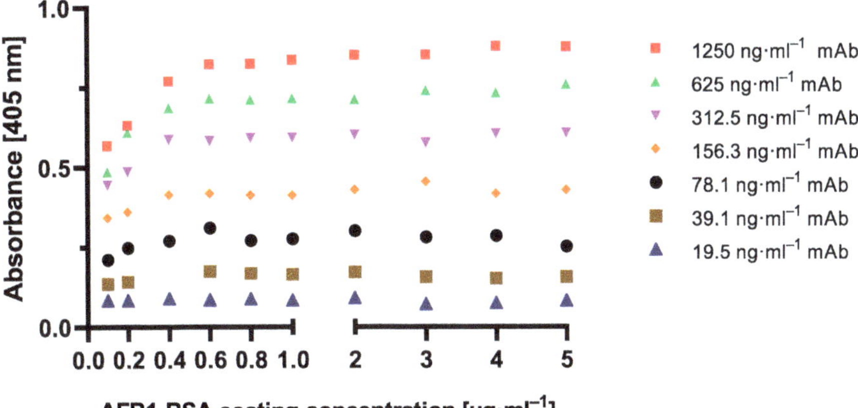

Figure A1. ELISA checkerboard titration determining suitable aflatoxin B1-BSA (AFB1-BSA) coating concentrations ranging from 0.1 µg·mL^{-1} to 5 µg·mL^{-1} and monoclonal antibody (mAb) AFB1_002 concentrations ranging from 19.5 ng·mL^{-1} to 1250 ng·mL^{-1}. Absorbance was measured at 405 nm by indirect readout after using a mouse-specific secondary antibody conjugated to horseradish peroxidase and application of respective substrate ($n = 1$).

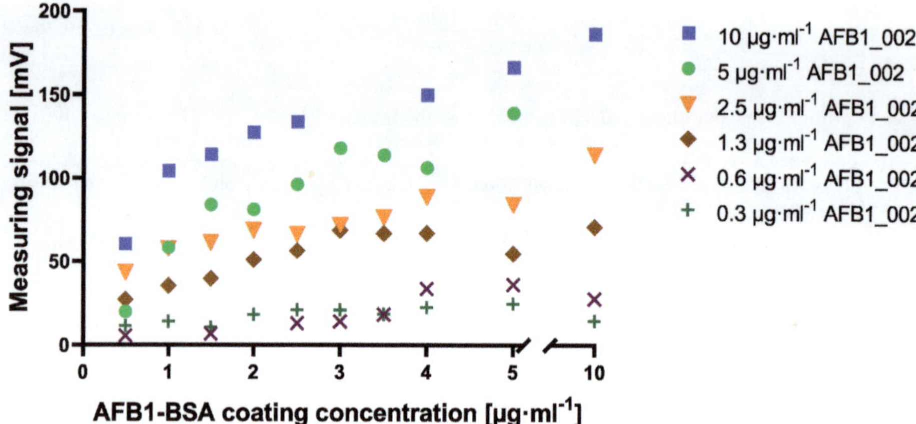

Figure A2. Magnetic immunodetection checkerboard titration determining suitable aflatoxin B1-BSA (AFB1-BSA) coating concentrations and aflatoxin B1-specific biotinylated monoclonal antibody AFB1_002 concentrations. The readout was done by detecting 700 nm streptavidin-functionalized magnetic particles coupled to biotinylated AFB1_002 using FMMD ($n = 1$).

References

1. Park, D.L.; Njapau, H.; Boutrif, E. *Minimizing Risks Posed by Mycotoxins Utilizing the HACCP Concept*; Food, Nutrition and Agriculture: Rome, Italy, 1999; pp. 49–55.
2. Eskola, M.; Kos, G.; Elliott, C.T.; Hajšlová, J.; Mayar, S.; Krska, R. Worldwide contamination of food-crops with mycotoxins: Validity of the widely cited 'FAO estimate' of 25. *Crit. Rev. Food Sci. Nutr.* **2019**, 1–17. [CrossRef]
3. Blount, W.P. Turkey "X" Disease. *Turkeys* **1961**, *9*, 55–58, 61, 77.
4. Bennett, J.W.; Klich, M. Mycotoxins. *Clin. Microbiol. Rev.* **2003**, *16*, 497–516. [CrossRef]
5. Kocabas, C.N.; Sekerel, B.E. Does systemic exposure to aflatoxin B(1) cause allergic sensitization? *Allergy* **2003**, *58*, 363–365. [CrossRef]
6. Azziz-Baumgartner, E.; Lindblade, K.; Gieseker, K.; Rogers, H.S.; Kieszak, S.; Njapau, H.; Rubin, C. Case-control study of an acute aflatoxicosis outbreak, Kenya, 2004. *Environ. Health Perspect.* **2005**, *113*, 1779–1783. [CrossRef]
7. Probst, C.; Njapau, H.; Cotty, P.J. Outbreak of an acute aflatoxicosis in Kenya in 2004: Identification of the causal agent. *Appl. Environ. Microbiol.* **2007**, *73*, 2762–2764. [CrossRef]
8. Tillett, T. Carcinogenic crops: Analyzing the effect of aflatoxin on global liver cancer rates. *Environ. Health Perspect.* **2010**, *118*, A258. [CrossRef]
9. Aguilar, F.; Hussain, S.P.; Cerutti, P. Aflatoxin B1 induces the transversion of G–>T in codon 249 of the p53 tumor suppressor gene in human hepatocytes. *Proc. Natl. Acad. Sci. USA* **1993**, *90*, 8586–8590. [CrossRef]
10. Klich, M.A. Aspergillus flavus: The major producer of aflatoxin. *Mol. Plant Pathol.* **2007**, *8*, 713–722. [CrossRef]
11. Engin, A.B.; Engin, A. DNA damage checkpoint response to aflatoxin B1. *Environ. Toxicol. Pharmacol.* **2019**, *65*, 90–96. [CrossRef]
12. Rushing, B.R.; Selim, M.I. Aflatoxin B1: A review on metabolism, toxicity, occurrence in food, occupational exposure, and detoxification methods. *Food Chem. Toxicol.* **2019**, *124*, 81–100. [CrossRef] [PubMed]
13. Stadler, D.; Berthiller, F.; Suman, M.; Schuhmacher, R.; Krska, R. Novel analytical methods to study the fate of mycotoxins during thermal food processing. *Anal. Bioanal. Chem.* **2019**, *412*, 9–16. [CrossRef] [PubMed]
14. Krska, R.; Schubert-Ullrich, P.; Molinelli, A.; Sulyok, M.; MacDonald, S.; Crews, C. Mycotoxin analysis: An update. *Food Addit. Contam.* **2008**, *25*, 152–163. [CrossRef]
15. Hajslova, J.; Zachariasova, M.; Cajka, T. Analysis of multiple mycotoxins in food. *Methods Mol. Biol.* **2011**, *747*, 233–258. [PubMed]

16. Anfossi, L.; Baggiani, C.; Giovannoli, C.; D'Arco, G.; Giraudi, G. Lateral-flow immunoassays for mycotoxins and phycotoxins: A review. *Anal. Bioanal. Chem.* **2013**, *405*, 467–480. [CrossRef] [PubMed]

17. Li, P.; Zhang, Z.; Hu, X.; Zhang, Q. Advanced hyphenated chromatographic-mass spectrometry in mycotoxin determination: Current status and prospects. *Mass Spectrom. Rev.* **2013**, *32*, 420–452. [CrossRef]

18. Wang, Y.K.; Yan, Y.X.; Ji, W.H.; Wang, H.A.; Li, S.Q.; Zou, Q.; Sun, J.H. Rapid simultaneous quantification of zearalenone and fumonisin B1 in corn and wheat by lateral flow dual immunoassay. *J. Agric. Food Chem.* **2013**, *61*, 5031–5036. [CrossRef]

19. Dzantiev, B.B.; Byzova, N.A.; Urusov, A.E.; Zherdev, A.V. Immunochromatographic methods in food analysis. *TRAC-Trend Anal. Chem.* **2014**, *55*, 81–93. [CrossRef]

20. Sulyok, M.; Stadler, D.; Steiner, D.; Krska, R. Validation of an LC-MS/MS-based dilute-and-shoot approach for the quantification of > 500 mycotoxins and other secondary metabolites in food crops: Challenges and solutions. *Anal. Bioanal. Chem.* **2020**, *412*, 2607–2620. [CrossRef]

21. Renaud, J.B.; Miller, J.D.; Sumarah, M.W. Mycotoxin Testing Paradigm: Challenges and Opportunities for the Future. *J. AOAC Int.* **2019**, *102*, 1681–1688. [CrossRef]

22. Pereira, V.L.; Fernandes, J.O.; Cunha, S.C. Mycotoxins in cereals and related foodstuffs: A review on occurrence and recent methods of analysis. *Trends Food Sci. Technol.* **2014**, *36*, 96–136. [CrossRef]

23. Xie, L.; Chen, M.; Ying, Y. Development of Methods for Determination of Aflatoxins. *Crit. Rev. Food Sci. Nutr.* **2016**, *56*, 2642–2664. [CrossRef]

24. Meyer, M.H.; Hartmann, M.; Krause, H.J.; Blankenstein, G.; Mueller-Chorus, B.; Oster, J.; Keusgen, M. CRP determination based on a novel magnetic biosensor. *Biosens. Bioelectron.* **2007**, *22*, 973–979. [CrossRef]

25. Hong, H.B.; Krause, H.J.; Song, K.B.; Choi, C.J.; Chung, M.A.; Son, S.W.; Offenhäusser, A. Detection of two different influenza A viruses using a nitrocellulose membrane and a magnetic biosensor. *J. Immunol. Methods* **2011**, *365*, 95–100. [CrossRef]

26. Rettcher, S.; Jungk, F.; Kühn, C.; Krause, H.J.; Nölke, G.; Commandeur, U.; Schröper, F. Simple and portable magnetic immunoassay for rapid detection and sensitive quantification of plant viruses. *Appl. Environ. Microbiol.* **2015**, *81*, 3039–3048. [CrossRef]

27. Achtsnicht, S.; Neuendorf, C.; Faßbender, T.; Nölke, G.; Offenhäusser, A.; Krause, H.J.; Schröper, F. Sensitive and rapid detection of cholera toxin subunit B using magnetic frequency mixing detection. *PLoS ONE* **2019**, *14*, e0219356. [CrossRef]

28. Meyer, M.H.; Stehr, M.; Bhuju, S.; Krause, H.J.; Hartmann, M.; Miethe, P.; Keusgen, M. Magnetic biosensor for the detection of Yersinia pestis. *J. Microbiol. Methods* **2007**, *68*, 218–224. [CrossRef]

29. Krause, H.; Wolters, N.; Zhanga, Y.; Offenhäusser, A.; Miethe, P.; Meyer, M.H.F.; Hartmann, M.; Keusgen, M. Magnetic particle detection by frequency mixing for immunoassay applications. *J. Magn. Magn. Mater.* **2007**, *311*, 436–444. [CrossRef]

30. Lee, H.M.; Song, S.O.; Cha, S.H.; Wee, S.B.; Bischoff, K.; Park, S.W.; Cho, M.H. Development of a monoclonal antibody against deoxynivalenol for magnetic nanoparticle-based extraction and an enzyme-linked immunosorbent assay. *J. Vet. Sci.* **2013**, *14*, 143–150. [CrossRef]

31. Xuan, Z.; Ye, J.; Zhang, B.; Li, L.; Wu, Y.; Wang, S. An Automated and High-Throughput Immunoaffinity Magnetic Bead-Based Sample Clean-Up Platform for the Determination of Aflatoxins in Grains and Oils Using UPLC-FLD. *Toxins* **2019**, *11*, 583. [CrossRef]

32. Achtsnicht, S.; Schönenborn, K.; Offenhäusser, A.; Krause, H.J. Measurement of the magnetophoretic velocity of different superparamagnetic beads. *J. Magn. Magn. Mater.* **2018**, *477*, 244–248. [CrossRef]

33. Achtsnicht, S.; Tödter, J.; Niehues, J.; Telöken, M.; Offenhäusser, A.; Krause, H.J.; Schröper, F. 3D Printed Modular Immunofiltration Columns for Frequency Mixing-Based Multiplex Magnetic Immunodetection. *Sensors* **2019**, *19*, 148. [CrossRef] [PubMed]

 *toxins*

Article

Development of an Immunofluorescence Assay Module for Determination of the Mycotoxin Zearalenone in Water

Borbála Gémes [1], Eszter Takács [1], Patrik Gádoros [2], Attila Barócsi [2], László Kocsányi [2], Sándor Lenk [2], Attila Csákányi [3], Szabolcs Kautny [3], László Domján [3], Gábor Szarvas [3], Nóra Adányi [4], Alexei Nabok [5], Mária Mörtl [1] and András Székács [1,*]

[1] Agro-Environmental Research Centre, Institute of Environmental Sciences, Hungarian University of Agriculture and Life Sciences, Herman O. út 15, H-1022 Budapest, Hungary; gemes.borbala.leticia@uni-mate.hu (B.G.); takacs.eszter84@uni-mate.hu (E.T.); mortl.maria@uni-mate.hu (M.M.)

[2] Department of Atomic Physics, Budapest University of Technology and Economics, Műegyetem rkp. 3, H-1111 Budapest, Hungary; gadorosp@eik.bme.hu (P.G.); barocsi@eik.bme.hu (A.B.); kocsanyi@eik.bme.hu (L.K.); lenk@eik.bme.hu (S.L.)

[3] Optimal Optik Ltd., Dayka Gábor u. 6/B, H-1118 Budapest, Hungary; attila_csakanyi@optimal-optik.hu (A.C.); szabolcs_kautny@optimal-optik.hu (S.K.); domjan@optimal-optik.hu (L.D.); gabor_szarvas@optimal-optik.hu (G.S.)

[4] Food Science Research Centre, Institute of Food Sciences, Hungarian University of Agriculture and Life Sciences, Herman O. út 15, H-1022 Budapest, Hungary; adanyi.nora@uni-mate.hu

[5] Materials and Engineering Research Institute, Sheffield Hallam University, Howard Street, Sheffield S1 1WB, UK; a.nabok@shu.ac.uk

* Correspondence: szekacs.andras@uni-mate.hu; Tel.: +36-1-796-0400

check for **updates**

Citation: Gémes, B.; Takács, E.; Gádoros, P.; Barócsi, A.; Kocsányi, L.; Lenk, S.; Csákányi, A.; Kautny, S.; Domján, L.; Szarvas, G.; et al. Development of an Immunofluorescence Assay Module for Determination of the Mycotoxin Zearalenone in Water. *Toxins* **2021**, *13*, 182. https://doi.org/10.3390/toxins13030182

Received: 10 February 2021
Accepted: 24 February 2021
Published: 2 March 2021

Publisher's Note: MDPI stays neutral with regard to jurisdictional claims in published maps and institutional affiliations.

Abstract: Project Aquafluosense is designed to develop prototypes for a fluorescence-based instrumentation setup for in situ measurements of several characteristic parameters of water quality. In the scope of the project an enzyme-linked fluorescent immunoassay (ELFIA) method has been developed for the detection of several environmental xenobiotics, including mycotoxin zearalenone (ZON). ZON, produced by several plant pathogenic *Fusarium* species, has recently been identified as an emerging pollutant in surface water, presenting a hazard to aquatic ecosystems. Due to its physico-chemical properties, detection of ZON at low concentrations in surface water is a challenging task. The 96-well microplate-based fluorescence instrument is capable of detecting ZON in the concentration range of 0.09–400 ng/mL. The sensitivity and accuracy of the analytical method has been demonstrated by a comparative assessment with detection by high-performance liquid chromatography and by total internal reflection ellipsometry. The limit of detection of the method, 0.09 ng/mL, falls in the low range compared to the other reported immunoassays, but the main advantage of this ELFIA method is its efficacy in combined in situ applications for determination of various important water quality parameters detectable by induced fluorimerty—e.g., total organic carbon content, algal density or the level of other organic micropollutants detectable by immunofluorimetry. In addition, the immunofluorescence module can readily be expanded to other target analytes if proper antibodies are available for detection.

Keywords: zearalenone; mycotoxin; competitive immunoassay; fluorescence detection; high-performance liquid chromatography; total internal reflection ellipsometry

Key Contribution: An in situ detection module with a dynamic detection range between 0.09 and 400 ng/mL was developed for zearalenone. The immunofluorescence method and module, as well as the instrument prototype, constitute a part of the Aquafluosense modular instrument family for determination of characteristic water parameters and contaminants.

1. Introduction

Natural mycotoxin contamination has been identified as an emerging problem in agriculture. These toxic secondary metabolites produced by some fungal species are often found in food and feed (especially in grains) and cause high risk for food- and feed-borne intoxication in both humans and livestock [1]. The wide range of their negative effects includes, e.g., genotoxic, cytotoxic, mutagenic, and teratogenic effects [2]. Among all the toxic filamentous fungi species, *Aspergillus*, *Fusarium*, and *Penicillium* are important genera, producing regularly detected and widely studied toxins including aflatoxins, ochratoxin A, deoxynivalenol, T-2 toxin, fumonisin, and zearalenone (ZON) [3]. Mycotoxins found in human urine, indicate the possibility of chronic exposure [4]. In addition, a northward migration of toxicogenic plant pathogenic fungi has been reported assumedly triggered by climate change [5–7]. ZON is a frequently occurring mycotoxin, produced by species of the *Fusarium* genus, including *F. graminearum*, *F. culmorum*, *F. semitectum*, *F. cerealis*, and *F. equiseti* [8]. Its most known impact on human health is endocrine disruption: ZON and derivatives trigger estrogen-like effects in mammals causing alteration in hormone-mediated processes, e.g., the production of follicle-stimulating hormone (FSH) and luteinizing hormone (LH), and reductions in the number of Leydig and granulosa cells [9]. The potential health and economic impacts of ZON necessitate its routine monitoring in food and commodities, which have led to the development and validation of analytical methods in recent decades.

1.1. Mycotoxins as Pollutants in Surface Waters

In addition to food- and feed-borne intoxication, humans can also be affected through exposures via surface water contamination. Various phytopathogenic fungi, including *Fusarium* species, have been demonstrated to be capable of continuing to produce their secondary metabolites in water [10,11], and this process has been indicated to be a potential route of human exposure to mycotoxins [12]. Numerous investigations have reported the presence and input pathways of the toxin in surface or groundwater [13–17]. Mycotoxins may occur in surface waters by direct fungal contamination, by leaching from infested soil as water runoff from agriculture, by washing out from contaminated agricultural commodities such as cereals, oil, forage, feed etc. [18,19], or by mycotoxin biosynthesis in water by fungi [20,21]. In turn, mycotoxins have been considered as emerging surface water contaminants of diffuse and point source occurrence [22], through which fungal contaminants are considered emerging evidence-based threats for drinking water quality safety regulations [23]. Thus, water contamination by aflatoxins B2 and G2 were detected in water in Southern England at levels of 0.1–1.7 ng/mL [24]; aflatoxins B1 and B2, fumonisin B3, and ochratoxin A were detected at concentrations up to 0.035 ng/mL in the Tagus river Portugal [25]; phytoestrogens and mycotoxins were monitored in agricultural stream basins in the United States in Iowa, with occasional occurrence of deoxynivalenol above 0.1 ng/mL level [26]; aflatoxins B2, B1, and G1, as well as ochratoxin A were detected at levels of no toxicological risk up to 0.0007 ng/mL in bottled water in Portugal [27]; fumonisins were detected at up to 0.048 ng/mL levels in aqueous environmental samples in Poland [28]. Along with other mycotoxins, ZON and its metabolites also appear to be water contaminants. ZON was in found in surface waters, groundwater, and wastewater in Poland at levels up to 0.081 ng/mL, originating from cereal crops [14,15], and ZON, along with the metabolites zearalanone, α-zearalenol, β-zearalenol, α-zearalanol, and β-zearalanol, was detected in surface waters in Brazil at levels up to 4.12 ng/mL [29]. In total, 32 of 159 surface water samples collected in central Illinois have been positively tested, and 10 of them were above limit of quantification (LOQ) with concentrations between 0.002 and 0.006 ng/mL [30]. ZON also has been found in nine samples collected from eight Portuguese rivers and creeks ranging between 0.006 and 0.083 ng/mL [17], and appeared in the drainage water of a *F. graminearum* infected field in Switzerland with higher concentrations in the summer vegetation periods in a two-year field experiment [31]. Lower but detectable (0.002–0.005 ng/mL) concentrations were found in the Tiber river in Italy [32].

Moreover, the appearance of mycotoxins in the aquatic environment can adversely influence entire ecosystems. Thus, *Fusarium* mycotoxins, including ZON, can exert hormonal (estrogenic), hepatotoxic or genotoxic effects on fish [33,34]. Nonetheless, maximal residue levels (MRLs) to ensure compliance with the tolerable daily intake for humans (µg/kg body weight) by the European Union legislation [35] have been set only for food and feed, e.g., MRLs from 20 µg/kg in processed maize-based foods for infants and young children up to 200 µg/kg in unprocessed maize [36]. There is also a commission recommendation for ZON (and other mycotoxins) in products intended as animal feed as recommended by two scientific opinions of the European Food Safety Authority [37,38], but no declared maximum level for drinking water or surface waters have been established yet.

1.2. Analytical Methods for Zearalenone Determination

There are numerous well-established methods for quantification of ZON, varying in their technical detail (e.g., sample preparation, principle of the analytical procedure) according to the complexity of the target matrix and other circumstances. Analytical methods for detection include colorimetric and fluorescence-based strategies (e.g., enzyme-linked immunosorbent assay, ELISA), chromatographic methods, and enzyme-linked oligonucleotide assays [39]. Traditional chromatographic separation, e.g., high-performance liquid chromatography (HPLC) [40–42], thin layer chromatography (TLC) [43–45], and liquid or gas chromatography coupled with mass spectroscopy (LC or GC MS) [46–49] have low limit of detection (LOD) and limit of quantification (LOQ) values but, usually, due to the complex sample preparation they are time consuming technologies requiring special instrumentation. Immunoanalytical techniques are cost-effective and suitable for rapid monitoring with detecting multiple samples at the same time [50–53]. Within this category, ELISA is the most prevailing method [54–61]. Through the advancement of the immunoanalytical techniques, analytical sensitivities increased and LOD and LOQ values dropped to the same level as those for chromatographic methods. Among immunoanalytical methods, immunosensors represent innovative and more sensitive analytical determination techniques than microplate- or immunostrip-based detection [62–68].

Project Aquafluosense (NVKP_16-1-2016-0049) [69] aims to develop a new water analysis system for natural and artificial waters, allowing complex, systematic and in situ fluorescence-based assessment and monitoring of water quality. The modular instrument family developed for main parameters (chlorophyll-a content, chemical and biochemical oxygen demands, total organic carbon, polycyclic aromatic hydrocarbon, and certain agricultural pollutant contents) can be individually configured for target tasks at each monitoring point. Within the project, we aimed to develop an enzyme-linked fluorescent immunoassay (ELFIA) module for monitoring and quantification of ZON.

2. Results and Discussion

2.1. Determination of Zearalenone by Autofluorescence

Detection capability of ZON by its own induced fluorescence was assessed in direct fluorescence (autofluorescence) measurements in water. A fluorescence intensity spectral map and a calibration curve are presented on Figure 1. Fluorescence is generated by the optical excitation of electrons, which emit fluorescent light when they return to their ground state from their excited state. As a loss of vibrational energy inevitably occurs during this process, the emission spectrum is shifted to longer wavelengths than the excitation wavelength (Stokes shift). Figure 1 depicts the relevant wavelength pairs of excitation and emission of fluorescence spectra, as well as the dependence of intensity of the emitted light (fluorescence) on the concentration of ZON. Excitation mapping was carried out by scanning emission intensities as a function of excitation intensities between 250 and 830 nm wavelengths depicting emission intensity in a color scale from blue to red (Figure 1a). On the basis of the fluorescence spectral map, the optimized peak for ZON measurement by autofluorescence was obtained with excitation at 280 nm wavelength and emission detection at 520 nm wavelengths. The dependence of the emitted light at these parameters on the concentration of ZON in the aqueous sample was also tested and was found to follow a sigmoidal (logistic) regression (Figure 1b). Based on the sigmoid curve for autofluorescence, an LOD value of 11.5 µg/mL was determined.

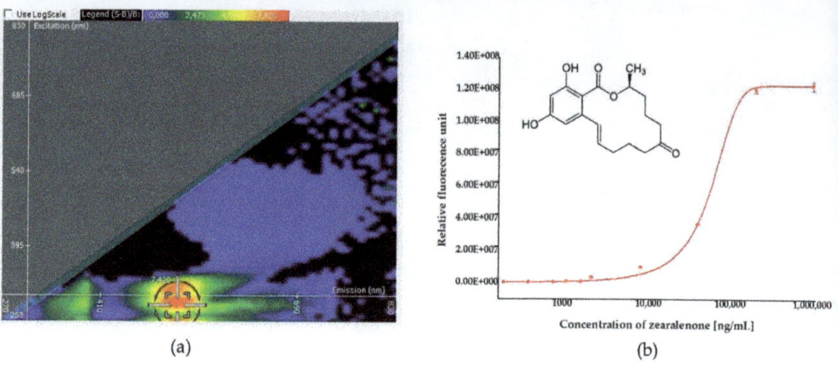

(a)   (b)

Figure 1. Results of zearalenone (ZON) quantification by autofluorescence. (**a**) A fluorescence spectral map of ZON in phosphate buffer saline and the optimized peak (in the range of a red patch, the middle point of cross-hair indicating optimal detection conditions) with 280 and 520 nm wavelengths for excitation (ordinate) and emission (abscissa), respectively. (**b**) A calibration curve obtained in a concentration range between 175 and 1,000,000 ng/mL of ZON and the chemical structure of ZON (insert).

2.2. Enzyme-Linked Fluorescent Immunoassay (ELFIA)

2.2.1. Titration and Inhibition of the Antiserum

Efficacy of the immunization was monitored by titration of the two rabbit antisera against the coating antigen, ZON conjugated to bovine serum albumin (BSA) (ZON–BSA) between 1:50 and 1:12,200 dilution factors. Microplates were coated with the BSA conjugate at concentrations of 1 to 5 µg/mL in coating buffer. Serum titers, defined as the serum dilution that binds 50% of the antigen under the given conditions, were determined for sera obtained from rabbits (rabbit 1 and rabbit 2). Only slight differences occurred between the efficacy of the two antisera: titer values were 1:828 and 1:448 for antisera from rabbit 1 and rabbit 2, respectively (Figure 2a). In the subsequent immunofluorescence assay experiments, antiserum from rabbit 1, showing somewhat higher affinity to the antigen, was applied. Accuracy and reproducibility of the measurements are better the nearer they are to the titer value, and thus antiserum was applied at a dilution factor of 1:1000 in the ELFIA tests.

To avoid the risk of saturatization or weak signal detection in assays, optimizations of the coating antigen concentration and serum dilution factor were performed by checkboard titration. The coating antigen, ZON-6′-carboxymethyloxime–BSA conjugate, was applied at concentrations in the range of 0.3125–2.5 µg/mL against the antiserum from rabbit 1 at dilutions in the range of 1:3375–1:1000 dilution factor. All combinations were investigated uninhibited and under inhibition by 3.2 ng/mL of ZON, as well. The coating antigen concentration and the antiserum dilution factor consistently influenced the analytical parameters (Figure 2b). The analytical signal (relative fluorescence unit, RFU) increased with increasing concentrations of the coating antigen and decreased with increasing dilution of the serum. The addition of ZON at a concentration of 3.2 ng/mL resulted in an average 40.0% ± 0.1% inhibition of the assay signal.

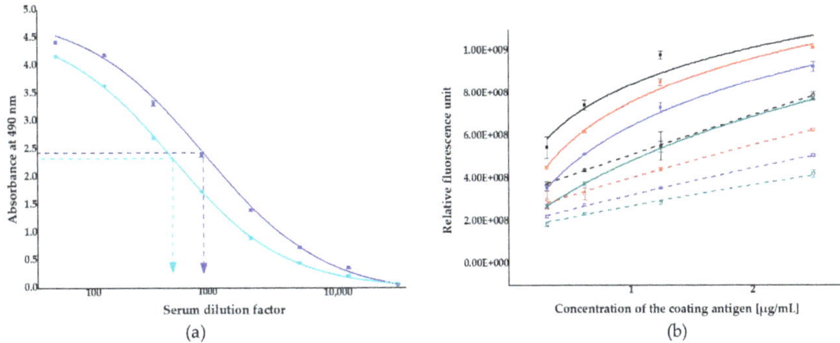

Figure 2. Analytical characterization of antisera collected from two 3-month old female New-Zealand white rabbits: (**a**) titer curves of antisera from rabbit 1 (■) and rabbit 2 (■) (dilution factor range of 1:50–1:12,200) using a zearalenone-6′-carboxymethyloxime-bovine serum albumin conjugate as a coating antigen at 5 µg/mL; blocked with 1% gelatin in phosphate buffer saline; (**b**) checkboard titration of the antiserum from rabbit 1 using the coating antigen at concentrations in the range of 0.3125–2.5 µg/mL and the serum at various dilution factors (solid symbols, solid lines): 1:1000 (■), 1:1500 (■), 1:2250 (■), 1:3375 (■). Titration was also performed under the same conditions with the serum inhibited by 3.2 ng/mL of zearalenone at various dilution factors (hollow symbols, slashed lines): 1:1000 (□), 1:1500 (□), 1:2250 (□), and 1:3375 (□).

2.2.2. Immunoassay

Indirect competitive ELFIAs were performed to established ZON calibration curves and to determine the LOD. The detection range was investigated in a concentration series of 0.004 pg/mL–2 µg/mL ZON in assay buffer. Matrix effects were determined by comparing calibration curves obtained in assay buffer and in surface water samples. No significant differences were determined among curves ($p > 0.05$), thus it has been concluded that determination of ZON in surface water can be performed without modification in sample preparation. Calibration curves and LODs were determined using both absorbance and fluorescence signals (Figure 3). For comparability of the two detection modes, assay signals are represented as relative values (signals ratios to maximal signal levels). ZON at a concentration of 2000 ng/mL and above reached its full inhibition potential on the surface binding of the antibodies. This occurs because at this concentration the avidity of the primary antibody is saturated by ZON molecules in the solution, and therefore, further increases in ZON concentration cannot push the immunocomplexation equilibrium any further—ZON has reached its full capacity to block binding of the antibody to the coating antigen ZON–BSA conjugate on the surface of the immunoplate. The average relative analytical signal corresponding to the maximal assay signal produced by the uninhibited serum was set to the upper plateau level of the sigmoid standard curve. The average relative analytical signal corresponding to full inhibition of the serum was considered as the lower plateau level of the sigmoid standard curve. Analytical parameters of calibration curves are summarized in Table 1.

LOD values were calculated for the two analytical detection modes of resorufin as a chromophore product. Thus, LOD = 0.25 and 0.09 ng/mL were determined for visual absorbance and fluorescence detections, respectively. Detection by fluorescence provided a wider and steeper dynamic range, thus ELFIA proved to be a 2.8-fold more sensitive method for ZON than the corresponding ELISA. It has to be noted that absorbance detection of resorufin by the application of QuantaRed Enhanced Chemifluorescent HRP Substrate Kit with HRP enzyme reaction (Thermo Fisher Scientific Inc., Waltham, MA, USA) provided 3.4-fold lower LOD than that of o-phenylenediamine dihydrochloride (OPD) as chromophore in a similar colorimetric ELISA (LOD for OPD = 0.85 ng/mL).

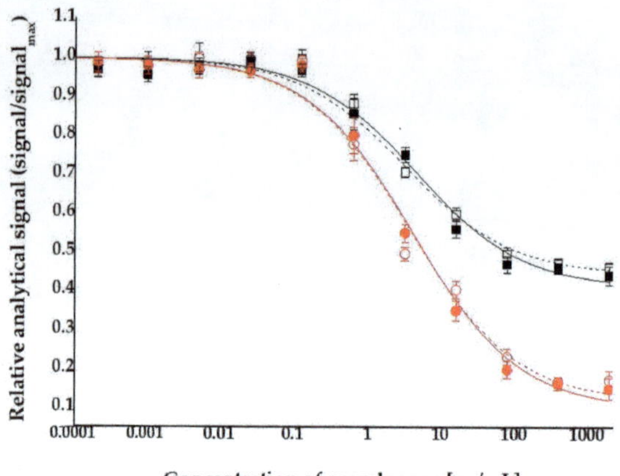

Concentration of zearalenone [ng/mL]

Figure 3. Competitive indirect calibration curves for zearalenone obtained in assay buffer (hollow marker, dashed lines) and in surface water from river Danube (solid marker, solid lines) determined by absorbance (■, □) and fluorescence (●, ○), detected at 576 and 593 nm wavelengths, respectively.

Table 1. Statistical parameters of the logistic mathematical fitting using the Rodbard equation [70] in the immunoassay format using assay signals by resorufin as a chromophore product with detection of absorbance and fluorescence.

Equation for fitting:

$$y = \frac{A_1 - A_2}{1 + \left(\frac{x}{x_0}\right)^p} + A_2 \quad ^1$$

Adjusted R^2: 0.990 (absorbance)
0.988 (fluorescence)

	Parameter	Value $\pm$ Standard Deviation
Absorbance	A_1	0.98 $\pm$ 0.02
	A_2	0.45 $\pm$ 0.01
	x_0	2.86 $\pm$ 0.38
	p	0.79 $\pm$ 0.12
Fluorescence	A_1	0.99 $\pm$ 0.01
	A_2	0.16 $\pm$ 0.03
	x_0	2.41 $\pm$ 0.27
	p	0.83 $\pm$ 0.19

^{1} Description of the equation parameters—A_1: upper plateau, A_2: lower plateau, x_0: 50% inhibition, p: curve slope at the inflexion (IC$_{50}$).

2.2.3. Effects of Light Source Intensity

Calibration curves were determined in the induced fluorescence method at different light source intensities. LED power can be digitally set between 0.001 µW and 4.63 mW in 256 nonequidistant steps at 532 nm wavelength, and it was tuned to provide assay signals between 20 and 30 RFU as the suggested least detectable value and 4095 RFU as the highest readable signal by the instrument. Correspondingly, the LED power values investigated ranged between 1.1 and 314 µW (1.1, 100.2, 169.4, 256 and 314 µW). As a background, 2000 ng/mL ZON solution was applied that triggered total inhibition of the antiserum. For 1.1 µW LED power there were no differences in the RFU values among different dilutions of ZON and for 314 µW LED power the RFU values reached the maximum (4095) at 0.64 ng/mL ZON concentration. Thus, effects of the LED power on ZON quantification were determined at values of 100.2, 169.4 and 256 µW. Recorded background signals were 24.7 $\pm$ 1.1, 27.9 $\pm$ 0.8 and 43.7 $\pm$ 1.0, while maximum (uninhibited) fluorescence levels were 1447, 2303 and 3131 for 100.2, 169.4, 256 µW, respectively. The analytical parameters of the calibration curves were determined with the optimized immunoassay system (see Section 4.4.3). In the data evaluation process, RFUs were corrected with the background. Among analytical parameters, IC_{50} values determined from the calibration curves in a concentration range of 0.0256–2000 ng/mL ZON were compared, and were found to be 2.52 ± 0.24, 3.04 ± 0.31 and 3.68 ± 0.29 ng/mL ZON for LED power values of 100.2, 169.4, 256 µW, respectively, while the LOD values did not appear to be significantly affected by the intensity of the light source.

2.2.4. Cross-Reactivity of the Antisera with Zearalenone Derivatives

Inhibition of the antiserum by metabolites and structural analogues of ZON was also determined under optimized assay conditions. ZON is metabolized mostly through hydrolysis mainly to β- and α-zearalenol in yeast and ovine species, respectively [71,72]. Thus, IC_{50} values by ZON metabolites and their reduced derivatives (α- and β zearalenol, α- and β-zearalanol, zearalanone) were determined in the immunoassay by absorbance and fluorescence and relative cross-reactivities (considering inhibition by ZON as 100%) are listed in Table 2. The results indicate that the antibodies appear to be most sensitive to the presence of the unsaturation in the resorcyclic lactone ring and exhibit lower affinity to hydroxy metabolites. Stereoconfiguration of the hydroxyl group also occurs to influence recognition.

Table 2. Percentage cross-reactivity (CR%) of the antiserum with zearalenone and its derivatives determined by fluorescence and absorbance.

Mycotoxin	Detection Mode			
	Fluorescence		Absorbance	
	IC_{50} (ng/mL) [1]	CR% [2]	IC_{50} (ng/mL)	CR% [1]
zearalenone	2.20 ± 0.31	100	2.73 ± 0.35	100
α-zearalenol	10.42 ± 0.24	21.1 ± 3.0	10.94 ± 1.28	20.1 ± 2.6
β-zearalenol	8.74 ± 0.90	25.2 ± 3.6	8.65 ± 0.84	25.4 ± 3.3
zearalanone	8.56 ± 0.74	25.7 ± 3.6	8.24 ± 0.82	26.7 ± 3.4
α-zearalanol	35.36 ± 2.86	6.2 ± 0.9	35.63 ± 3.05	6.2 ± 0.8
β-zearalanol	200.7 ± 12.32	1.1 ± 0.2	250.02 ± 21.68	0.9 ± 0.1

[1] IC_{50}: half maximal inhibitory concentration; [2] CR%: Cross-reactivity defined as the percentage ratio of the IC_{50} values of zearalenone and of the given derivative.

2.2.5. Analytical Detection Capability Compared to Other Immunoanalytical Methods

The analytical performance of the above ELFIA method was compared to that of other immunoanalytical methods (immunoassays, immunosensors) reported in the scientific literature. Detection capabilities of the immunoanalytical methods are listed in Table 3. The immunoanalytical methods reported are mostly developed to be used for crop commodities (maize, wheat, barley, rice), and LODs are specified in the method descriptions as detectable ZON concentrations in the commodity (e.g., µg/kg). For this comparison, Table 3 enlists LODs in the final diluted extract according to the method specifications published. In these assays, organic solvent extracts in aqueous acetonitrile or methanol (MeOH) were used for ZON determination, but the solvent was diluted to 0.1% or below during dilution with the assay buffer to reach the detection range. LODs ranged between 10 ng/mL by an IgY-based ELISA [57] down to 0.002 pg/mL by an an optical waveguide ligthmode spectroscopy immunosensor [67]. As seen from the analytical sensitivity data, the current ELFIA method is located in the middle range of the reported procedures regarding the LOD values. Analytical detection performance can be possibly improved by using more specific antibodies, but the main advantage of the ELFIA method lies not primarily in its sensitivity, but in its utility in combined in situ application in the determination of other water quality parameters detectable by induced fluorimetry—e.g., total organic carbon content, algal density or other organic microcontaminants (herbicide glyphosate or pharmaceutical carbamazepin). In addition, the immunofluorescence module can be easily extended to other target analytes if proper antibodies are available for their detection.

Table 3. Analytical performance characteristics of various immunoanalytical methods for the determination of zearalenone.

Analytical Method	LOD [1] (ng/mL)	IC$_{50}$ [2] Detection Range (ng/mL)	Matrix	Organic Solvent Content in the Sample Extract	Reference
ELISA [3]	10	40 / 10–200	maize	10% AcCN [4]	[57]
Radioimmunoassay	5	NR [5] / 0.25–10	human serum	-	[61]
ELISA	1	3 / 0.5–50	wheat, maize	10% MeOH [6]	[55]
SPR (Sensor) [7]	0.56	5	wheat	16% MeOH	[65]
DPV (Sensor) [8]	0.25	NR	beer, wine	20% AcCN	[64]
ELISA	0.24	0.855 [9]	maize	14% MeOH	[54]
ELISA	0.15 (PBS) / 0.23 (maize)	1.13 (PBS) / 1.4 (maize)	maize	8% MeOH	[60]
FLISA [10]	0.10	0.95	maize flour	14% MeOH	[56]
ELFIA [11]	0.09	2.4	water	0.2% MeOH	this study
ELISA	0.05	NR	wheat	10% MeOH	[58]
ELISA	0.02	0.18	maize flour	14% MeOH	[56]
SPR (Sensor)	0.01	NR	NR	9% AcCN	[63]
PW PI (Sensor) [12]	0.01	NR	water	10% MeOH	[66]
CPG-Based Immunosensor [13]	0.007	0.087	wheat	0.21% MeOH; 0.2% AcCN	[68]
ELISA (Coupled with IAC) [14]	0.002	0.02	maize	10% AcCN	[59]
OWLS (Sensor) [15]	2×10^{-6}	0.014	maize	AcCN	[54]

[1] LOD: limit of detection; [2] IC$_{50}$: half maximal inhibitory concentration; [3] ELISA: enzyme-linked immunosorbent assay; [4] AcCN: acetonitrile; [5] NR: not reported; [6] MeOH: methanol; [7] SPR: surface plasmon resonance; [8] DPV: differential pulse voltammetry; [9] PBS: phosphate buffer saline; [10] FLISA: fluorescence-linked immunosorbent assay; [11] ELFIA: enzyme-linked fluorescent immunoassay; [12] PW PI: planar waveguide operating as a polarization interferometer; [13] CPG: controlled pore glass; [14] IAC: immunoaffinity column; [15] OWLS: optical waveguide lightmode spectroscopy.

2.3. High-Performance Liquid Chromatography (HPLC)

Concentrations of ZON were also determined by HPLC instrumental analysis on the basis of peak areas in the chromatograms at the corresponding retention time (6.12 min) with excellent linear calibration characteristics in three parallel measurements. Peak areas determined at 236 nm for ZON concentrations between 10 and 2000 ng/mL were applied in linear regression, where regression coefficient of the concentration dependence was 0.999 in all measurements. Calibration curves with standard solutions were also investigated in MeOH:water = 7:3 and MeOH:phosphate buffer saline (PBS) = 1:1, where slope coefficients of linear regression were 26.90 ± 0.06 and 25.92 ± 0.06 for MeOH:water and MeOH:PBS, respectively. Peak purity was assessed by ratios of signal intensities (peak areas) recorded at 236 and 274 nm. These values for standard solutions were 2.15. Relative SDs established for different concentration levels for three parallel injections ranged between 0.65% and 1.76%. The LOD of the method was determined to be 10 ng/mL for ZON. A chromatogram and linear calibration (average of peak areas from three parallel measurements and their SDs) are presented in Figure 4.

2.4. Total Internal Reflection Ellipsometry

The immunoassay setup has also been applied in a system based on detection by total internal reflection ellipsometry (TIRE), where two parameters Ψ and Δ are related, respectively, to the amplitude and phase shift of p- and s-components of the polarized light detected [73]. Since variations in the refractive index and the thickness of the adsorbed layers cause 10 times higher values of Δ than of Ψ, $\Delta(\lambda)$ spectra were used in the TIRE method as a sensor response. A typical series of $\Delta(\lambda)$ spectra for ZON competitive immunoassays and the respective calibration curve of the assay signal in TIRE (δd corresponding to the shift in the adsorbed layer thickness vs. the concentration of ZON) obtained by sigmoidal fitting are depicted in Figure 5. The response is similar to that shown in Figure 2, where the highest concentration of ZON yields the lowest response, which is typical for competitive immunoassays. On the basis of the standard calibration curve, a LOD of 0.01 ng/mL for ZON was determined.

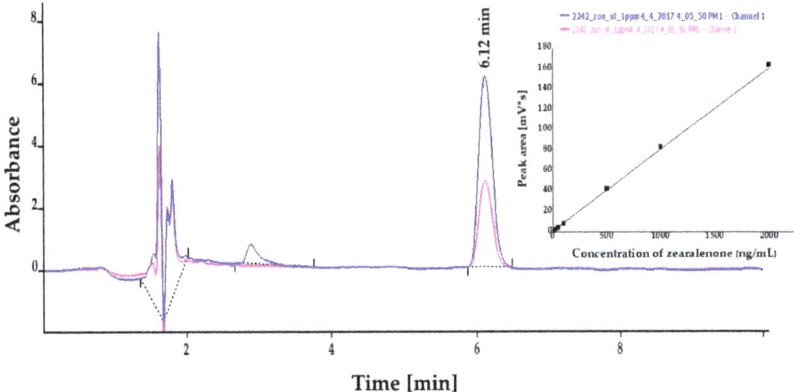

Figure 4. High-performance liquid chromatography (HPLC) chromatogram of zearalenone (ZON) at 1 µg/mL concentration dissolved in methanol:phosphate buffer saline (1:1). Linear calibration (average of peak area from three parallel measurements and their SDs) of ZON in a concentration range of 10–2000 ng/mL determined at 236 nm by high-performance liquid chromatography coupled with UV detection (insert).

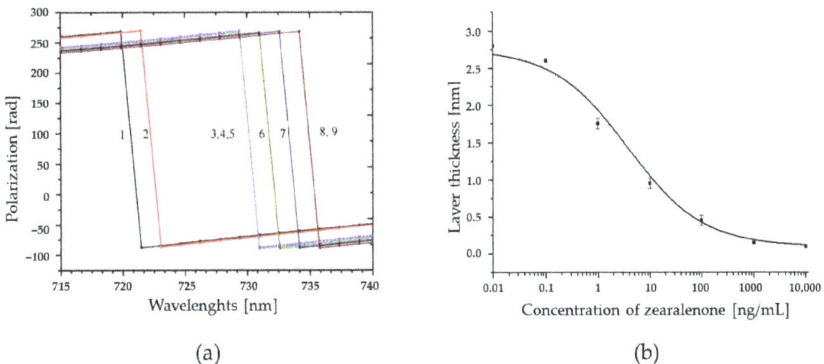

Figure 5. A competitive immunoassay for zearalenone (ZON) carried out by detection via total internal reflection ellipsometry (TIRE). (**a**) A typical set of $\Delta(\lambda)$ spectra measured on bare Au surface (1), polyallylamine hydrochloride (2) ZON–bovine serum albumin conjugate (3), bovine serum albumin (4), Ab-ZON of from preincubated mixtures containing ZON: 100 (5), 10 (6), 1 (7) and 0.1 ng/mL (8). (**b**) Changes in the adsorbed layer thickness versus the concentration of ZON (in the mixture with Ab-ZON) obtained by fitting the TIRE data.

3. Conclusions

Within project Aquafluosense, successful development resulted in a modular instrumentation setup for fluorescence-based determination of several characteristic parameters of water quality. The application of fluorescence, as an analytical signal in an enzyme-linked immunoassay format, results in a method of improved sensitivity with a lower LOD value than in the colorimetric ELISA (0.09 and 0.25 ng/mL, respectively). Moreover, resorufin-based determination proved lower LOD than application of OPD in colorimetric assay (LOD_{OPD} = 0.85 ng/mL). This benefit of this is that it allows determination of lower pollutant concentrations in surface water, which contributes to more effective monitoring. The detection results were validated by HPLC instrumental analysis and by a TIRF immunosensor method. Although the sensor technology provided orders of magnitude lower LOD than the immunofluorescence method developed, the great advantage of the latter is that it makes in situ determination possible, and the 96-well microplate format used in the immunofluorescence determination prototype allows an assay capacity of 25 samples in parallel in triplicates (with standard curves of seven calibration points). An in situ detection module with a dynamic detection range between 0.4 and 400 ng/mL was developed for ZON. The immunofluorescence method and the instrument prototype constitute a part of the Aquafluosense modular instrument family for determination of characteristic water parameters and contaminants.

4. Materials and Methods

4.1. Materials and Reagents

Organic chemicals and solvents, mycotoxin ZON and its derivatives, goat antirabbit immunoglobulin conjugated to horseradish peroxidase (HRP) as secondary antibody, and salts for buffers were purchased from Sigma-Aldrich Inc. (St. Louise, MO, USA). The purity of standard solutions was $\geq 98\%$. Immunoassays were carried out in high-capacity 96-well microplates (Nunc, Roskilde, Denmark) for colorimetric assay and in low profile 96-well microplates with white wells for increased fluorescence (Bio-Rad Laboratories, Hercules CA, USA). QuantaRed Enhanced Chemifluorescent HRP Substrate Kit was used as the last step in immunoassays. Surface water samples were obtained from river Danube at Budapest and lakes Velencei at Agárd and Balaton at Tihany (sampling site GPS latitude and longitude and coordinates—Danube: 47.517519, 19.045519; lake Velencei: 47.200922, 18.578361; lake Balaton: 46.914043, 17.893401).

4.2. Instrumentation

RFUs were determined by the prototype of a novel instrumentation developed in project Aquafluosense and realized in a modular setup (Figure 6). The instrument was developed partly (motor, optics, sample holder) at Optimal Optik Ltd. (Budapest, Hungary) and partly (detector electronics) at the Budapest University of Technology and Economics (Budapest, Hungary), and was designed to fit the 96-well ELISA microplate format using a self-designed, 3D-printed holder for ELFIA (Figure 6). The samples were illuminated in a dual head configuration with a high-power LED (Cree XPEBGR-L1-0000-00F01 with 520 to 535 nm minimum to maximum dominant wavelength range) in each head. The emitted fluorescence is measured in a dichroic beam path with silicon photodiodes (PIN-25D, OSI Optoelectronics) having large active area (d = 27.9 mm). The necessary high-spectral blocking and contrast was achieved by a combination of dichroic (Semrock FF562-Di03, edge: 562 nm) and bandpass optical filters on both the excitation (Semrock FF01-531/40-25, peak: 531 nm, width: 40 nm) and emission (Semrock FF01-593/40-25, peak: 593 nm, width: 40 nm) paths. The photodetector signal was coupled to a 2-stage amplifier unit (1st stage: OPA129 electrometer preamplifier, Texas Instruments; 2nd stage: AD620 instrumentation amplifier, Analog Devices) and then fed to a 12-bit analog-to-digital converter (Analog Devices AD7864-2 with 0 to 5 V unipolar input range) yielding 4095 resolvable RFUs. Gain and offset of the 2nd stage and the LED optical power were controlled by 256-stage (8-bit) digital potentiometers. As the LED power adjustment was nonequidistant, the optical power-control number curve was calibrated by a FieldMaxII-TO (Coherent) power meter with an OP-2 VIS sensor head set to a nominal wavelength of 532 nm. The instrument was equipped with stepping motors to move the detector heads over the 96-well microplates which provides fast and effective determination of individual RFUs in each microplate well. The instrument development is currently in the experimental phase; further decision about possible commercialization will be made by the project consortium.

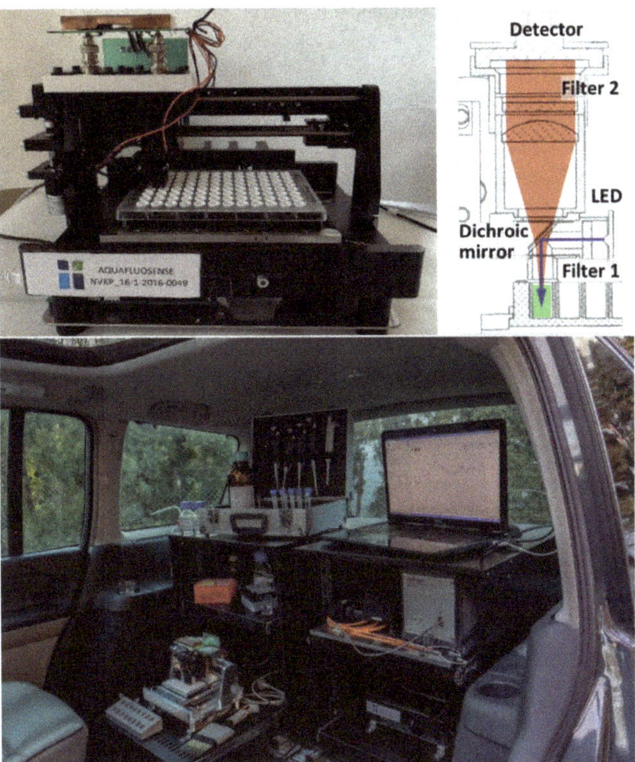

Figure 6. The immunofluorescence module developed in project Aquafluosense and appropriate for zearalenone determination (**top left**). The optical path in the detector head (**top right**). The modular instrumental setup during on site operation in a laboratory motor vehicle (**bottom**).

4.3. Determination of Zearalenone by Autofluorescence

Fluorescence spectra of ZON were recorded on a SpectraMax iD3 Multi-Mode Microplate Reader (Molecular Devices, San Jose, CA, USA) by scanning the excitation wavelengths between 250 and 830 nm and emission wavelengths between 270 and 850 nm, with step sizes of 10 nm, where the emission wavelength must be a minimum of 20 nm greater than excitation. A spectrum map was established by RFUs measured for a solution of ZON in PBS at a concentration of 2000 ng/mL with a corresponding RFU for PBS at each point as a background. An optimized peak on the basis of the fluorescence spectral map obtained was applied to establish a calibration curve (in the concentration range of 0.6–2000 ng/mL) and a LOD of ZON based on autofluorescence.

4.4. Enzyme-Linked Immunofluorescence Assay

4.4.1. Hapten Synthesis and Conjugation

The corresponding hapten, ZON-6′-carboxymethyloxime was converted from ZON by the method of Thouvenot and Morfin [61]. The reaction mixture containing 300 mg of ZON dissolved in dry pyridine and 600 mg of carboxymethoxylamine was stirred overnight at room temperature (22 ± 0.5 °C), then evaporated and the residue was taken up in 50 mL of slightly alkaline (pH = 8) water. The aqueous phase was extracted with ethyl acetate (4 × 100 mL). The organic phase was dried over sodium sulfate, then separated and evaporated to afford 156 mg of the product. The process of conversion of ZON to the corresponding hapten was followed by thin layer chromatography using hexane–ethyl acetate (1:2) as an eluent. BSA and conalbumin (CONA) were applied as carrier proteins. The hapten was conjugated to these proteins through amide bonds [55], using the active ester method for conjugation. Thus, 125 mg of the hapten ZON-6′-carboxymethyloxime was dissolved in 6.2 mL of dry tetrahydrofurane (THF), then 24 mg of *N*-hydroxy-succinimide and 73 mg of *N*,*N*′-dicyclohexylcarbodiimide were added. The mixture was stirred for 2 h at room temperature and the precipitation formed (dicyclohexylurea) was filtered off. In the mixture of 15.5 mL of water and 0.9 mL of THF, 150 mg of the proteins were dissolved in two separate batches. To these solutions, 3.1 mL of the above THF solution of the hapten active ester was added dropwise. The mixture was stirred for 24 h at 4 °C, and then the products (hapten–protein conjugate) were dialyzed against water at 4 °C for 1 week. Conjugation was monitored by UV spectroscopy; conjugates were lyophilized and stored at −20 °C until the analytical measurements.

4.4.2. Serum Preparation

Two 3-month old female New Zealand white rabbits were immunized intradermally with the CONA-hapten conjugate (immunogen). Rabbit immunisation was performed under the supervision of the Ethics Committee of Research on Animals (Food Science Research Institute, Institute of Food Sciences, Hungarian University of Agriculture and Life Sciences, Budapest, Hungary) and under the authorisation and inspection by the Government Office for Pest County in Hungary (Official permit for animal testing # PE/EA/45-6/2020, last date of approval: 21 February 2020). Initial immunization was carried out with 0.1 mg immunogen dissolved in PBS and emulsified in Freund's complete adjuvant (1:1 volume fraction). Then, injection of 0.15 mg of the immunogen in PBS and Freund's incomplete adjuvant (1:1 volume fraction) was given (the first three booster injections at 3-week intervals, with the subsequent one at a 1-month interval). One week after each immunization, rabbits were bled and after coagulation of the blood (4 °C overnight) the sera were centrifuged (2400 g, 15 min) and purified by gel chromatography on PD-10 desalting columns.

4.4.3. Immunoassay

Plates were coated with 1 µg/mL ZON-6′-carboxymethyloxime–BSA conjugate (ZON–BSA) in carbonate buffer (15 mM Na_2CO_3, 35 mM $NaHCO_3$, pH = 9.6) overnight (~8 h) at 4 °C. The unbound conjugate was washed 4 times with PBS with 2% Tween20 (137 mM NaCl, 2.7 KCl, 10 mM $Na_2HPO_4 \times 2H_2O$, pH = 7.4). Blocking was carried out with 150 µL/well 1% gelatin in PBS, for 1.5 h of incubation at 37 °C. After washing, competition was initiated by adding 50 µL/well of both ZON analytical standard and purified rabbit anti-ZON serum (in PBS buffer with 5% Tween20, dilution 1:1000). An analytical ZON standard stock solution (1 mg/mL ZON in MeOH) was used in serial dilution (0.004 pg/mL–2 µg/mL). Thus, MeOH content in the actual ELFIA measurements was 0.2% or below. After 1 h of incubation at 37 °C and four washes, 100 µL/well goat antirabbit IgG–HRP (horseradish peroxidase, dilution 1:7500) conjugate was added as secondary antibody and incubated for 1 h at 37 °C. Unbound secondary antibodies were washed out 4 times with PBS and 100 µL/well working solution of QuantaRed Enhanced Chemifluorescent HRP Substrate Kit were added (content of the working solution: QuantaRed ADHP concentrate, QuantaRed Enhancer Solution, and QuantaRed stable Peroxide in 1:50:50 (*v/v*) proportion). The kit contains 10-acetyl-3,7-dihydroxyphenoxazine (ADHP), a nonfluorescent compound that is dehydrogenated (oxidized) by HRP to resorufin, a highly fluorescent reaction product, which can also be measured in a colorimetric plate reader. After 5 min of incubation, the enzymatic activity was stopped by 10 µL of QuantaRed Stop Solution. Both absorbance and fluorescence were measured at 576 and at 593 nm wavelengths, respectively. After the colorimetric assay, the liquid phase was transferred with an 8-channel pipette into a white, low profile 96-well microplate where fluorescence was determined. Absorbances were read by SpectraMax iD3 Multi-Mode Microplate Reader (Molecular Devices, San Jose, CA, USA) at 576 nm wavelength, while relative fluorescence signals were determined by the prototype fluorimeter developed in project Aquafluosense [69,74] (see Section 4.2). Standard curves were obtained in different surface water samples as well, for evaluation of matrix effects in ZON determination from natural water bodies. Statistical analysis of standard curves was performed by comparison of IC_{50} values by one-way analysis of variance followed by post hoc Tukey test at a significance level of 0.05.

The immunoassay developed in this study can be performed in situ by the instrument developed and built in a laboratory motor vehicle. Coating the assay microplates by the ZON–BSA conjugate can be carried out under laboratory conditions (method in Section 4.4.3) prior the measurement, and in situ determination can be performed on pre-coated microplates in the mobile equipment. Thus, the total immunoassay performance time is 2.5 h. Since the immunofluorescence instrument is equipped with stepping motor units, fluorescence determination on an entire plate requires approximately 2 min. A possibility of further shortening the procedure is potentially through performing both the coating and blocking steps in the laboratory a day before the in situ determination. In this case, the coating step can be applied at room temperature for 1 h and the blocking step with 1% gelatin in PBS for 1.5 h. Coated and blocked microplates can be stored at 4 °C until measurement; however, it is necessary to prevent evaporation in the wells by covering the microplate with parafilm.

4.5. High-Performance Liquid Chromatography (HPLC)

A calibration curve and the LOD for ZON were determined also by HPLC measurements on a Younglin YL9100 HPLC system equipped with a YL9150 autosampler (YL Instruments, Anyang, Korea). Compounds were separated on a column (150 × 4.6 mm i.d., 5 μm) at 30 °C containing C18 stationary phase. The PerfectSil 100 ODS-3 column was manufactured by MZ-Analysentechnik GmbH (Mainz, Germany). UV detector signals were recorded at $\lambda_1 = 236$ nm and $\lambda_2 = 274$ nm. Eluent flow rate was 1.0 mL/min with isocratic elution for 10 min (30:70 = A:B eluents, A = 90% water:10% MeOH, B = MeOH). The retention time of ZON under the current conditions was 6.12 min. LOD, defined as an analyte concentration corresponding to a signal level of signal/noise ratio of 3, was determined with standard solutions. For establishing an analytical standard curve, a stock solution of ZON at a concentration of 1.0 g/mL was prepared in MeOH. Calibration curves were obtained with 7 standard solutions between 10 and 2000 ng/mL in MeOH:water = 7:3) and MeOH:PBS (1:1).

4.6. Total Internal Reflection Ellipsometry (TIRE)

Total internal reflection ellipsometry (TIRE) experiments were carried out on an M-2000 automatic spectroscopic ellipsometer (J.A. Woollam Co., Lincoln, NE, USA) operating in the 370–1000 nm range using glass-based sensor chips fabricated in the laboratory by vacuum evaporation. Sensor surfaces were prepared by thermal evaporation of layers of chromium (Cr)—3 nm thick and gold (Au)—25 nm on standard microscopic glass slides (BK-7). The Cr layer improves the adhesion of gold to the glass surface. The Au surface was modified with mercaptoethyl sodium sulfonate to enhance the negative surface charge. The ellipsometer was equipped with a 68° trapezoidal prism which allowed coupling the light beam at total internal reflection conditions to the gold film on the glass slide. The 0.2 mL reaction cell with the inlet and outlet tubes was attached underneath to the gold surface, allowing the injection of the required chemicals to perform binding reactions. The ellipsometry spectral scans were performed in a standard Trisma/HCl buffer solution (pH = 7.5) after completing each adsorption (binding) stage. For a competitive immunoassay, a ZON-6′-carboxymethyloxime–BSA conjugate (ZON–BSA) was electrostatically immobilized on the Au surface via a polyallylamine hydrochloride layer. In order to block all the remaining binding sites, an additional adsorption of BSA was carried out. Then a mixture of ZON-specific antiserum and solutions of free ZON (at a concentration range of 0.01 ng/mL–10 μg/mL) were injected into the cell with the intermediate rinsing with buffer. The mixtures were preincubated for 5 min before injection. A series of Δ spectra were recorded after binding of ZON to the antibodies immobilized on the chip surface.

Author Contributions: Conceptualization, A.S., L.K., N.A. and A.N.; methodology, A.B., G.S., M.M., S.L., A.C., S.K. and E.T.; software, A.B., P.G. and L.D.; formal analysis and investigation, B.G., M.M., A.N. and E.T.; data curation, A.S., E.T., A.N. and M.M.; writing—original draft preparation, B.G., A.S. and E.T.; writing—review and editing, A.S., L.K. and A.B.; visualization, B.G., E.T., M.M., A.S. and A.N.; supervision, A.S.; project administration, A.S., L.K., A.B. and N.A.; funding acquisition, A.S., L.K., A.B., A.N. and L.D. All authors have read and agreed to the published version of the manuscript.

Funding: This research was funded by the Hungarian National Research, Development and Innovation Office within the National Competitiveness and Excellence Program, project NVKP_16-1-2016-0049 "In situ, complex water quality monitoring by using direct or immunofluorimetry and plasma spectroscopy" (Aquafluosense 2017-2021).

Institutional Review Board Statement: The study was conducted according to the guidelines of the Declaration of Helsinki, and approved by the Ethics Committee of Research on Animals of the Food Science Research Institute, Institute of Food Sciences, Hungarian University of Agriculture and Life Sciences, Budapest, Hungary (protocol code PE/EA/45-6/2020, last date of approval 21 February 2020).

Informed Consent Statement: Not applicable.

Data Availability Statement: The data presented in this study are available on request from the corresponding author. The data are not publicly available due to privacy reasons.

Acknowledgments: The authors express their appreciation to Csilla Magor at the Agro-Environmental Research Institute, National Agricultural Research and Innovation Centre, Budapest, Hungary, and to Ali Madlool Al-Jawdah at the Materials and Engineering Research Institute, Sheffield Hallam University, Sheffield, UK, for their technical contribution in the HPLC determination and the TIRF experiments, respectively.

Conflicts of Interest: The authors declare no conflict of interest.

References

1. Khaneghah, A.M.; Fakhri, Y.; Gahruie, H.H.; Niakousari, M.; Sant'Ana, A.S. Mycotoxins in cereal-based products during 24 years (1983–2017): A global systematic review. *Trends Food Sci. Technol.* **2019**, *91*, 95–105. [CrossRef]
2. Cimbalo, A.; Alonso-Garrido, M.; Font, G.; Manyes, L. Toxicity of mycotoxins in vivo on vertebrate organisms: A review. *Food Chem. Toxicol.* **2020**, *137*, 111161. [CrossRef]
3. Krska, R.; Molinelli, A. Mycotoxin analysis: State-of-the-art and future trends. *Anal. Bioanal. Chem.* **2006**, *387*, 145–148. [CrossRef]
4. Rubert, J.; Soriano, J.M.; Mañes, J.; Soler, C. Rapid mycotoxin analysis in human urine: A pilot study. *Food Chem. Toxicol.* **2011**, *49*, 2299–2304. [CrossRef]
5. Van Der Fels-Klerx, H.; De Rijk, T.; Booij, C.; Goedhart, P.; Boers, E.; Zhao, C.; Waalwijk, C.; Mol, H.; Van Der Lee, T. Occurrence of *Fusarium* head blight species and *Fusarium* mycotoxins in winter wheat in the Netherlands in 2009. *Food Addit. Contam. Part A* **2012**, *29*, 1716–1726. [CrossRef]
6. Valverde-Bogantes, E.; Bianchini, A.; Herr, J.R.; Rose, D.J.; Wegulo, S.N.; Hallen-Adams, H.E. Recent population changes of *Fusarium* head blight pathogens: Drivers and implications. *Can. J. Plant Pathol.* **2020**, *42*, 315–329. [CrossRef]
7. Valencia-Quintana, R.; Milić, M.; Jakšić, D.; Šegvić Klarić, M.; Tenorio-Arvide, M.G.; Pérez-Flores, G.A.; Bonassi, S.; Sánchez-Alarcón, J. Environment changes, aflatoxins, and health issues, a review. *Int. J. Environ. Res. Public Health* **2020**, *17*, 7850. [CrossRef]
8. Dänicke, S.; Winkler, J. Invited review: Diagnosis of zearalenone (ZEN) exposure of farm animals and transfer of its residues into edible tissues (carry over). *Food Chem. Toxicol.* **2015**, *84*, 225–249. [CrossRef] [PubMed]
9. Zheng, W.; Feng, N.; Wang, Y.; Noll, L.; Xu, S.; Liu, X.; Lu, N.; Zou, H.; Gu, J.; Yuan, Y.; et al. Effects of zearalenone and its derivatives on the synthesis and secretion of mammalian sex steroid hormones: A review. *Food Chem. Toxicol.* **2019**, *126*, 262–276. [CrossRef] [PubMed]
10. Kelley, J.; Paterson, R.R.M. Comparisons of ergosterol to other methods for determination of *Fusarium graminearum* biomass in water as a model system. In Proceedings of the 22nd European Culture Collections' Organization Meeting, Braga, Portugal, 17–19 September 2003; Lia, N., Smith, D., Eds.; Micoteca da Universidade do Minho: Braga, Portugal, 2003; pp. 235–239. Available online: https://repositorium.sdum.uminho.pt/bitstream/1822/55312/1/document_3404_1.pdf (accessed on 8 February 2021).
11. Picardo, M.; Filatova, D.; Nuñez, O.; Farré, M. Recent advances in the detection of natural toxins in freshwater environments. *TrAC Trends Anal. Chem.* **2019**, *112*, 75–86. [CrossRef]
12. Russell, R.; Paterson, M. Zearalenone production and growth in drinking water inoculated with *Fusarium graminearum*. *Mycol. Prog.* **2007**, *6*, 109–113. [CrossRef]
13. Bucheli, T.D.; Wettstein, F.E.; Hartmann, N.; Erbs, M.; Vogelsang, S.; Forrer, H.-R.; Schwarzenbach, R.P. *Fusarium* mycotoxins: Overlooked aquatic micropollutants? *J. Agric. Food Chem.* **2008**, *56*, 1029–1034. [CrossRef] [PubMed]
14. Gromadzka, K.; Waśkiewicz, A.; Goliński, P.; Świetlik, J. Occurrence of estrogenic mycotoxin—Zearalenone in aqueous environmental samples with various NOM content. *Water Res.* **2009**, *43*, 1051–1059. [CrossRef]

15. Waśkiewicz, A.; Gromadzka, K.; Bocianowski, J.; Pluta, P.; Goliński, P. Zearalenone Contamination of the aquatic environment as a result of its presence in crops/Pojava mikotoksina u vodenom okolišu zbog njihove prisutnosti u usjevima. *Arch. Ind. Hyg. Toxicol.* **2012**, *63*, 429–435. [CrossRef]

16. Jarošová, B.; Javůrek, J.; Adamovský, O.; Hilscherová, K. Phytoestrogens and mycoestrogens in surface waters—Their sources, occurrence, and potential contribution to estrogenic activity. *Environ. Int.* **2015**, *81*, 26–44. [CrossRef]

17. Laranjeiro, C.S.; Da Silva, L.J.G.; Pereira, A.M.; Pena, A.; Lino, C.M. The mycoestrogen zearalenone in Portuguese flowing waters and its potential environmental impact. *Mycotoxin Res.* **2017**, *34*, 77–83. [CrossRef]

18. Al-Gabr, H.M.; Zheng, T.; Yu, X. Fungi contamination of drinking water. *Rev. Environ. Contam. Toxicol.* **2013**, *228*, 121–139. [CrossRef]

19. Hartmann, N.; Erbs, M.; Wettstein, F.E.; Schwarzenbach, R.P.; Bucheli, T.D. Quantification of estrogenic mycotoxins at the ng/L level in aqueous environmental samples using deuterated internal standards. *J. Chromatogr. A* **2007**, *1138*, 132–140. [CrossRef]

20. Criado, M.V.; Pinto, V.E.F.; Badessari, A.; Cabral, D. Conditions that regulate the growth of moulds inoculated into bottled mineral water. *Int. J. Food Microbiol.* **2005**, *99*, 343–349. [CrossRef] [PubMed]

21. Pereira, V.; Fernandes, D.; Carvalho, G.; Benoliel, M.; Romão, M.S.; Crespo, M.B. Assessment of the presence and dynamics of fungi in drinking water sources using cultural and molecular methods. *Water Res.* **2010**, *44*, 4850–4859. [CrossRef] [PubMed]

22. Kolpin, D.W.; Schenzel, J.; Meyer, M.T.; Phillips, P.J.; Hubbard, L.E.; Scott, T.-M.; Bucheli, T.D. Mycotoxins: Diffuse and point source contributions of natural contaminants of emerging concern to streams. *Sci. Total. Environ.* **2014**, 669–676. [CrossRef] [PubMed]

23. Babič, M.N.; Gunde-Cimerman, N.; Vargha, M.; Tischner, Z.; Magyar, D.; Veríssimo, C.; Sabino, R.; Viegas, C.; Meyer, W.; Brandão, J. Fungal contaminants in drinking water regulation? A tale of ecology, exposure, purification and clinical relevance. *Int. J. Environ. Res. Public Health* **2017**, *14*, 636. [CrossRef]

24. Paterson, R.; Kelley, J.; Gallagher, M. Natural occurrence of aflatoxins and *Aspergillus flaws* (Link) in water. *Lett. Appl. Microbiol.* **1997**, *25*, 435–436. [CrossRef]

25. Oliveira, B.R.; Mata, A.T.; Ferreira, J.P.; Crespo, M.T.B.; Pereira, V.J.; Bronze, M.R. Production of mycotoxins by filamentous fungi in untreated surface water. *Environ. Sci. Pollut. Res.* **2018**, *25*, 17519–17528. [CrossRef]

26. Kolpin, D.W.; Hoerger, C.C.; Meyer, M.T.; Wettstein, F.E.; Hubbard, L.E.; Bucheli, T.D. Phytoestrogens and Mycotoxins in Iowa Streams: An examination of underinvestigated compounds in agricultural basins. *J. Environ. Qual.* **2010**, *39*, 2089–2099. [CrossRef]

27. Mata, A.; Ferreira, J.; Oliveira, B.; Batoréu, M.; Crespo, M.B.; Pereira, V.; Bronze, M. Bottled water: Analysis of mycotoxins by LC–MS/MS. *Food Chem.* **2015**, *176*, 455–464. [CrossRef]

28. Waśkiewicz, A.; Bocianowski, J.; Perczak, A.; Goliński, P. Occurrence of fungal metabolites—Fumonisins at the ng/L level in aqueous environmental samples. *Sci. Total. Environ.* **2015**, *524-525*, 394–399. [CrossRef] [PubMed]

29. Emídio, E.S.; Da Silva, C.P.; De Marchi, M.R.R. Estrogenic mycotoxins in surface waters of the Rico Stream micro-basin, São Paulo, Brazil: Occurrence and potential estrogenic contribution. *Eclética Química J.* **2020**, *45*, 23–32. [CrossRef]

30. Maragos, C.M. Zearalenone occurrence in surface waters in central Illinois, USA. *Food Addit. Contam. Part B* **2012**, *5*, 55–64. [CrossRef] [PubMed]

31. Hartmann, N.; Erbs, M.; Forrer, H.-R.; Vogelgsang, S.; Wettstein, F.E.; Schwarzenbach, R.P.; Bucheli, T.D. Occurrence of Zearalenone on *Fusarium graminearum* infected wheat and maize fields in crop organs, soil, and drainage water. *Environ. Sci. Technol.* **2008**, *42*, 5455–5460. [CrossRef]

32. Laganà, A.; Bacaloni, A.; De Leva, I.; Faberi, A.; Fago, G.; Marino, A. Analytical methodologies for determining the occurrence of endocrine disrupting chemicals in sewage treatment plants and natural waters. *Anal. Chim. Acta* **2004**, *501*, 79–88. [CrossRef]

33. Pietsch, C. Zearalenone (ZEN) and its influence on regulation of gene expression in carp (*Cyprinus carpio* L.) liver tissue. *Toxins* **2017**, *9*, 283. [CrossRef] [PubMed]

34. Khezri, A.; Herranz-Jusdado, J.G.; Ropstad, E.; Fraser, T.W. Mycotoxins induce developmental toxicity and behavioural aberrations in zebrafish larvae. *Environ. Pollut.* **2018**, *242*, 500–506. [CrossRef] [PubMed]

35. European Food Safety Authority (EFSA). Scientific Opinion on the risks for public health related to the presence of zearalenone in food. *EFSA J.* **2011**, *9*, 2197. [CrossRef]

36. European Commission. Commission regulation (EC) No 1881/2006 of 19 December 2006. Setting maximum levels for certain contaminants in foodstuffs. *Off. J. Eur. Union* **2006**, 364–365.

37. EFSA Panel on Contaminants in the Food Chain (CONTAM). Scientific Opinion on the risks for human and animal health related to the presence of modified forms of certain mycotoxins in food and feed. *EFSA J.* **2014**, *12*. [CrossRef]

38. EFSA Panel on Contaminants in the Food Chain (CONTAM); Knutsen, H.; Alexander, J.; Barregård, L.; Bignami, M.; Brüschweiler, B.; Ceccatelli, S.; Cottrill, B.; DiNovi, M.; Edler, L.; et al. Risks for animal health related to the presence of zearalenone and its modified forms in feed. *EFSA J.* **2017**, *15*, e04851. [CrossRef]

39. Caglayan, M.O.; Şahin, S.; Üstündağ, Z. Detection strategies of zearalenone for food safety: A review. *Crit. Rev. Anal. Chem.* **2020**, *67*, 1–20. [CrossRef]

40. Turner, G.V.; Phillips, T.D.; Heidelbaugh, N.D.; Russell, L.H. High Pressure liquid chromatographic determination of zearalenone in chicken tissues. *J. Assoc. Off. Anal. Chem.* **1983**, *66*, 102–104. [CrossRef]

41. Sebaei, A.S.; Gomaa, A.M.; Mohamed, G.G.; El-Di, F.N. Simple validated method for determination of deoxynivalenol and zearalenone in some cereals using high performance liquid chromatography. *Am. J. Food Technol.* **2012**, *7*, 668–678. [CrossRef]

42. Majerus, P.; Graf, N.; Krämer, M. Rapid determination of zearalenone in edible oils by HPLC with fluorescence detection. *Mycotoxin Res.* **2009**, *25*, 117–121. [CrossRef] [PubMed]

43. Scott, P.M.; Lawrence, J.W.; van Walbeek, W. Detection of mycotoxins by thin-layer chromatography: Application to screening of fungal extracts. *App. Environ. Microbiol.* **1970**, *20*, 839–842. [CrossRef]

44. Larionova, D.A.; Goryacheva, I.Y.; Van Peteghem, C.; De Saeger, S. Thin-layer chromatography of aflatoxins and zearalenones with β-cyclodextrins as mobile phase additives. *World Mycotoxin J.* **2011**, *4*, 113–117. [CrossRef]

45. Soares, L.M.V. Multi-toxin TLC methods for aflatoxins, ochratoxin a, zearalenone and sterigmatocystin in foods. In *Analysis of Plant Waste Materials*; Springer International Publishing: Geneva, Switzerland, 1992; Volume 13, pp. 227–238.

46. Thouvenot, D.; Morfin, R. Quantitation of zearalenone by gas-liquid chromatography on capillary glass columns. *J. Chromatogr. A* **1979**, *170*, 165–173. [CrossRef]

47. Yan, Z.; Wang, L.; Wang, J.; Tan, Y.; Yu, D.; Chang, X.; Fan, Y.; Zhao, D.; Wang, C.; De Boevre, M.; et al. A QuEChERS-based liquid chromatography-tandem mass spectrometry method for the simultaneous determination of nine zearalenone-like mycotoxins in pigs. *Toxins* **2018**, *10*, 129. [CrossRef]

48. De Santis, B.; Debegnach, F.; Gregori, E.; Russo, S.; Marchegiani, F.; Moracci, G.; Brera, C. Development of a LC-MS/MS Method for the multi-mycotoxin determination in composite cereal-based samples. *Toxins* **2017**, *9*, 169. [CrossRef]

49. Jodlbauer, J.; Zöllner, P.; Lindner, W. Determination of zeranol, taleranol, zearalenone, α- and β-zearalenol in urine and tissue by high-performance liquid chromatography-tandem mass spectrometry. *Chromatographia* **2000**, *51*, 681–687. [CrossRef]

50. Pestka, J.J.; Abouzied, M.N. Immunological assays for mycotoxin detection. *Food Technol.* **1995**, *49*, 120–128.

51. Maragos, C.M.; Kim, E.-K. Detection of zearalenone and related metabolites by fluorescence polarization immunoassay. *J. Food Prot.* **2004**, *67*, 1039–1043. [CrossRef]

52. Hao, K.; Suryoprabowo, S.; Song, S.; Liu, L.; Kuang, H. Rapid detection of zearalenone and its metabolite in corn flour with the immunochromatographic test strip. *Food Agric. Immunol.* **2017**, *29*, 498–510. [CrossRef]

53. Renaud, J.B.; Miller, J.D.; Sumarah, M.W. Mycotoxin testing paradigm: Challenges and opportunities for the future. *J. AOAC Int.* **2019**, *102*, 1681–1688. [CrossRef]

54. Dong, G.; Pan, Y.; Wang, Y.; Ahmed, S.; Liu, Z.; Peng, D.; Yuan, Z. Preparation of a broad-spectrum anti-zearalenone and its primary analogues antibody and its application in an indirect competitive enzyme-linked immunosorbent assay. *Food Chem.* **2018**, *247*, 8–15. [CrossRef]

55. Liu, M.T.; Ram, B.P.; Hart, L.P.; Pestka, J.J. Indirect enzyme-linked immunosorbent assay for the mycotoxin zearalenone. *Appl. Environ. Microbiol.* **1985**, *50*, 332–336. [CrossRef]

56. Liu, N.; Nie, D.; Zhao, Z.; Meng, X.; Wu, A. Ultrasensitive immunoassays based on biotin–streptavidin amplified system for quantitative determination of family zearalenones. *Food Control.* **2015**, *57*, 202–209. [CrossRef]

57. Pichler, H.; Krska, R.; Szekacs, A.; Grasserbauer, M. An enzyme-immunoassay for the detection of the mycotoxin zearalenone by use of yolk antibodies. *Anal. Bioanal. Chem.* **1998**, *362*, 176–177. [CrossRef]

58. Radová, Z.; HajšLová, J.; Králová, J.; Papoušková, L.; Sýkorová, S. Analysis of Zearalenone in wheat using high-performance liquid chromatography with fluorescence detection and/or enzyme-linked immunosorbent assay. *Cereal Res. Commun.* **2001**, *29*, 435–442. [CrossRef]

59. Tang, X.; Li, X.; Li, P.; Zhang, Q.; Li, R.; Zhang, W.; Ding, X.; Lei, J.; Zhang, Z. Development and application of an immu-noaffinity column enzyme immunoassay for mycotoxin zearalenone in complicated samples. *PLoS ONE* **2014**, *9*, e85606.

60. Thongrussamee, T.; Kuzmina, N.; Shim, W.-B.; Jiratpong, T.; Eremin, S.; Intrasook, J.; Chung, D.-H. Monoclonal-based enzyme-linked immunosorbent assay for the detection of zearalenone in cereals. *Food Addit. Contam. Part A* **2008**, *25*, 997–1006. [CrossRef] [PubMed]

61. Thouvenot, D.; Morfin, R.F. Radioimmunoassay for zearalenone and zearalanol in human serum: Production, properties, and use of porcine antibodies. *Appl. Environ. Microbiol.* **1983**, *45*, 16–23. [CrossRef]

62. Chun, H.S.; Choi, E.H.; Chang, H.-J.; Choi, S.-W.; Eremin, S.A. A fluorescence polarization immunoassay for the detection of zearalenone in corn. *Anal. Chim. Acta* **2009**, *639*, 83–89. [CrossRef]

63. Van Der Gaag, B.; Spath, S.; Dietrich, H.; Stigter, E.; Boonzaaijer, G.; Van Osenbruggen, T.; Koopal, K. Biosensors and multiple mycotoxin analysis. *Food Control.* **2003**, *14*, 251–254. [CrossRef]

64. Goud, Y.K.; Kumar, S.V.; Hayat, K.; Gobi, V.K.; Song, H.; Kim, K.-H.; Marty, J.L. A highly sensitive electrochemical immunosensor for zearalenone using screen-printed disposable electrodes. *J. Electroanal. Chem.* **2019**, *832*, 336–342. [CrossRef]

65. Hossain, M.Z.; Maragos, C.M. Gold nanoparticle-enhanced multiplexed imaging surface plasmon resonance (iSPR) detection of *Fusarium* mycotoxins in wheat. *Biosens. Bioelectron.* **2018**, *101*, 245–252. [CrossRef] [PubMed]

66. Nabok, A.; Al-Jawdah, A.; Gémes, B.; Takács, E.; Székács, A. An optical planar waveguide-based immunosensors for determination of *Fusarium* mycotoxin zearalenone. *Toxins* **2021**, *13*, 89. [CrossRef] [PubMed]

67. Székács, I.; Adányi, N.; Szendrő, I.; Székács, A. Direct and Competitive optical grating immunosensors for determination of *Fusarium* mycotoxin zearalenone. *Toxins* **2021**, *13*, 43. [CrossRef]

68. Urraca, J.L.; Benito-Peña, E.; Pérez-Conde, C.; Moreno-Bondi, M.C.; Pestka, J.J. Analysis of zearalenone in cereal and swine feed samples using an automated flow-through immunosensor. *J. Agric. Food Chem.* **2005**, *53*, 3338–3344. [CrossRef]

69. Aquafluosense. komplex vízminősítést in situ megvalósító, közvetlen és immunfluoreszcencián, valamint optikai és lézeres plazma-színképelemzésen alapuló, moduláris, érzékelő- és műszercsalád kifejlesztése, továbbá az alkalmazási területek kutatása. Available online: http://aquafluosense.hu (accessed on 8 February 2021).

70. Rodbard, D.; Hutt, D.M. Statistical analysis of radioimmunoassays and immunoradiometric (labeled antibody) assays: A generalized, weighted, iterative, least-squares method for logistic curve fitting. In Proceedings of the Symposium on RIA and Related Procedures in Medicine, Proceedings of the Internationaé Atomic Energy Agency, Vienna, Austria, 1974; p. 165.

71. Miles, C.O.; Erasmuson, A.F.; Wilkins, A.L.; Towers, N.R.; Smith, B.L.; Garthwaite, I.; Scahill, B.G.; Hansen, R.P. Ovine metabolism of zearalenone to α-zearalanol (Zeranol). *J. Agric. Food Chem.* **1996**, *44*, 3244–3250. [CrossRef]

72. Megharaj, M.; Garthwaite, I.; Thiele, J.H. Total biodegradation of the oestrogenic mycotoxin zearalenone by a bacterial culture. *Lett. Appl. Microbiol.* **1997**, *24*, 329–333. [CrossRef]

73. Nabok, A.; Tsargorodskaya, A. The method of total internal reflection ellipsometry for thin film characterisation and sensing. *Thin Solid Film.* **2008**, *516*, 8993–9001. [CrossRef]

74. Csősz, D.; Lenk, S.; Barócsi, A.; Csőke, T.L.; Klátyik, S.; Lázár, D.; Berki, M.; Adányi, N.; Csákányi, A.; Domján, L.; et al. Sensitive fluorescence instrumentation for water quality assessment. In Proceedings of the Optical Sensors and Sensing Congress (ES, FTS, HISE, Sensors), San Jose, CA, USA, 25–27 June 2019; OSA: Washington, DC, USA, 2019.

Article

An Optical Planar Waveguide-Based Immunosensors for Determination of *Fusarium* Mycotoxin Zearalenone

Alexei Nabok [1,*], Ali Madlool Al-Jawdah [1,2], Borbála Gémes [3], Eszter Takács [3] and András Székács [3]

1   Materials and Engineering Research Institute, Sheffield Hallam University, Howard Street, Sheffield S1 1WB, UK; aalimadlol@yahoo.com
2   College of Sciences, Babylon University, P.O. Box 4, Hilla 51002, Iraq
3   Agro-Environmental Research Institute, National Research and Innovation Centre, Herman Ottó út 15, H-1022 Budapest, Hungary; szekacs.andras@akk.naik.hu (A.S.); gemes.borbala.leticia@akk.naik.hu (B.G.); akacs.eszter@akk.naik.hu (E.T.)
*   Correspondence: a.nabok@shu.ac.uk; Tel.: +44-7854-805603

Abstract: A planar waveguide (PW) immunosensor working as a polarisation interferometer was developed for the detection of mycotoxin zearalenone (ZON). The main element of the sensor is an optical waveguide consisting of a thin silicon nitride layer between two thicker silicon dioxide layers. A combination of a narrow waveguiding core made by photolithography with an advanced optical set-up providing a coupling of circular polarised light into the PW via its slanted edge allowed the realization of a novel sensing principle by detection of the phase shift between the p- and s-components of polarised light propagating through the PW. As the p-component is sensitive to refractive index changes at the waveguide interface, molecular events between the sensor surface and the contacting sample solution can be detected. To detect ZON concentrations in the sample solution, ZON-specific antibodies were immobilised on the waveguide via an electrostatically deposited polyelectrolyte layer, and protein A was adsorbed on it. Refractive index changes on the surface due to the binding of ZON molecules to the anchored antibodies were detected in a concentration-dependent manner up to 1000 ng/mL of ZON, allowing a limit of detection of 0.01 ng/mL. Structurally unrelated mycotoxins such as aflatoxin B1 or ochratoxin A did not exert observable cross-reactivity.

Keywords: mycotoxin; zearalenone; planar waveguide sensor; polarisation interferometer; label-free detection; limit of detection

Key Contribution: A planar waveguide immunosensor operating as a polarisation interferometer showing rapid response; high sensitivity and selectivity for detecting the *Fusarium* mycotoxin zearalenone in a wide concentration range from 1000 to 0.01 ng/mL was developed.

Citation: Nabok, A.; Al-Jawdah, A.M.; Gémes, B.; Takács, E.; Székács, A. An Optical Planar Waveguide-Based Immunosensors for Determination of *Fusarium* Mycotoxin Zearalenone. *Toxins* **2021**, *13*, 89. https://doi.org/10.3390/toxins 13020089

Received: 30 December 2020
Accepted: 21 January 2021
Published: 25 January 2021

Publisher's Note: MDPI stays neutral with regard to jurisdictional claims in published maps and institutional affiliations.

1. Introduction

Label-free optical biosensor techniques based on evanescent field effects are of increasing interest for agro-environmental safety in monitoring the quality of food and animal feed [1]. Evanescent waves (or fields) created at interfaces between two transparent media having different refractive indices (RIs) in optical devices such as waveguides propagate along the interfaces with their intensity decaying rapidly away from the interfaces. The evanescent field can penetrate into the medium of the lower RI to a distance of approximately 200 nm [2], indicating that the evanescence phenomenon is sensitive to RI changes not only in the waveguide but also in its immediate vicinity, and therefore can undergo characteristic modulations due to the changes in the medium RI caused by molecular interactions taking place near that surface. While these interactions can be detected in real time and with outstandingly high sensitivity, the use of molecular size bio-receptors (antibodies, aptamers, or molecularly imprinted polymers) in the biosensor set-up introduces the required specificity to the target analytes for analytical determination. Several waveguide-based biosensor formats exist for the optical detection of various mycotoxin molecules such as zearalenone (ZON).

Label-free optical immunosensor techniques include reflectance-based methods, such as total internal reflection ellipsometry (TIRE) [3], and grating-based methods, such as optical waveguide light-mode spectroscopy (OWLS) [4]. However, these methods typically rely on lab-based equipment and as a consequence are often incapable of fulfilling the current demands of portable biosensors suitable for in-field analysis, as the laboratory benchtop instrument cannot be moved to and operated at the site of sampling. One of the most promising directions in label-free biosensing is based on the use of the most sensitive optical technique of interferometry. Several successful developments of interferometric biosensor devices were accomplished recently [5]; they include dual beam interferometers [6], ring-resonators [7], and Mach-Zehnder (MZ) interferometers [6,8]. Biosensors based on MZ interferometers combining a high sensitivity of detection and portable design [6–10] were the most impressive with a pinnacle achievement being a monolithic silicon-based MZ biosensor combining in one chip the light source, multichannel biosensor with microfluidic sample delivery, photodetectors, and signal processing electronics [11,12]. Such biosensors are particularly suitable for in-field or point-of-need use. Thus, optical immunosensors based on planar waveguide (PW) technology are gaining attention [12] and have been developed for mycotoxins such as deoxynivalenol, ochratoxin A, ZON, and T-2 [13–17] and other aquatic toxins including sweet water and marine algal toxins such as microcystins, okadaic acid, domoic acid, and cylindrospermopsin involved in direct toxicity [18–21], saxitoxin involved in indirect toxicity through the food chain [22], as well as a microbial toxin tetrodotoxin [23]. In the current work, a PW immunosensor based on a novel principle of polarisation interferometry was adapted for the detection of ZON.

2. Results and Discussion

2.1. Planar Waveguide Biosensor Design and Testing

The detection principle of the planar waveguide (PW) biosensor acting as a polarisation interferometer (PI) is similar to Mach–Zehnder (MZ) interferometers, but instead of two optical arms in the MZ biosensor, the p- and s-polarisations of light were used as parallel parameters in this set-up. The PW being the main element of the developed biosensor was devised on an Si wafer and consisted of a 200 nm thick silicon nitride (Si_3N_4) ore layer (having a RI n = 2.01) placed between two much thicker (3 μm) cladding silicon oxide (SiO_2) layers of lower RI (n = 1.46). Such design allowed the propagation of a single mode electromagnetic (EM) wave through the waveguide by multiple internal reflection; the large difference in RIs between the Si_3N_4 core and the SiO_2 cladding resulted in light propagation at a steep angle of 47°, creating a large number of internal reflections of light (up to 3000 reflections/mm) along the PW.

As shown in Figure 1, the polarised 630 nm light from a laser diode (1) was coupled into waveguide (4) via a slant edge, which was polished at a 47° angle to provide a 90° incidence angle and therefore maximal efficiency of coupling. The light was converted to circular polarisation using a λ/4 plate (2) and focused on the slant edge using a lens (3). The outcoming light is going through a polariser (7), which converts the changes in the EM wave polarisation into modulation of its intensity, and collected by a charge-coupled device (CCD) array photodetector (8). The waveguide (4) with the dimensions of 25 × 8 mm is held between two pieces of black nylon with the upper piece forming an 8 × 2 × 6 mm (≈0.1 mL) cell (6) sealed against the top side of the waveguide and equipped with inlet and outlet tubes enabling injecting different chemicals into the cell. In the earlier versions of the set-up, the top layer of SiO_2 is etched away by injecting 1:10 diluted hydrofluoric acid into the cell to form the sensing window. Later on, in the advanced experimental set-up, both the waveguiding core and sensing window were formed by photolithography.

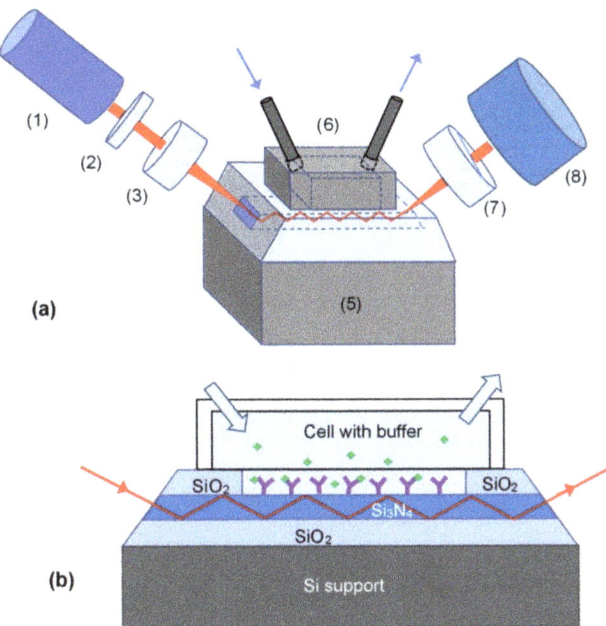

Figure 1. (**a**) The planar waveguide (PW) biosensor experimental set-up: laser diode (1), $\lambda/4$ plate (2), collimating lens (3), PW (4) on Si wafer support (5), reaction cell (6) with inlet and outlet tubes, polariser (7), and charge-coupled device (CCD) array (8); (**b**) Cross-section of waveguide section showing schematically the multiple reflections of light, the sensing window, the reaction cell, and antibodies immobilised on the PW surface binding zearalenone molecules.

The resulted set-up operates as a planar polarisation interferometer (PPI); the p-component of polarised light (lying in the plane of incidence) is affected by changes in the RI of the medium, while the s-polarised component (orthogonal to the plane of incidence) is almost invariant to the RI variation in the medium and subsequently used as a reference. Any changes in the medium RI in the sensing window including the variations of RI caused by molecular adsorption result in a multi-periodic sensor response cause by a variable phase shift between p- and s- polarisations of light, which could be converted by a polariser to a multiperiodic signal. In a way, the principle of PI is a logical expansion of the TIRE method [3], which is based on the detection of a phase shift between p- and s-components of polarised light, utilising a large number of reflections in the optical waveguides.

The experimental set-up for PI went through several stages of optimisation. Previously, the light from a fan-beam laser diode was coupled into the PW and was propagated over the entire width ($\approx$8 mm) of the waveguide. As the result of a modal dispersion of light across the waveguide, and therefore not equal conditions of light propagation (see Figure 2a), averaging of the light intensity over the entire width of the waveguide has led to losing the contrast of the interference pattern. To avoid that, the number of pixels for light averaging had to be limited. However, improved results were obtained with photolithography to form a narrow strip ($\approx$2 mm) of silicon nitride (Figure 2b). Another advantage of photolithography was the formation of a well-defined sensing window. The photographs of Figure 3 show the PW biosensor set-up (Figure 3a), the top views of the waveguide at different preparative stages in Figure 3b, e.g., a 24 $\times$ 6 mm chip with an SiO_2 and Si_3N_4 layer deposited (1), after etching of Si_3N_4 to form a narrow (2 mm) waveguide core (2), and the final structure with the sensing window (2 $\times$ 8 mm) etched in the top SiO_2 layer (3), and the waveguide inserted in the cell (Figure 3c). A Thorlabs LC100 camera was interfaced to a PC; the output signal acquisition was carried out using SPLICCO software (A Thorlabs GmbH, Bergkirchen, Germany).

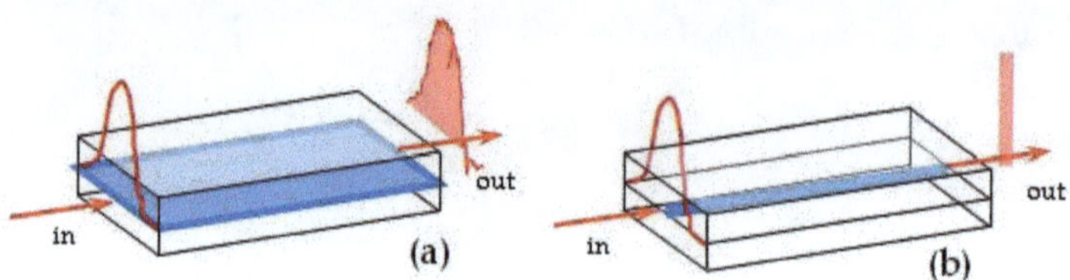

Figure 2. Propagation of light through the planar waveguide (PW) in a wide core (**a**) and narrow core (**b**) set-up.

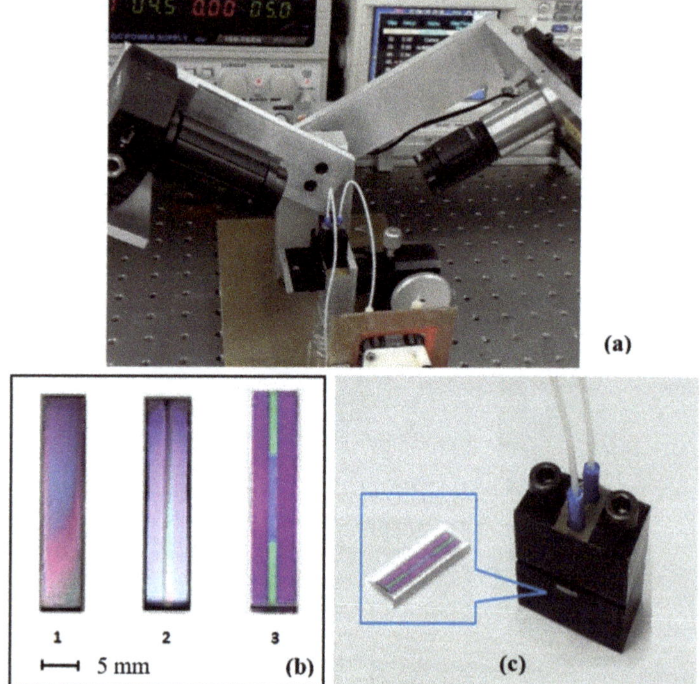

Figure 3. The upgraded planar polarisation interferometry device. Planar waveguide-based polarisation interferometry experimental set-up upgraded from the prototype (**a**); photolithography steps the waveguide preparation (**b**); the reaction cell with the waveguide inserted (**c**).

2.2. Testing the Waveguide Sensor

The testing of a PW sensor was performed by varying the RI of the liquid medium, i.e., by injecting into the cell initially filled with water aqueous solutions of NaCl of different concentrations having different RIs and recording the corresponding multi-periodic output waveforms [15,16]. The number of periods of signal oscillation were roughly estimated from these waveforms and presented in Table 1 along with corresponding changes in the RI, which allows estimating the refractive index sensitivity (RIS) of the PW sensors as:

$$RIS = \frac{2 \times \pi \times N}{\Delta n}$$

where N is the number of periods of oscillation, and Δn is the difference between the RI of aqueous sodium chloride (NaCl) and water i.e., $\Delta n = n_{NaCl} - n_{water}$, where $n_{water} = 1.332$. The obtained RIS values are also shown in Table 1.

Table 1. Evaluation of the refractive index sensitivity (RIS) of the planar waveguide sensor as a function of the number of periods of oscillation (N) and differences in the refractive index ($\triangle$n) in aqueous sodium chloride (NaCl) solutions at different concentrations (NaCl%).

NaCl%	N	$\triangle$n	No. of Periods	RIS (rad/RI unit)
2	1.3370	0.0050	3	3769.90
5	1.3395	0.0075	6	5026.55
8	1.3420	0.0100	10	6283.19
10	1.3460	0.0140	12.5	5610.00
15	1.3495	0.0175	16	5744.63
20	1.3610	0.0290	19	4166.57
		Average RIS = 5091.8 ± 787.5 rad/RI unit		

An average RIS around 5100 radians per RI unit was calculated and is quite remarkable, since it is much higher than in traditional optical methods, e.g., surface plasmon resonance or total internal reflection ellipsometry. Relatively large standard deviation values are due to a rough estimation of a number of periods of oscillations in the output waveforms.

2.3. Detection of Zearalenone by Planar Waveguide Immunoosensor

The detection of ZON was carried out using the experimental PW set-up described in Section 2.2, which has an RIS of approximately 5100 rad/RI unit. Experiments of detecting ZON were performed in a direct immunosensor format with specific antibodies immobilised electrostatically on the sensor surface via the layers poly-allylamine hydrochloride (PAH) and protein A (ProtA) in a following sequence: (i) deposition of PAH polycations carrying positive charge; (ii) deposition of ProtA being negatively charged at pH = 7, (iii) deposition of polyclonal antibodies to ZON via a biding site at the second constant domain to ProtA. In the above experiments, a very thin ($\approx$1 nm) layer of PAH [24] deposited on the waveguide surface yields a phase shift of about a $\frac{1}{2}$ period. The absorption of larger molecules of ProtA (42 kDa) causes a larger phase shift of about two periods, while much larger molecules of polyclonal antibodies to ZON (150–900 kDa) gave about 3.5 periods of phase change.

The detection of ZON was undertaken by sequential injections of ZON standards in increasing order of concentrations, e.g., 0.01, 0.1, 1, 10, 100, and 1000 ng/mL. Typical responses to injections of different concentrations of ZON are shown in Figure 4. As one can see, the number of periods of signal oscillation increases with the increase in ZON concentration. The accuracy of phase shift calculation was about 0.1 of a period or about 0.6 rad. However, the increase in concentration of ZON cannot lead to a limitless increase in the number oscillation periods of the output signal. The sequential injections of ZON at increasing concentration cause a gradual saturation of the ZON-specific antibodies. The saturation and even slight decrease of the individual phase shifts is due to the beginning saturation of the binding sites of the antibodies. A complete saturation of binding sites resulted in a very small phase shift due to non-specific binding. This corresponds to the exhaustion of the immobilised antibodies on the sensor surface. It has also to be noted that the washing out of non-specifically bound ZON molecules during purging with pure buffer solution through the cell caused a phase shift of about 1/4 of a period, which was the baseline (background) level of the experiments.

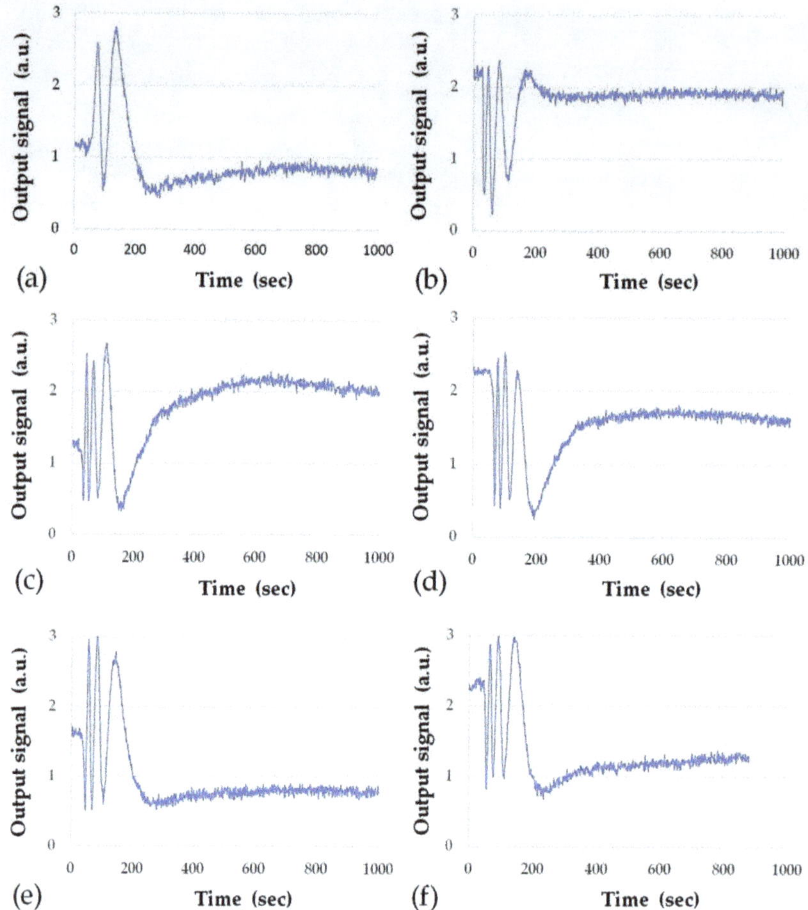

Figure 4. Typical sensor responses to injection of zearalenone (ZON) of different concentrations: 0.01 ng/mL (**a**), 0.1 ng/mL (**b**), 1 ng/mL (**c**), 10 ng/mL (**d**), 100 ng/mL (**e**), and 1000 (ng/mL) (**f**).

As seen in Figure 4, the minimal detection concentration of ZON (limit of detection, LOD) was 0.01 ng/mL, which is an order of magnitude lower than the results obtained earlier using TIRE in a similar direct immunoassay format and have the same LOD as in TIRE measurements in more sensitive competitive assay format [3]. Cross-reactivities (CRs) of the immobilised antibodies, which were carried out with aflatoxin B1 (AFB1) and ochratoxin A (OTA), did not show any signal oscillations and thus no phase shifts, even at a concentration of 1000 ng/mL. This demonstrated the high specificity of the antibodies towards ZON.

2.4. Comparison of ZON Detection with PW and ELISA Methods

The analysis of the enzyme-linked immunosorbent assays (ELISA) detection of ZON in the optimised direct immunoassay with respective specific antibodies and 3,3′,5,5′-tetramethylbenzidine (TMB) as a chromophore showed a saturation curve in a semi-logarithmic plot, while increments in the accumulated phase shift due to various concentrations of ZON could also be plotted. Such signal saturation and accumulated responses for the injections of ZON are given in Figure 5.

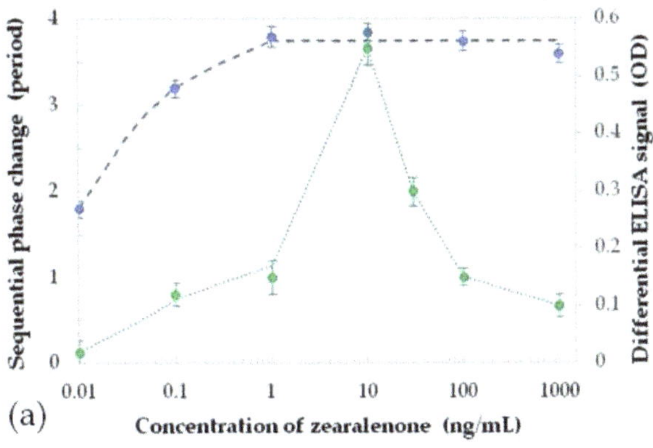

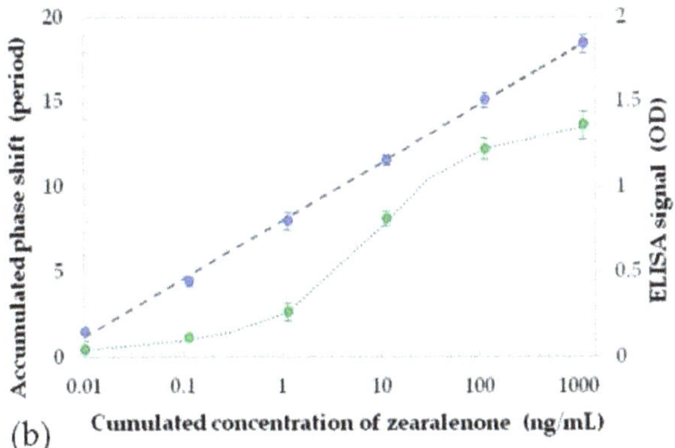

Figure 5. Dependence of the assay signals in the PW sensor (phase shifts, blue symbols, and blue slashed line) and the enzyme-linked immunosorbent assays (ELISA) (optical densities, green symbols, and green dotted line) caused by injections of zearalenone (ZON) in different concentrations. Total responses against the accumulated concentration of ZON (**a**); differential responses among subsequent injections (**b**).

A semi-logarithmic graph (Figure 5a) indicates that the sequential number of periods of phase change initially increases with increasing ZON concentrations; then, from 1 ng/mL ZON, the response becomes constant (approximately four periods per one order of magnitude change in ZON concentration). This constant increase in the phase shift steadily continues as the concentration of ZON is increased to 1000 ng/mL. Therefore, as seen, injections of 1, 10, 100, and 1000 ng/mL of ZON, respectively, resulted in the same increase in the analytical signal (number of periods of phase change). Since the biosensing test was performed by consecutive injections of ZON, the total phase shifts for each concentration can be calculated by adding the responses of individual injections and subtracting the background of non-specific binding. Such dependence of a total phase shift against the accumulated concentration of ZON (Figure 5b) indicates that the sensor response is very close to linear, which means that the saturation of binding centres was not achieved. The low LOD could be estimated as 0.01 ng/mL by extrapolation of the linear dependence to the triple noise level being approximately 1.8 rad.

The results of ELISA control measurements are presented in Figure 5 for comparative assessment. As the ELISA was performed in a competitive assay format, its calibration resulted in a decreasing sigmoid curve, showing that the ratio of bound antibodies to the immobilised conjugate of ZON to bovine serum albumin (BSA) onto the surface decreases with increasing concentrations of ZON. Thus, the higher the concentration of ZON is in the sample, the lower the analytical signal becomes. For a straight comparison with the direct immunosensor format, this sigmoid standard curve was inverted by showing the ratio of the unbound antibodies, and this proportion shows an increasing trend as a function of ZON concentration with a saturation of the binding sites on Figure 5b. The differential diagram of that sigmoid standard curve, plotted in Figure 5a, shows a bell-shaped curve for the incremental signals with a maximum at 10 ng/mL and a decreasing rate of signal increases between 10 and 1000 ng/mL, indicating the saturation of the antibody binding sites available on the surfaces of the wells of the microplates. This means that near the peak of the bell-shaped curve (corresponding to the half maximal inhibitory concentration or IC_{50} value i.e., the inflection point of the sigmoid curve), the increase in the analytical signal is the highest. Towards the right and left tails (the lower and upper plateaus of the sigmoid curve), the differences in the analytical signal between the measured ZON concentrations decrease.

The actual smallest concentration detected in these experiments was 0.01 ng/mL, corresponding to 1.5 periods of phase change in the PW sensor, while approximately 0.8 ng/mL was achieved in the ELISA test. The PW sensor baseline was established by injecting pure buffer solution (with no toxin content), which gave a response of a quarter-of-a-period of phase change related to small changes in RI due to washing out non-specifically bound ZON molecules. The intercept of the baseline with the initial linear slope of PW sensor response gives a value of 0.004 ng/mL (Figure 5b). This indicates that the PW sensor is capable of detecting at least two orders of magnitude lower ZON concentrations that the corresponding competitive ELISA using the same antibodies, and it represents an outstanding sensitivity for optical biosensors operating in the direct immunoassay format. The performance of the immunoreagents in other immunoassay formats indicates proper robustness for application e.g., in surface water. Moreover, the sensitivity can be increased even further by the use of a reference channel, e.g., the section of the waveguide without immobilised antibodies.

3. Conclusions

The development of a PW biosensor for the detection of ZON was achieved in several steps. A prototype of a PW biosensor operating as a PI was devised using a silicon nitride waveguide layer sandwiched between two silicon oxide layers. The prototype was upgraded to a narrow core waveguide and the sensing window made by photolithography and the improved optical system, which resulted in a better quality of signal and therefore a higher refractive index sensitivity. The biosensor was characterised for its refractive index sensitivity under standardised conditions by creating spectrograms of the signal waveforms in response to injections of aqueous NaCl solutions of different concentrations The sensor surface was functionalised using electrostatically deposited PAH and ProtA on the sensor surface to entrap ZON-specific antibodies. Immunosensing of ZON was carried out by recording the PW sensor responses caused by sequential injections of ZON solutions at increasing concentrations in the range of 0.01 and 1000 ng/mL. The application of monoclonal antibodies could provide better analyte specificity (specific recognition of ZON); therefore, it is a possible direction of further development in this biosensor approach to apply monoclonal antibodies. In this study, advantages of polyclonal antibodies were utilised, namely, that polyclonal antibodies are often of higher affinity than monoclonals. The quantification of ZON in the direct immunosensor was plotted as a saturation curve with increasing ZON concentrations, the corresponding incremental built-up of the sensor signal in relation to ZON concentrations was calculated, and the LOD of the methods was determined to be below 0.01 ng/mL.

The developed PW experimental set-up is still a benchtop type because of other electronic equipment used; the dimensions of the optical assembly is only about $10 \times 20 \times 20$ cm and it could be scaled down further down using miniature optical components. We are planning to develop a portable hand-held biosensor including the signal processing electronics, which will be suitable for in-field use.

4. Materials and Methods

4.1. Reagents and Instrumentation

All chemical reagents, including PAH, ProtA, mycotoxin standards, BSA, 2-amino-2-(hydroxymethyl)propane-1,3-diol (tris(hydroxymethyl)aminomethane, Tris) buffer and biochemicals were purchased from Sigma-Aldrich Co. (Dorset, UK or Budapest, HU), unless indicated otherwise. ZON-specific antibodies and protein conjugates were obtained at the Agro-Environmental Research Institute, National Research and Innovation Centre, Budapest, Hungary, as described before [4]. Immunoassays were performed in a Spectra-Max iD3 Multi-Mode Microplate Reader (Molecular Devices, San Jose, CA, USA) using high-capacity 96-well microplates (Nunc, Roskilde, Denmark).

4.2. Planar Waveguide Immunosensor Design

The PW structures were devised by standard microelectronic processes as a thin (200 nm) layer of Si_3N_4 sandwiched between much thicker (3 µm) layers of SiO_2. The 200 nm core layer thickness was required to accommodate a single mode electromagnetic wave propagating along the waveguide [25,26]. Due to the large difference in RIs between the Si_3N_4 core ($n = 2.01$) and the SiO_2 cladding ($n = 1.46$), the light propagates at an angle of $47°$ (corresponding to an angle of total internal reflection) and with a consequent 3000 reflections/mm approximately. The use of a slant edge of the waveguide cut at $47°$ was an optimal light coupling solution, which provided sufficient light intensity propagating through the waveguide.

In the experimental PI set-up, 650 nm light from a laser diode was focused with a lens to a narrow (less than a millimetre) spot on a slant edge of the waveguide and collected on the other side with a CCD array. A polarising element in front of a CCD camera allows the visualisation of a phase shift between p- and s-components of polarised light. The reaction cell equipped with the inlet and outlet tubes is sealed against a sensing window, which was etched in the top SiO_2 layer. The surface of Si_3N_4 in the sensing window could be coated with a biosensing layer. Any changes in RI and/or thickness of this sensing layer affect mostly the p-component of polarised light (while the s-component acts as a reference), thus resulting in a multi-periodic output signal.

4.3. Functionalisation of the Sensor Surface

The immunoreagents were immobilised onto the biosensor transducer, as depicted in Figure 6, using PAH as a polycation electrostatically deposited on the biosensor surface. A thin (10–20 nm) layer of SiO_2 was left of the surface of the Si_3N_4 core to provide a negative surface charge of OH^- for binding PAH polycations; this was achieved using ellipsometry thickness measurements to calibrate the etching time. The coating layer was deposited by incubating the sensor for 20 min with a 1 mg/mL aqueous solution of PAH. After removal of the solution, the sensor was rinsed three times with de-ionised water, and it was coated by incubating the sensor with a 0.01 mg/mL solution of ProtA in 35 mM Tris-HCl buffer (pH = 7.5) for 15 min. After being rinsed three times with Tris-HCl buffer, incubation with polyclonal antibodies to ZON at a concentration of 1 µg/mL in Tris-HCl buffer was carried out for 15 min. Finally, after final rinsing the cell three times with Tris-HCl buffer, biosensing tests were performed by injecting ZON aliquots at increasing concentrations into the cell and recording the PW sensor responses.

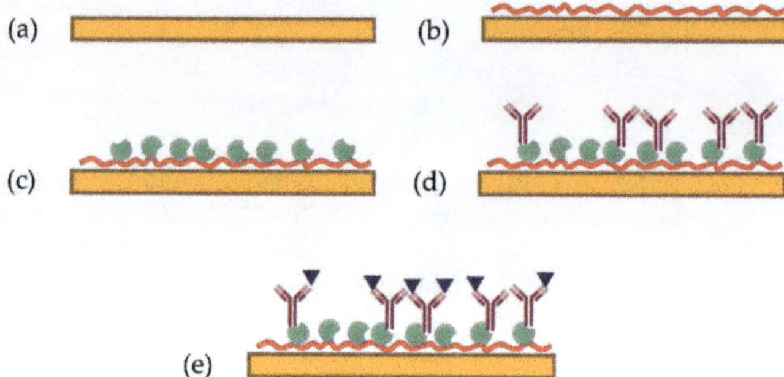

Figure 6. Steps in functionalisation of the sensor surface, and detection of zearalenone (ZON) with the functionalised sensor. Bare planar waveguide sensor (**a**); electrostatic immobilisation of a poly-allylamine hydrochloride (PAH) layer (**b**); binding protein A (Prot A) to PAH (**c**); oriented anchoring of ZON-specific antibodies by Prot A (**d**), detection of ZON by its specific binding to the anchored antibodies (**e**).

4.4. Planar Waveguide Immunosensor Assay

The detection of ZON was carried out in direct assay with specific polyclonal antibodies (Ab) immobilised electrostatically on the waveguide surface via a polycation layer of PAH followed by electrostatic binding of protA, which have a binding site to immunoglobulins (IgG or IgM) following the procedure described in detail earlier [16] (see Section 4.3). Then, the detection of ZON was carried out by sequential injections of ZON solutions with progressively increasing concentration of ZON of 0.01, 0.1, 1, 10, 100, and 1000 ng/mL in phosphate buffer. Typical sensor responses to the injection of different concentrations of ZON were recorded and evaluated. Each injection was followed by purging the cell three times with 1 mL of pure buffer solution in order to remove unbound mycotoxin molecules. Before each series of measurements, the cell was thoroughly cleaned in ethanol and de-ionised water. Negative tests were carried out by injecting structurally unrelated mycotoxins, e.g., AFB1 and OTA at concentrations up to 1000 ng/mL.

4.5. Enzyme-Linked Immunosorbent Assay

Enzyme-linked immunosorbent assays (ELISAs) were carried out in 96-well microplates (see Section 4.1) in a competitive immunoassay format. Microplate wells were coated with 1 µg/mL ZON–BSA conjugate in carbonate buffer (15 mM Na_2CO_3, 35 mM $NaHCO_3$, pH = 9.6) for overnight ($\approx$8 h) at 4 °C. The unbound conjugate was washed out 4 times with phosphate buffer saline (PBS) (137 mM NaCl, 2.7 KCl, 10 mM Na_2HPO_4 × $2H_2O$, pH = 7.4) with 2% Tween20. Blocking was carried out with 150 µL/well 1% gelatine in PBS for 1.5 h of incubation at 37 °C. After washing, competition was performed by adding 50 µL/well of both ZON analytical standard and purified rabbit anti-ZON serum (in PBS buffer with 5% Tween20, dilution 1:1000). After 1 h of incubation at 37 °C and four times washing, 100 µL/well goat anti-rabbit IgG-HRP (horseradish peroxidase, dilution 1:7500) conjugate were added as secondary antibody and incubated for 1 h at 37 °C. Unbound secondary antibodies were washed out 4 times with PBS, and enzymatic activity was measured using 100 µL/well substrate solution containing 1.3 mM hydrogen peroxide as a substrate and 0.42 mM TMB as a chromophore in 100 mM citrate buffer (pH = 6.0). After 10 min of incubation, the enzymatic activity was stopped by 50 µl of 4 M sulphuric acid (H_2SO_4), and absorbance was measured at a wavelength of 450 nm.

Author Contributions: Conceptualisation, A.N. and A.S.; methodology, A.N.; validation, A.M.A.-J., B.G. and E.T.; formal analysis, A.M.A.-J. and B.G.; investigation, A.N.; writing—original draft preparation, A.M.A.-J., B.G. and E.T.; writing—review and editing, A.N. and A.S.; visualisation, A.M.A.-J. and B.G.; supervision, A.N. and A.S.; project administration, A.S.; funding acquisition, A.N. and A.S. All authors have read and agreed to the published version of the manuscript.

Funding: This research was funded by the Hungarian National Research, Development and Innovation Office, project NVKP_16-1-2016-0049 "*In situ*, complex water quality monitoring by using direct or immunofluorimetry and plasma spectroscopy" and TKP2020-NKA 24 "Tématerületi kiválóság program" in the Thematic Excellence Programme 2020.

Institutional Review Board Statement: Not applicable.

Informed Consent Statement: Not applicable.

Data Availability Statement: The data presented in this study are available on request from the corresponding author. The data are not publicly available due to privacy reasons.

Acknowledgments: A.M.A.-J. would like to thank Babylon University, College of Sciences, Hilla, Iraq for sponsoring his PhD project at Sheffield Hallam University, UK.

Conflicts of Interest: The authors declare no conflict of interest.

References

1. Adányi, N.; Majer-Baranyi, K.; Székács, A. Evanescent field effect based nanobiosensors for agro-environmental and food safety. In *Nanobiosensors: Nanotechnology in the Agri-Food Industry*; Grumezescu, A.M., Ed.; Elsevier: Amsterdam, The Netherlands, 2017; Volume 8, pp. 429–474. [CrossRef]
2. Horváth, R.; Lindvold, L.R.; Larsen, N.B. Reverse-symmetry waveguides: Theory and fabrication. *Appl. Phys. B* **2002**, *74*, 383–393. [CrossRef]
3. Nabok, A.; Tsargorodskaya, A.; Mustafa, M.K.; Székács, I.; Starodub, N.F.; Székács, A. Detection of low molecular weight toxins using an optical phase method of ellipsometry. *Sens. Actuators B* **2011**, *154*, 232–237. [CrossRef]
4. Székács, I.; Adányi, N.; Szendrő, I.; Székács, A. Direct and competitive optical grating immunosensors for determination of *Fusarium* mycotoxin zearalenone. *Toxins* **2021**, *13*, 43. [CrossRef] [PubMed]
5. Al-Rubaye, A.; Nabok, A.; Abu-Ali, H.; Szekacs, A.; Takacs, E. LSPR/TIRE bio-sensing platform for detection of low molecular weight toxins. In Proceedings of the IEEE Sensors, Glasgow, UK, 29 October–1 November 2017. [CrossRef]
6. Al Rubaye, A.G.; Nabok, A.; Catanante, G.; Marty, J.-L.; Takacs, E.; Szekacs, A. Label-free optical detection of mycotoxins using specific aptamers immobilized on gold nanostructures. *Toxins* **2018**, *10*, 291. [CrossRef] [PubMed]
7. Nabok, A. Comparative studies on optical biosensors for detection of bio-toxins, In *Advanced Sciences and Technologies in Security Applications, Biosensors for Security and Bioterrorism Applications*; Nikolelis, D.P., Nikoleli, G.P., Eds.; Springer: Cham, Switzerland, 2016; pp. 491–508. [CrossRef]
8. Lechuga, L.M. Optical biosensors. *Comp. Anal. Chem.* **2005**, *44*, 209–250. [CrossRef]
9. Berney, H.; Oliver, K. Dual polarization interferometry size and density characterization of DNA immobilisation and hybridisation. *Biosens. Bioelectron.* **2005**, *21*, 618–626. [CrossRef] [PubMed]
10. Escorihuela, J.; González-Martínez, M.Á.; López-Paz, J.L.; Puchades, R.; Maquieira, Á.; Gimenez-Romero, D. Dual-polarization interferometry: A novel technique to light up the nanomolecular world. *Chem. Rev.* **2015**, *115*, 265–294. [CrossRef] [PubMed]
11. Sun, Y.; Fan, X. Optical ring resonators for biochemical and chemical sensing. *Anal. Bioanal. Chem.* **2011**, *399*, 205–211. [CrossRef]
12. Kozma, P.; Kehl, F.; Ehrentreich-Foerster, E.; Stamm, C.; Bier, F.F. Integrated planar optical waveguide interferometer biosensors: A comparative review. *Biosens. Bioelectron.* **2014**, *58*, 287–307. [CrossRef]
13. Tittlemier, S.A.; Roscoe, M.; Drul, D.; Blagden, R.; Kobialka, C.; Chan, J.; Gaba, D. Single laboratory evaluation of a planar waveguide-based system for a simple simultaneous analysis of four mycotoxins in wheat. *Mycotoxin Res.* **2013**, *29*, 55–62. [CrossRef]
14. Nabok, A.; Al-Jawdah, A.M.; Tsargorodska, A. Development of planar waveguide-based immunosensor for detection of low molecular weight molecules such as mycotoxins. *Sens. Actuat. B* **2017**, *247*, 975–980. [CrossRef]
15. Al-Jawdah, A.; Nabok, A.; Jarrah, R.; Holloway, A.; Tsargorodska, A.; Takacs, E.; Szekacs, A. Mycotoxin biosensor based on optical planar waveguide. *Toxins* **2018**, *10*, 272. [CrossRef] [PubMed]
16. Al-Jawdah, A.; Nabok, A.; Abu-Ali, H.; Catanante, G.; Marty, J.-L.; Szekacs, A. Highly sensitive label-free in vitro detection of aflatoxin B1 in an aptamer assay using optical planar waveguide operating as a polarization interferometer. *Anal. Bioanal. Chem.* **2019**, *411*, 7717–7724. [CrossRef] [PubMed]
17. Nabok, A.; Al-Rubaye, A.G.; Al-Jawdah, A.M.; Tsargorodska, A.; Marty, J.-L.; Catanante, G.; Szekacs, A.; Takacs, E. Novel optical biosensing technologies for detection of mycotoxins. *Optics Laser Technol.* **2019**, *109*, 212–221. [CrossRef]
18. Devlin, S.; Meneely, J.P.; Greer, B.; Greef, C.; Lochhead, M.J.; Elliott, C.T. Next generation planar waveguide detection of microcystins in freshwater and cyanobacterial extracts, utilising a novel lysis method for portable sample preparation and analysis. *Anal Chim. Acta* **2013**, *769*, 108–113. [CrossRef]
19. McNamee, S.E.; Elliott, C.T.; Greer, B.; Lochhead, B.; Campbell, K. Development of a planar waveguide microarray for the monitoring and early detection of five harmful algal toxins in water and cultures. *Environ. Sci. Technol.* **2014**, *48*, 13340–13349. [CrossRef]

20. Murphy, C.; Stack, E.; Krivelo, S.; McPartlin, D.A.; Byrne, B.; Greef, C.; Lochhead, M.J.; Husar, G.; Devlin, S.; Elliott, C.T.; et al. Detection of the cyanobacterial toxin, microcystin-LR, using a novel recombinant antibody-based optical-planar waveguide platform. *Biosens. Bioelectron.* **2015**, *67*, 708–714. [CrossRef]
21. Bickman, S.R.; Campbell, K.; Elliott, C.; Murphy, C.; O'Kennedy, R.; Papst, P.; Lochhead, M.J. An innovative portable biosensor system for the rapid detection of freshwater cyanobacterial algal bloom toxins. *Environ. Sci. Technol.* **2018**, *52*, 11691–11698. [CrossRef]
22. Meneely, J.P.; Campbell, K.; Greef, C.; Lochhead, M.; Elliott, C.T. Development and validation of an ultrasensitive fluorescence planar waveguide biosensor for the detection of paralytic shellfish toxins in marine algae. *Biosens. Bioelectron.* **2013**, *41*, 691–697. [CrossRef]
23. Reverté, L.; Campàs, M.; Yakes, B.J.; Deeds, J.R.; Katikou, P.; Kawatsu, K.; Lochhead, M.; Elliott, C.T.; Campbell, K. Tetrodotoxin detection in puffer fish by a sensitive planar waveguide immunosensor. *Sens. Actuat. B* **2002**, *253*, 967–976. [CrossRef]
24. Nabok, A.V.; Hassan, A.K.; Ray, A.K. Optical and electrical characterisation of polyelectrolyte self-assebled thin films. *Mater. Sci Eng. C* **1999**, *8–9*, 505–508. [CrossRef]
25. Shirshov, Y.M.; Svechnikov, S.V.; Kiyanovskii, A.P.; Ushenin, Y.V.; Venger, E.F.; Samoylov, A.V.; Merker, R. A sensor based on the planar polarisation interferometer. *Sens. Actuat. A* **1998**, *68*, 384–387. [CrossRef]
26. Shirshov, Y.M.; Snopok, B.A.; Samoylov, A.V.; Kiyanovskij, A.P.; Venger, E.F.; Nabok, A.V.; Ray, A.K. Analysis of the response of planar interferometer to molecular layer formation: Fibrinogen adsorption on silicon nitride surface. *Biosens. Bioelectron.* **2001**, *16*, 381–390. [CrossRef]

toxins

Article

Direct and Competitive Optical Grating Immunosensors for Determination of *Fusarium* Mycotoxin Zearalenone

Inna Székács [1], Nóra Adányi [2], István Szendrő [3] and András Székács [4],*

1   Nanobiosensorics Group, Institute for Technical Physics and Material Science, Centre for Energy Research, Konkoly-Thege M. út 29-33, H-1121 Budapest, Hungary; szekacs@mfa.kfki.hu
2   Food Science Research Institute, National Agricultural Research and Innovation Centre, Herman O. út 15, H-1022 Budapest, Hungary; adanyi.nora@eki.naik.hu
3   MicroVacuum Ltd., Kerékgyártó u. 10, H-1147 Budapest, Hungary; istvan.szendro@microvacuum.com
4   Agro-Environmental Research Institute, National Research and Innovation Centre, Herman Ottó út 15, H-1022 Budapest, Hungary
*   Correspondence: szekacs.andras@akk.naik.hu; Tel.: +36-30-927-2297

Abstract: Novel optical waveguide lightmode spectroscopy (OWLS)-based immunosensor formats were developed for label-free detection of *Fusarium* mycotoxin zearalenone (ZON). To achieve low limits of detection (LODs), both immobilised antibody-based (direct) and immobilised antigen-based (competitive) assay setups were applied. Immunoreagents were immobilised on epoxy-, amino-, and carboxyl-functionalised sensor surfaces, and by optimising the immobilisation methods, standard sigmoid curves were obtained in both sensor formats. An outstanding LOD of 0.002 pg/mL was obtained for ZON in the competitive immunosensor setup with a dynamic detection range between 0.01 and 1 pg/mL ZON concentrations, depending on the covalent immobilisation method applied. This corresponds to a five orders of magnitude improvement in detectability of ZON relative to the previously developed enzyme-linked immonosorbent assay (ELISA) method. The selectivity of the immunosensor for ZON was demonstrated with structural analogues (α-zearalenol, α-zearalanol, and β-zearalanol) and structurally unrelated mycotoxins. The method was found to be applicable in maize extract using acetonitrile as the organic solvent, upon a dilution rate of 1:10,000 in buffer. Thus, the OWLS immunosensor method developed appears to be suitable for the quantitative determination of ZON in aqueous medium. The new technique can widen the range of sensoric detection methods of ZON for surveys in food and environmental safety assessment.

Keywords: mycotoxin; zearalenone; immunosensor; optical waveguide lightmode spectroscopy; label-free detection

Key Contribution: A real time, label-free OWLS immunosensor with a dynamic detection range between 0.01 and 1 pg/mL and an outstanding limit of detection of 0.002 pg/mL was developed and validated in maize extract for zearalenone. The novel immunosensor showed a five orders of magnitude improvement in analytical sensitivity for zearalenone compared to ELISA.

Citation: Székács, I.; Adányi, N.; Szendrő, I.; Székács, A. Direct and Competitive Optical Grating Immunosensors for Determination of *Fusarium* Mycotoxin Zearalenone. *Toxins* **2021**, *13*, 43. https://doi.org/10.3390/toxins13010043

Received: 1 December 2020
Accepted: 5 January 2021
Published: 8 January 2021

Publisher's Note: MDPI stays neutral with regard to jurisdictional claims in published maps and institutional affiliations.

1. Introduction

Zearalenone (ZON) is a mycotoxin produced by several *Fusarium* species, most frequently by *F. graminearum*, and is commonly found in maize and also in wheat, barley, sorghum, and rye throughout various countries of the world, causing substantial human exposure [1]. ZON and its metabolites have oestrogenic activity in several species [2–5] accompanied by hepatotoxicity, haematotoxicity, immunotoxicity, and genotoxicity [6–8]. No uniform regulations have been imposed for this toxin in different countries. Tolerance levels in grains and grain products have been set in several countries at a concentration range of 20 to 1000 µg/kg [9], e.g., 20 to 200 µg/kg in unprocessed and processed cereal products in the EU [10]. Data evaluation on the most sensitive animal species—swine—and comparing with humans, a tolerable daily intake for ZON has been set as 0.25 μg/kg body weight [11,12].

Common analytical methods for identifying and quantifying mycotoxins include thin-layer chromatography (TLC) [13,14] or high-pressure TLC [15], laser fluorimetry [16], gas chromatography (GC) [14] often coupled with mass spectrometry (GC-MS) [17,18], high-performance liquid chromatography (HPLC) [19–21] with standardised sample preparation [22,23], ultra-performance liquid chromatography (UPLC) [24], and capillary electrophoresis [25,26]. HPLC methods have become the most widespread for mycotoxin analysis. These methods are sensitive and accurate but require extensive sample preparation steps, well-trained personnel, and expensive instrumentation. Therefore, just as for other mycotoxins, and on the basis of the historical radioimmunoassay method [27], immunochemical methods, e.g., enzyme-linked immunosorbent assays (ELISAs), have been developed and utilised for rapid screening of ZON [14,19,28–33]. These immunoassays were further amplified with fluorescent quantum dots [34–36], magnetic nanoparticles [37], or helical carbon nanotubes [38]. Alternatively, antibodies [24,36,39–44] or molecularly imprinted polymers [45–47] could be applied for affinity chromatography or pre-column sample purification prior to chromatographic analyses (HPLC, UPLC). Similarly, nanoparticle-assisted lateral flow immunochromatographic strips [48,49] were devised, occasionally with surface-enhanced Raman scattering detection [50]. Recently, micro- and nanoarray immunoassays were reported in microplate-based [51] or microfluidic sensor-based [52] setups. A cut-off level of 100 μg/kg was established (4 min) for ZON and T2 toxin in a gel-based immunoassay [53]. Fluorescence polarisation immunoassays allowed for a detection range for ZON of 150–1000 μg/kg and a limit of detection (LOD) of 137 μg/kg, and required less than 2 min per sample to carry out [54]. A magnetic nanotag-based immunoassay [55] and a multiplexed quantum dot immunochromatographic assay [56] allowed the parallel detection of ZON in the presence of other mycotoxins. Label-free biosensors on the basis of antibodies [57–60], aptamers [58,61–67], or molecularly imprinted polymers [68–71] as recognition elements have also been developed with various signal amplification and detection routes involved, and the range of sensoric detection techniques is expanding [72]. Thus, a surface plasmon resonance (SPR) biosensor has been developed for the simultaneous detection of four mycotoxins, with an LOD below 0.2 ng/mL for ZON [73], a gold nanoparticle-amplified imaging SPR (iSPR) biosensor allowed an LOD for ZON of 59.2 pg/mL in multiplex mycotoxin determination [74], a method of total internal reflection ellipsometry (TIRE) allowed detection of ZON at concentrations as low as 0.1 ng/mL [75], and electrochemical sensors resulted in LODs of 0.15–0.25 pg/mL [42,60]. The immunosensors developed allow rapid quantitative determination of the target compounds in plant samples and in environmental matrices, mainly in ground water.

Immunosensors based on the technique of optical waveguide lightmode spectroscopy (OWLS) have been applied with success to detect different molecules, and gained importance in environmental and food analysis [59,76,77]. In the current study, an OWLS immunosensor has been developed for the determination of ZON in maize samples. Different chemical methods for functionalisation and accordingly for immobilisation were compared regarding analytical sensitivity and sensor stability. Upon optimisation, the novel immunosensor was used for the detection of ZON contamination in maize and the results were compared to ELISA measurements to demonstrate the outstanding applicability of the method in complex food matrices and assumedly, in environmental samples as well.

2. Results and Discussion

OWLS immunosensors were devised both in the direct (immobilised antibody) and competitive (immobilised antigen conjugate)-based formats. Immobilisation of the protein reactants has been carried out by several chemical routes utilising hydroxyl groups on the sensor surface converted into epoxy or amino functionalities, further reacted with appropriate chemical reagents for covalent immobilisation of the protein immunoreagents (Figure 1).

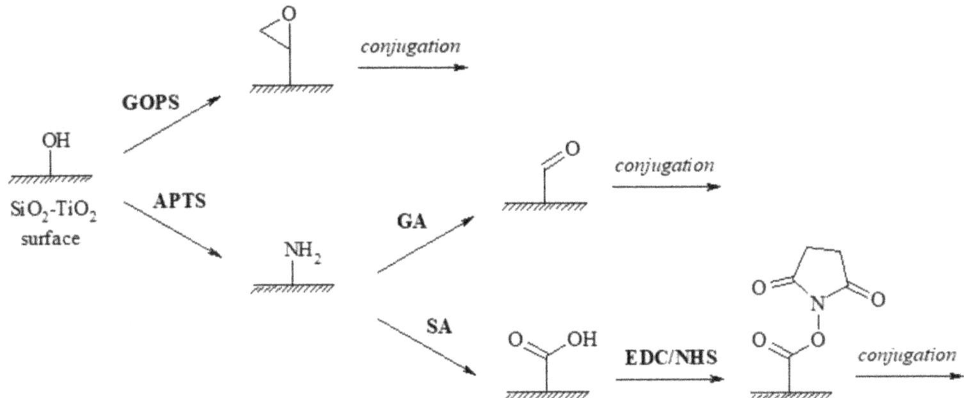

Figure 1. Functionalisation of the sensor surface with epoxy, amino, or carboxyl functional groups. GOPS: γ-glycidoxypropyl-trimethoxysilane; APTS: (3-aminopropyl) triethoxysilane; GA: glutaraldehyde; SA: succinic anhydride; EDC: 1-ethyl-3-(3-dimethylaminopropyl) carbodiimide; NHS: N-hydroxysuccinimide.

2.1. Direct Immunosensor

Immunoglobulin (IgG) fractions purified from ZON-specific rabbit antisera obtained against a conjugate of ZON to conalbumin (ovotransferrin, CONA) as an immunogen were used in an immobilised antibody-based (direct) immunosensor format. The main characteristics that determined achievable assay signals were the quality and concentration (dilution) of the ZON-specific antibodies. Using a dilution of the IgG purified from the serum of 1:2000 for immobilisation by all three methods, the epoxy-functionalised sensor surface modified with γ-glycidoxypropyl-trimethoxysilane (GOPS), as well as the amino-functionalised sensor surface modified with (3-aminopropyl)triethoxysilane (APTS), and glutaraldehyde (GA) or succinic anhydride with 1-ethyl-3-(3-dimethylaminopropyl)carbodiimide and N-hydroxysuccinimide (SA/EDC-NHS), standard calibration curves were obtained for ZON determination by applying ZON onto the immobilised antibodies on the sensor surface at various concentrations up to 100 μg/mL (Figure 2).

The highest sensor signals were obtained by APTS/GA modification, followed by APTS/SA-EDC/NHS, while immobilisation with GOPS provided the lowest assay signals. Detection sensitivity, characterised with the analyte (ZON) 50% effective concentrations (EC_{50}) corresponding to the half-maximal signal level, indicated EC_{50} values of 3.6 ± 0.2, 2.2 ± 0.6, and 1.1 ± 0.1 μg/mL for the GOPS, APTS/GA, and APTS/SA-EDC/NHS modifications, respectively. Signal intensities and statistics indicated that immobilisation on the epoxy-modified surface (GOPS) provided lower binding efficacy and reproducibility than that on amino-modified surfaces (APTS) with homo-bifunctional cross-linking (GA) with further modification to carboxyl groups (SA/EDC-NHS). Nonetheless, the lowest detectable ZON concentrations in these setups were in all three cases above 500 ng/mL, which is not sufficiently sensitive for analysis of real samples.

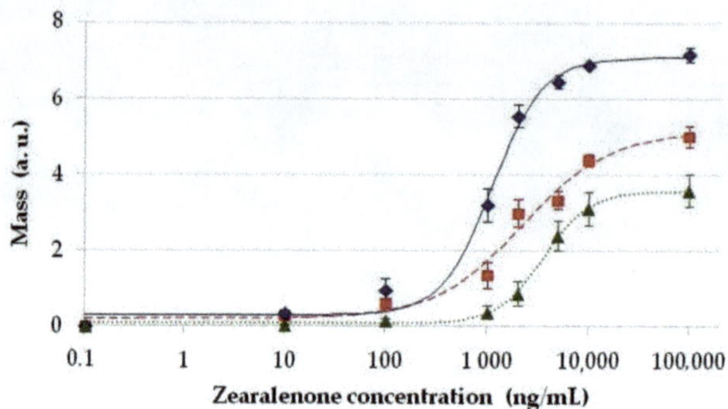

Figure 2. Standard calibration curves for zearalenone (ZON) determination by the direct optical waveguide lightmode spectroscopy (OWLS) immunosensor format. Sensor signals proportional to relative surface mass (ng mm^{-2}) on the OWLS sensor, expressed in arbitrary units (a.u.), as a function of concentration of ZON applied in the calibration standard samples in the sensor format with ZON-specific serum immobilised on amino- and epoxy-modified sensor surfaces using (3-aminopropyl) triethoxysilane and glutaraldehyde (APTS/GA) (■, red dashed line), (3-aminopropyl) triethoxysilane, succinic anhydride and 1-ethyl-3-(3-dimethylaminopropyl) carbodiimide with N-hydroxysuccinimide (APTS/SA/EDC-NHS) (♦, blue solid line), and γ-glycidoxypropyl-trimethoxysilane (GOPS) (▲, green dotted line).

2.2. Competitive Immunosensor

2.2.1. Serum Titration

The polyclonal IgG fraction purified ZON-specific rabbit antisera obtained against ZON-CONA as an immunogen were titrated in the OWLS immunosensor setup using a protein-heterologous conjugate to bovine serum albumin (BSA) at a ZON-BSA concentration of 10 µg/mL as a sensor surface antigen. Purified antisera were injected onto this sensor surface at increasing concentrations (decreasing dilutions) to assess the binding affinity of the antibodies. Typical titration curves are shown in Figure 3, indicating the peak signals obtained in the flow-through system in 3.4 min upon injection, decreasing peak intensities with increasing serum dilution and optimal dilution (the highest dilution still allowing distinguishable signal) at a serum dilution of 1:2000. The use of more concentrated serum for further competitive measurements results in deteriorated method sensitivity, while lower antibody concentrations allow for less stable sensor performance.

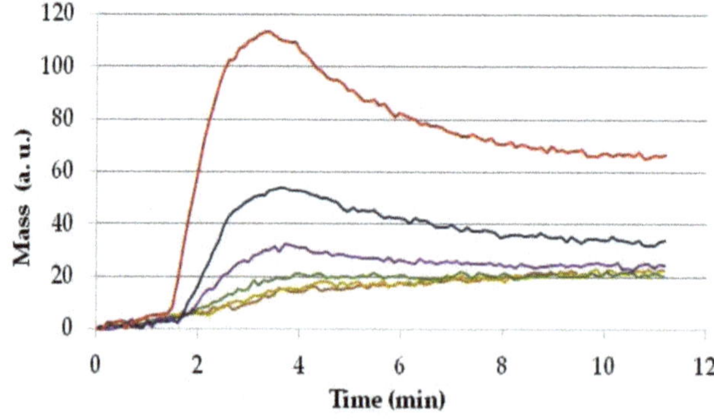

Figure 3. Optimisation of serum dilution by recording sensor responses to polyclonal antiserum at various dilutions using a sensor surface modified with 10 µg/mL of zearalenone conjugate to bovine serum albumin. Serum dilutions at 1:500 (red line), 1:1000 (blue line), 1:2000 (purple line), 1:4000 (green line), 1:8000 (yellow line), and 1:16,000 (brown line).

The determination of the amount of polyclonal IgG applied is essential in both the direct and indirect (competitive) measurements, particularly in the latter as it is a rather sensitive equilibrium. As seen in Figure 3, when the IgG is applied at small concentrations (high dilutions, e.g., 1:8000 or 1:16,000), antibodies poorly saturate the sensor surface, and small, unstable sensor responses are obtained. On the contrary, in the case of high IgG concentrations (dilutions of 1:500 or 1:1000), although the signal obtained is well measurable (exceeding 100 or 50 arbitrary units, respectively), the surface becomes saturated, and it loses its sensitivity during the measurement of standards and samples. For the measurements, we chose an IgG concentration that is high enough to provide well-measurable signals (at least 20 arbitrary units). On the other hand, the IgG concentration should not be too high, so that the system remains sensitive enough to detect standards containing low amounts of the antigen. Taking the height and shape of the signals into consideration, in the case of competitive measurement of ZON, the dilution of the antibody solution was chosen to be 1:2000.

2.2.2. Competitive Immunosensor Setups with Different Surfaces Modifications

The protein-heterologous conjugate of ZON (ZON-BSA) was used as a sensor surface antigen at various concentrations between 2 and 20 µg/mL with the above three immobilisation methods. Upon serum titration, surface coating conditions were optimised for immunosensor sensitivity i.e., analytical standard curves were obtained using concentration series of ZON and recording its inhibitory effect on antibody binding to the treated immunosensor surfaces (Figure 4).

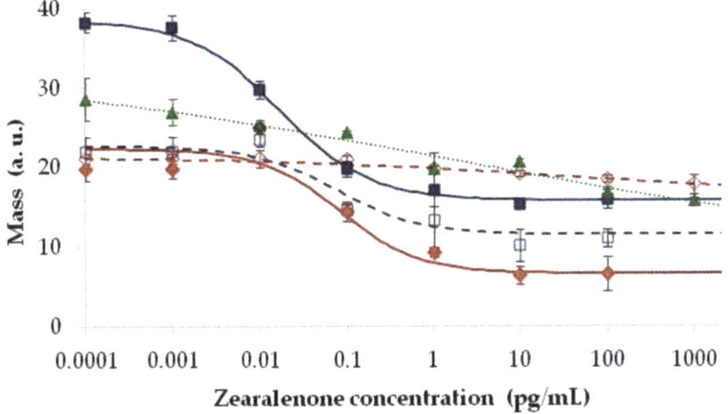

Figure 4. Standard calibration curves for zearalenone (ZON) determination by the competitive optical waveguide lightmode spectroscopy (OWLS) immunosensor format. Sensor signals proportional to relative surface mass (ng mm^{-2}) on the OWLS sensor, expressed in arbitrary units, as a function of concentration of ZON applied in the calibration standard samples in the sensor format with ZON conjugate to bovine serum albumin (ZON-BSA) immobilised on amino- and epoxy-modified sensor surfaces with (3-aminopropyl) triethoxysilane and glutaraldehyde (APTS/GA) using 10 µg/mL ZON-BSA (■, blue solid line) and 5 µg/mL ZON-BSA (□, blue dashed line); (3-aminopropyl) triethoxysilane, succinic anhydride and 1-ethyl-3-(3-dimethylaminopropyl) carbodiimide with N-hydroxysuccinimide (APTS/SA/EDC-NHS) using 10 µg/mL ZON-BSA (♦, red solid line), 5 µg/mL ZON-BSA (◊, red dashed line), and γ-glycidoxypropyl-trimethoxysilane (GOPS) (▲, green dotted line).

The highest and stable sensor signals were obtained when ZON-BSA was at 10 µg/mL concentration, above which further improvement in assay signals could not be obtained, and excess of the antigen even caused less reproducible or deteriorated signals. Similarly to the direct sensor format, immobilisation on the epoxy-modified surface (GOPS) was found to be of limited utility in the competitive sensor format as well. Although addition of ZON resulted in concentration-dependent sensor signals, nonetheless, this means of immobilisation was improper for analytical purposes due to the lack of proper regression possibility with the four-parameter logistic fitting. Immobilisation on amino- (APTS/GA) or carboxyl-modified surfaces (APTS/SA/EDC-NHS) allowed for better quantitative detection possibilities, with dynamic detection ranges of 0.001–1 pg/mL when the antisera were used in 1:2000 dilution. Immobilisation with GA appeared to be applicable at coating levels with ZON-BSA both at 10 and 5 µg/mL, while the SA/EDC-NHS method resulted in a standard sigmoid curve only at a coating level of 10 µg/mL. In these cases, linear detection ranges were found to be similar with good reproducibility. Antisera obtained from different rabbits (under the same immunisation protocol), although showing somewhat different titration characteristics, provided similar results in the competitive formats, indicating that slight differences in serum composition did not have inhibitory activity. The highest sensor signals were obtained here also by APTS/GA modification. Detection sensitivity showed outstandingly low EC_{50} values in the range of 0.017–0.083 pg/mL, corresponding to at least six orders of magnitude improvement in detection range compared to the direct immunosensor. The dynamic detection range of ZON was found in the 0.010–1 pg/mL ZON concentrations, and an LOD for ZON of 0.002 pg/mL was obtained in the APTS/GA modification using the ZON-BSA conjugate as the surface coating antigen at 10 µg/mL. As the competitive immunosensors based on amino- and carboxyl-modified surfaces provided similar ranges of detection, due to the better reproducibility and longer shelf-life of the latter, the sensor setup using 10 µg/mL of ZON-BSA as the surface coating antigen immobilised with the APTS/SA/EDC-NHS method, as well as ZON-specific antibodies at 1:2000 dilution, was chosen to be used for practical purposes.

2.3. Immunosensor Specificity

Immunosensor specificity was tested on the optimised immunosensor setup (see above) by measuring EC_{50} values obtained with ZON derivatives and structurally unrelated mycotoxins, and cross-reactivities (CRs), defined as a percentage ratio between the EC_{50} values of ZON and the given compound, were calculated. Among the structurally unrelated compounds tested, aflatoxin B1 and ochratoxin did not cause a decrease in the OWLS sensor signal up to 1000 ng/mL concentration in the diluted standards. Among the compounds tested, only α-zearalenol, α-zearalanol, and β-zearalanol showed significant CRs (Table 1) with ZON in the competitive immunosensor format. These are major reductive metabolites of ZON in mammals, but are also formed to a lesser extent in plants as well [12]; therefore, the potential presence of these metabolites should be also considered upon positive detection of ZON in commodities by the current immunosensor method. These CR values are in good agreement with the corresponding values reported for our ELISA system for ZON [29]; however, the detection sensitivity of the current OWLS immunosensor exceeds that of the ELISA by five orders of magnitude. Such outstanding improvement in the detection range of an OWLS immunosensor compared to the corresponding ELISA has been reported [59,78].

Table 1. Percentage of cross-reactivity (CR%) of the competitive OWLS immunosensor and the corresponding ELISA method [29] with zearalenone and its derivatives.

Compound	OWLS Sensor		ELISA	
	IC_{50} (pg/mL)	CR% [1]	IC_{50} (ng/mL)	CR% [1]
zearalenone	14.3	100	14.1	100
α-zearalenol	56.5	25.2	50.1	28.2
α-zearalanol	111.0	12.8	199.5	7.1
β-zearalanol	526.5	2.7	1259.0	1.1

[1] Cross-reactivity defined as the percentage ratio of the 50% inhibitory concentration (IC_{50}) values of zearalenone and of the given derivative.

2.4. Method Validation in Commodity Matrix

The optimised competitive immunosensor was applied to determine ZON concentrations in maize commodity. For this purpose, maize samples spiked with ZON at concentration levels of 0–10 µg/kg were extracted in acetonitrile/water (6:4), and were analysed by the competitive OWLS immunosensor and ELISA. These analyses aimed to assess matrix effects by the maize extract on the one hand, and were also targeted to investigate whether the two analytical methods detect the same ZON concentrations, identical to the nominal values, on the other hand. To assess possible matrix effects on immunosensor performance, the aqueous extracts were diluted 1:100 to 1:10,000 in 42 mM 2-amino-2-(hydroxymethyl)-1,3-propanediol (Tris) buffer (pH 7.4). Figure 5 shows analytical standard curves obtained by the optimised competitive immunosensor setup in diluted maize extracts, and demonstrates that matrix effects are diluted out at 1:10,000. ZON concentrations detected by the competitive immunosensor indicated analytical recoveries at initial ZON concentrations between 5 ng/kg and 10 µg/kg, carried out in triplicates, were found to be 84% and 124%, mostly suitable for practical use. It has to be noted, however, that the maximal recovery value fell by 3.3% out of the acceptable recovery range of the European legislation performance criteria for ZON detection set to be 60–120% and 70–120% for ZON concentration at or below 50 µg/kg and above 50 µg/kg, respectively [79].

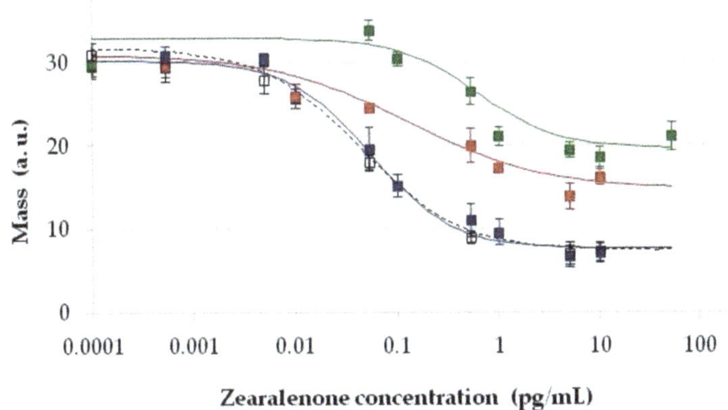

Figure 5. Standard calibration curves for zearalenone (ZON) determination by the competitive optical waveguide lightmode spectroscopy (OWLS) immunosensor format in maize extract at various dilutions. The amino-modified sensor surface with ZON conjugate to bovine serum albumin (ZON-BSA) immobilised at 10 µg/mL on amino-modified sensor surfaces by (3-aminopropyl) triethoxysilane, succinic anhydride and 1-ethyl-3-(3-dimethylaminopropyl) carbodiimide with N-hydroxysuccinimide (APTS/SA/EDC-NHS) using ZON-specific serum at 1:2000 dilution. Maize extracts were applied at dilutions of 1:100 (■, green line), 1:1000 (■, red line), and 1:10,000 (■, blue line); ZON calibration curve in 42 mM 2-amino-2-(hydroxymethyl)-1,3-propanediol (Tris) buffer (pH 7.4) (□, black slashed line).

ZON concentrations measured in maize extract by OWLS and ELISA methods were compared to each other as shown in Figure 6. Results indicate that concentrations detected by the two methods well correlated with each other in the 0.1–10 µg/kg range ($r^2 = 0.984$), and both methods are applicable. However, while the ELISA method required an extract dilution of 1:10 and detected ZON above 0.1 ng/mL, the OWLS immunosensor required an extract dilution of 1:10,000, but detected ZON above 0.01 pg/mL.

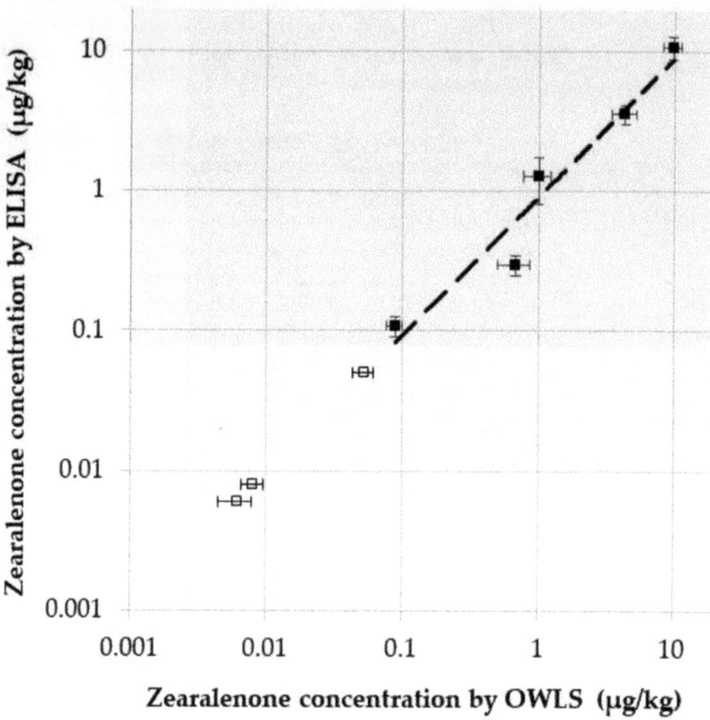

Figure 6. Determination of zearalenone (ZON) content in maize samples by optical waveguide lightmode spectroscopy (OWLS) immunosensor and enzyme-linked immunosorbent assay (ELISA). The sensing range for OWLS (>0.002 µg/kg) and for ELISA (>0.09 µg/kg) are indicated in blue and red, respectively. (Ordinate values for ZON concentrations below 0.09 µg/kg are virtual for visualisation—indicated by hollow rectangles).

3. Conclusions

To provide stable immunosensors for the detection of mycotoxin ZON, continuous flow OWLS sensor setups were established. To immobilize protein immunoreagents (ZON-specific antibodies or ZON-BSA conjugate), the immunosensor surface was modified by epoxy, amino, and carboxyl functional groups under laboratory conditions by optimised silanisation protocols. Epoxy functional groups allowed direct immobilisation of the proteins under alkaline conditions (pH = 9.5). Amino functional groups allowed direct immobilisation of the proteins with 2.5% GA, or could be converted to carboxylic acid functional groups by 0.2% SA and conjugate to proteins using a 1:1 mixture of 0.1 M NHS and 0.4 M EDC.

In the direct (immobilised antibody) format, immobilisation on epoxy-modified surfaces (GOPS) provided lower binding efficacy and reproducibility than that on amino- (APTS/GA) or carboxyl-modified surfaces (APTS/SA/EDC-NHS). However, detectable ZON concentrations fell in all three cases above 500 ng/mL, not being sufficient for practical purposes. In the competitive (immobilised antigen) format, immobilisation on epoxy-modified surfaces (GOPS) remained improper for use, not providing sigmoid analyte concentration dependence, but amino- or carboxyl-modified surfaces were found of high utility. Both methods (APTS/GA and APTS/SA/EDC-NHS) resulted in similar analytical detection levels (EC_{50} values in the range of 0.017–0.083 pg/mL) and linear detection ranges. Higher signal levels—therefore, greater signal decreases by inhibition—were achieved with amino-modified surfaces; however, carboxyl-modified surfaces allowed for more stable and reproducible results. The optimised competitive immunosensor using 10 µg/mL of ZON-BSA as the surface coating antigen immobilised by the APTS/SA/EDC-NHS method, as well as ZON-specific antibodies at 1:2000 dilution, was found to show excellent sensitivity and specificity to ZON, allowed an LOD of 0.002 pg/mL, and was found to be applicable to the determination of ZON in maize extracts. Detectable analyte concentrations in assay buffer were found to be five orders of magnitude lower by the immunosensor than by the related ELISA method, which, considering the sample preparation requirements, corresponds to a three orders of magnitude improvement for determination of ZON content in maize commodity. Such unique improvements in the analytical sensitivity of the OWLS technique compared to the corresponding ELISA method have previously been evidenced for the detection of other analytes, including a nearly three orders of magnitude enhancement for the endocrine biomarker protein vitellogenin [77] and a six orders of magnitude improvement for a herbicide active ingredient trifluralin [76]. Moreover, the current OWLS immunosensor represents substantial advancements compared to previous immunosensors for ZON, e.g., based on SPR [71], TIRE [73], and electrochemical detection with antibodies immobilised on gold nanoparticles embedded on multi-walled carbon nanotubes [42], to which the LOD of the current competitive OWLS immunosensor represents 30,000-, 5000-, and 75-fold improvements, respectively.

4. Materials and Methods

4.1. Reagents and Instrumentation

Chemical reagents, including γ-glycidoxypropyl-trimethoxysilane (GOPS) and (3-aminopropyl) triethoxysilane (APTS), mycotoxin standards, proteins, and biochemicals, were purchased from Sigma-Aldrich Kft. (Budapest, Hungary) unless indicated otherwise. OWLS immunosensor measurements were carried out on an OWLS 210 instrument and BioSense 3.8 software (MicroVacuum Ltd., Budapest, Hungary) using OWLS 2400 sensor chips with optical grating of 2400 lines per mm in the SiO_2-TiO_2 waveguide layer (MicroVacuum Ltd., Budapest, Hungary). Immunoassays were carried out in an iEMS MF microplate reader (LabSystems, Helsinki, Finland) using high-capacity 96-well microplates (Nunc, Roskilde, Denmark).

4.2. Immunogen and Antibody Production

Since ZON is low-molecular-weight hapten, it is non-immunogenic and should be conjugated to a protein carrier for immunisation. ZON was converted to the corresponding hapten, ZON-6'-carboxymethyloxime, and was conjugated to carrier proteins BSA and CONA by the method from the literature [26]. Polyclonal antibodies directed against ZON were produced in female, 3-month-old New Zealand white rabbits immunised periodically and intradermally with a standard mixture of 25 µg of the ZON-CONA conjugate immunogen per kg body weight and 50 µL of Freund's complete or incomplete adjuvant. Rabbit immunisation was carried out under the supervision of the Ethics Committee of Research on Animals (Food Science Research Institute, National Agricultural Research and Innovation Centre, Budapest, Hungary) and under the authorisation and inspection by the Government Office for Pest County in Hungary (Official permit for animal testing # PE/EA/45-6/2020, last date of approval: 21 February 2020). Serum from the whole blood obtained was centrifuged at 2400 g for 15 min, and its IgG fraction was purified by sodium-sulphate precipitation [80].

4.3. OWLS Immunosensor Measurements

OWLS sensoric determinations were carried out in a flow-through cell of the OWLS 210 instrument. The optical grating of the sensor surface is illuminated with a polarised He-Ne laser light (632.8 nm), and the sensor chip is rotated along its axis in a narrow angle range ($\pm 7°$). The laser beam is diffracted on the grating, and enters the waveguide at the characteristic incoupling angles, where it propagates by total internal reflection, and is detected by photodiodes at the ends of the waveguide layer. Incoupling of the incident laser beam occurs at two well-defined angles of incidence: one for transverse electric (TE) and one for transverse magnetic (TM) mode. Rotating the cuvette with ± 7 degrees, effective refractive indices (NTE and NTM) are monitored, and four characteristic photocurrent peaks (TE and TM peaks on both, positive and negative sides) can be detected at the incoupling angles αTE and αTM. Apparent incoupling angles can be measured with 10^{-4} degree accuracy, and signal resolution relative to the effective refraction index is $\Delta N \sim 10^{-6}$. The mass of the deposited material absorbed on the waveguide surface from the continuous-flow medium can be calculated from the effective refractive indices (NTE and NTM), expressed as thickness of the protein layer deposited (nm) or surface coverage (ng cm^{-2}). All determinations were carried out at room temperature in a flow-injection analyser system at a flow rate of 200 µl/min and with injection volumes of 200 µL.

4.3.1. Functionalisation of the Sensor Surface

The sensor surface was derivatised by several routes to form reactive functional groups for covalent immobilisation of the protein immunoreagents (ZON-specific antibodies or ZON-BSA conjugate) (Figure 1). Proteins were immobilised to the derivatised sensor chip in a flow-through system using a 42 mM Tris running buffer (pH 7.4). Reactive epoxy groups were formed on the sensor surface by heating the sensor chips to 60 °C in 10% GOPS in toluene for 20 hrs, followed by washing the chips with toluene and further heating to 100 °C for 1 hr. Epoxy functionalised surfaces allowed for direct anchoring of biomolecules carrying amino or hydroxy moieties by nucleophilic addition to them in alkaline medium (pH > 8.5) [81]. Thus, protein immunoreagents were injected onto the epoxylated sensor chips at 1–20 µg/mL concentrations in 0.2 M carbonate buffer at pH 9.5, followed by buffer exchange to 42 mM Tris buffer at pH 7.4, and removal of unbound proteins from the surface by injecting 0.1 M aqueous hydrochloric acid. Amino groups were formed on the surface of the sensor by treating the chips at 75 °C with 10% APTS at pH 3.0 for 4 hrs, followed by washing with distilled water and heat treatment at 95 °C for 6 hrs [81]. The amino functionality was activated with 2.5% aqueous glutaraldehyde (GA), allowing direct additive anchoring biomolecules carrying amino groups. The immobilisation reaction was carried out within the flow-through cuvette, by injecting GA into the flowing distilled water medium, followed by medium exchange to Tris buffer (42 mM, pH 7.4), subsequent injection of the proteins at 1–20 µg/mL concentration, and elution of the unbound reagent fraction by the injection of 0.1 M aqueous hydrochloric acid. Alternatively, amino groups were modified to carboxyl groups by derivatisation with succinic anhydride (SA), and were utilised for covalent attachment of biomolecules by the activated ester method using NHS with a dehydrating agent EDC. Carboxylation was carried out in separate vessels with 1% SA in dry ethanol at 25 °C for 1 hr, followed by drying the chips at 90 °C for 15 min. Active ester formation on the chip surface was carried out in a stopped flow mode by injecting a 1:1 solution of 0.2 M EDC and 0.05 M NHS, incubating at room temperature for 10 min, rinsing with Tris buffer (42 mM, pH 7.4), adding the protein-ZON conjugate at 1–20 µg/mL concentration in 10 mM sodium acetate buffer at pH 4.0, and incubating again at room temperature for 10 min. Alternatively, active ester formation could be carried out outside the OWLS 210 instrument, on a Petri dish, under similar reaction conditions. Residual active ester functional groups were finally deactivated by the injection of 1 M ethanolamine at pH 8.5 for 10 min.

4.3.2. Immunosensor Formats

Non-competitive and competitive detection formats were applied for developing OWLS immunosensors. The first format was based on the immobilisation of 2000-fold diluted polyclonal antibodies. Such immobilised antibodies capture their analyte (the antigen or similar immunoreactive compounds) from the sample; therefore, this format is often also termed a direct format, and the amount of antigen bound to the immobilised antibodies is proportional to the quantity of the antigen in the standard solutions. In the second format, 10 µg/mL of ZON-BSA conjugate was bound to the solid surface, then standards or samples were mixed in a 1:1 ratio with solutions containing known amounts of antibodies, and the mixture upon a short incubation was injected into the system. The amount of antibodies bound to the immobilised conjugates is inversely proportional to the quantity of the antigen in the standard solutions.

4.4. Sample Preparation

Ground maize samples were spiked with ZON at the concentration range of 5 ng/kg to 10 µg/kg (5, 10, 50, 100, and 500 ng/kg and 1, 5, and 10 µg/kg). One gram aliquots of the spiked samples were extracted with 10 mL of acetonitrile/water (6:4) as a solvent. The samples were stirred for 10 min and centrifuged on ultrafiltration membranes with a 100,000 nominal molecular weight limit at 5000 rpm for 10 min, and the filtrate was collected for OWLS and ELISA measurements. Upon sample preparation, all samples were stored at 4 °C until measurement and were diluted with 42 mM Tris buffer (pH 7.4) to the appropriate rate prior to analysis.

4.5. Determination of ZON by ELISA Method

To confirm the utility of the immunosensor, ZON content was also determined in a corresponding competitive ELISA system [29]. ELISA plates were coated with 5 µg/mL ZON-BSA conjugate, and inhibition of binding of the polyclonal antibody by ZON was measured using a commercial horseradish peroxidase labelled second (anti-rabbit IgG) antibody and a colorimetric immunoassay signal measured at 450 nm.

4.6. Data Analysis and Statistics

All determinations, except for the real time recordings in direct sensor titration experiments (Figure 3), were performed at least in triplicates, and error bars on the graphs represent the standard deviation (SD) of the replicates for each datum point. SDs were calculated as the square root of variance of the deviation of each datum point relative to the mean. Sigmoid calibration curves were obtained by logistic mathematical fitting using the Rodbard equation [82], which were also used for determination of the IC_{50} values. LOD values were defined as an analyte concentration corresponding to a signal that differs from the background level by 3 SDs of the background.

Author Contributions: Conceptualisation, I.S. (Inna Székács), N.A. and A.S.; methodology, I.S. (Inna Székács) and N.A.; software, I.S. (István Szendő); validation, I.S. (Inna Székács) and N.A.; formal analysis, N.A. and A.S.; investigation, I.S. (Inna Székács); resources, I.S. (István Szendő); data curation, I.S. (Inna Székács); writing—original draft preparation, I.S. (Inna Székács); writing—review and editing, N.A., I.S. (István Szendő) and A.S.; visualisation, A.S.; supervision, A.S.; project administration, A.S.; funding acquisition, N.A. and A.S. All authors have read and agreed to the published version of the manuscript.

Funding: This research was funded by the Hungarian National Research, Development and Innovation Office, project NVKP_16-1-2016-0049 "In situ, complex water quality monitoring by using direct or immunofluorimetry and plasma spectroscopy".

Institutional Review Board Statement: The study was conducted according to the guidelines of the Declaration of Helsinki, and approved by the Ethics Committee of Research on Animals of the Food Science Research Institute, National Agricultural Research and Innovation Centre, Budapest, Hungary (protocol code PE/EA/45-6/2020, last date of approval 21 February 2020).

Informed Consent Statement: Not applicable.

Data Availability Statement: The data presented in this study are available on request from the corresponding author. The data are not publicly available due to privacy reasons.

Conflicts of Interest: The authors declare no conflict of interest.

References

1. Maragos, C.M. Zearalenone occurrence and human exposure. *World Mycotoxin J.* **2010**, *3*, 369–383. [CrossRef]
2. Kiang, D.T.; Kennedy, B.J.S.; Pathre, V.; Mirocha, C.J. Binding characteristicsof zearalenone analogs to estrogen receptors. *Cancer Res.* **1978**, *38*, 3611–3615.
3. Kuiper, G.G.J.M.; Lemmen, J.G.; Carlsson, B.; Corton, J.C.; Safe, S.H.; Van der Saag, P.T.; Van der Burg, B.; Gustafsson, J.A. Interaction of estrogenic chemicals and phytoestrogens with estrogen receptor beta. *Endocrinology* **1998**, *139*, 4252–4263. [CrossRef]
4. Zinedine, A.; Soriano, J.M.; Moltó, J.C.; Mañes, J. Review on the toxicity, occurrence, metabolism, detoxification, regulations and intake of zearalenone: An oestrogenic mycotoxin. *Food Chem. Toxicol.* **2007**, *45*, 1–18. [CrossRef]
5. Turcotte, J.C.; Hunt, P.J.B.; Blaustein, J.D. Estrogenic effects of zearalenone on the expression of progestin receptors and sexual behavior in female rats. *Horm. Behav.* **2005**, *47*, 178–184. [CrossRef]
6. Hueza, I.M.; Raspantini, P.C.F.; Raspantini, L.E.R.; Latorre, A.O.; Górniak, S.L. Zearalenone, an estrogenic mycotoxin, is an immunotoxic compound. *Toxins* **2014**, *6*, 1080–1095. [CrossRef]
7. Marin, D.E.; Motiu, M.; Taranu, I. Food contaminant zearalenone and its metabolites affect cytokine synthesis and intestinal epithelial integrity of porcine cells. *Toxins* **2015**, *7*, 1979–1988. [CrossRef]
8. Zhou, H.; George, S.; Hay, C.; Lee, J.; Qian, H.; Sun, X. Individual and combined effects of aflatoxin B1, deoxynivalenol and zearalenone on HepG2 and RAW 264.7 cell lines. *Food Chem. Toxicol.* **2017**, *103*, 18–27. [CrossRef]
9. Food and Agricultural Organization (FAO). *Worldwide Regulations for Mycotoxins in Food and Feed in 2003*; Food and Agricultural Organization of the United Nations: Rome, Italy, 2004.
10. European Commission (EC). Commission Regulation (EC) No 1881/2006 of 19 December 2006 setting maximum levels for certain contaminants in foodstuffs. *Off. J. Eur. Union* **2006**, *L 364*, 5–24.
11. European Food Safety Authority (EFSA). Scientific Opinion on the risks for public health related to the presence of zearalenone in food. *EFSA J.* **2011**, *9*, 2197. [CrossRef]
12. European Food Safety Authority (EFSA). Scientific Opinion on the risks for animal health related to the presence of zearalenone and its modified forms in feed. *EFSA J.* **2017**, *15*, 4851.
13. Schaafsma, A.W.; Nicol, R.W.; Savard, M.E.; Sinha, R.C.; Reid, L.M.; Rottinghaus, G. Analysis of *Fusarium* toxins in maize and wheat using thin layer chromatography. *Mycopathologia* **1998**, *142*, 107–113. [CrossRef]
14. Josephs, R.D.; Shumacher, R.; Krska, R. International interlaboratory study for the determination of the *Fusarium* mycotoxins zearalenone and deoxynivalenol in agricultural commodities. *Food Addit. Contam.* **2001**, *18*, 417–430. [CrossRef]
15. Dawlatana, M.; Coker, R.D.; Nagler, M.J.; Blunden, G.; Oliver, W.O. An HPTLC method for the quantitative determination of zearalenone in maize. *Chromatographia* **1998**, *47*, 215–218. [CrossRef]
16. Diebold, G.J.; Karny, N.; Zare, R.N. Determination of zearalenone in corn by laser fluorimetry. *Anal. Chem.* **1979**, *51*, 67–69. [CrossRef] [PubMed]
17. Tanaka, T.; Teshima, R.; Ikebuchi, H.; Sawada, J.; Ichinoe, M. Sensitive enzyme-linked immunosorbent assay for the mycotoxin zearalenone in barley and Job's-tears. *J. Agric. Food Chem.* **1995**, *43*, 946–950. [CrossRef]
18. Kinani, S.; Bouchonnet, S.; Bourcier, S.; Porcher, J.M.; Aït-Aïssa, S. Study of the chemical derivatization of zearalenone and its metabolites for gas chromatography–mass spectrometry analysis of environmental samples. *J. Chromatogr. A* **2008**, *1190*, 307–315. [CrossRef]
19. Radová, Z.; Hajšlová, J.; Králová, J. Analysis of zearalenone in wheat using high-performance liquid chromatography with fluorescence detection and/or enzyme–linked immunosorbent assay. *Cereal Res. Commun.* **2001**, *29*, 435–442. [CrossRef]
20. Berthiller, F.; Schumacher, R.; Buttinger, G.; Krska, R. Rapid simultaneous determination of major type A- and B-trichothecenes as well as zearalenone in maize by high performance liquid chromatography-tandem mass spectrometry. *J. Chromatogr. A* **2005**, *1062*, 209–216. [CrossRef]
21. Ok, H.E.; Chung, S.H.; Lee, N.; Chun, H.S. Simple high-performance liquid chromatography method for the simultaneous analysis of aflatoxins, ochratoxin A, and zearalenone in dried and ground red pepper. *J. Food Prot.* **2015**, *78*, 1226–1231. [CrossRef]
22. Nakhjavan, B.; Ahmed, N.S.; Khosravifard, M. Development of an improved method of sample extraction and quantitation of multi-mycotoxin in feed by LC-MS/MS. *Toxins* **2020**, *12*, 462. [CrossRef] [PubMed]
23. Rausch, A.-K.; Brockmeyer, R.; Schwerdtle, T. Development and validation of a QuEChERS-based liquid chromatography tandem mass spectrometry multi-method for the determination of 38 native and modified mycotoxins in cereals. *J. Agric. Food Chem.* **2020**, *68*, 4657–4669. [CrossRef] [PubMed]
24. Sun, S.; Yao, K.; Zhao, S.; Zheng, P.; Wang, S.; Zeng, Y.; Liang, D.; Ke, Y.; Jiang, H. Determination of aflatoxin and zearalenone analogs in edible and medicinal herbs using a group-specific immunoaffinity column coupled to ultra-high-performance liquid chromatography with tandem mass spectrometry. *J. Chromatogr. B* **2018**, *1092*, 228–236. [CrossRef] [PubMed]
25. Holland, R.D.; Sepaniak, M.J. Qualitative analysis of mycotoxins using micellar electrokinetic capillary chromatography. *Anal. Chem.* **1993**, *65*, 1140–1146. [CrossRef]
26. Böhs, B.; Seidel, V.; Lindner, W. Analysis of selected mycotoxins by capillary electrophoresis. *Chromatographia* **1995**, *41*, 631–637. [CrossRef]
27. Thouvenot, D.; Morfin, R.F. Radioimmunoassay for zearalenone and zearalanol in human serum: Production, properties, and use of porcine antibodies. *Appl. Environ. Microbiol.* **1983**, *45*, 16–23. [CrossRef]

28. Liu, M.T.; Ram, B.P.; Hart, L.P.; Pestka, J.J. Indirect enzyme-linked immunosorbent assay for the mycotoxin zearalenone. *Appl. Environ. Microbiol.* **1985**, *50*, 332–336. [CrossRef]

29. Székács, A. Enzyme-linked immunosorbent assay for monitoring the *Fusarium* toxin zearalenone. *Food Technol. Biotechnol.* **1998**, *36*, 105–110.

30. Pichler, H.; Krska, R.; Székács, A.; Grasserbauer, M. An enzyme-immunoassay for the detection of the mycotoxin zearalenone by use of yolk antibodies. *Fresenius J. Anal. Chem.* **1998**, *362*, 176–177. [CrossRef]

31. Thongrussamee, T.; Kuzmina, N.S.; Shim, W.-B.; Jiratpong, T.; Eremin, S.A.; Intrasook, J.; Chung, D.-H. Monoclonal-based enzyme-linked immunosorbent assay for the detection of zearalenone in cereals. *Food Addit. Contam. A* **2008**, *25*, 997–1006. [CrossRef]

32. Tang, X.; Li, X.; Li, P.; Zhang, Q.; Li, R.; Zhang, W.; Ding, X.; Lei, J.; Zhang, Z. Development and application of an immunoaffinity column enzyme immunoassay for mycotoxin zearalenone in complicated samples. *PLoS ONE* **2014**, *9*, e85606. [CrossRef]

33. Dong, G.; Pan, Y.; Wang, Y.; Ahmed, S.; Liu, Z.; Peng, D.; Yuan, Z. Preparation of a broad-spectrum anti-zearalenone and its primary analogues antibody and its application in an indirect competitive enzyme-linked immunosorbent assay. *Food Chem.* **2018**, *247*, 8–15. [CrossRef]

34. Zhan, S.; Huang, X.; Chen, R.; Li, J.; Xiong, Y. Novel fluorescent ELISA for the sensitive detection of zearalenone based on H2O2-sensitive quantum dots for signal transduction. *Talanta* **2016**, *158*, 51–56. [CrossRef]

35. Beloglazova, N.V.; Speranskaya, N.V.; Wu, A.; Wang, Z.; Sanders, M.; Goftman, V.V.; Zhang, D.; Goryacheva, I.Y.; De Saeger, S. Novel multiplex fluorescent immunoassays based on quantum dot nanolabels for mycotoxins determination. *Biosens. Bioelectron.* **2014**, *62*, 59–65. [CrossRef] [PubMed]

36. Zhang, F.; Liu, B.; Sheng, W.; Zhang, Y.; Liu, Q.; Li, S.; Wang, S. Fluoroimmunoassays for the detection of zearalenone in maize using CdTe/CdS/ZnS quantum dots. *Food Chem.* **2018**, *255*, 421–428. [CrossRef] [PubMed]

37. Hendrickson, O.D.; Chertovich, J.O.; Zherdev, A.V.; Sveshnikov, P.G.; Dzantiev, B.B. Ultrasensitive magnetic ELISA of zearalenone with pre-concentration and chemiluminescent detection. *Food Control* **2018**, *84*, 330–338. [CrossRef]

38. Fang, D.; Zeng, B.; Zhang, S.; Dai, H.; Lin, Y. A self-enhanced electrochemiluminescent ratiometric zearalenone immunoassay based on the use of helical carbon nanotubes. *Microchim. Acta* **2020**, *187*, 303. [CrossRef]

39. Visconti, A.; Pascale, M. Determination of zearalenone in corn by means of immunoaffinity clean-up and high-performance liquid chromatography with fluorescence detection. *J. Chromatogr. A* **1998**, *815*, 133–140. [CrossRef]

40. Fazekas, B.; Tar, A. Determination of zearalenone content in cereals and feedstuffs by immunoaffinity column coupled with liquid chromatography. *J. AOAC Int.* **2001**, *84*, 1453–1459. [CrossRef]

41. Hao, K.; Suryoprabowo, S.; Song, S.; Liu, L.; Kuang, H. Rapid detection of zearalenone and its metabolite in corn flour with the immunochromatographic test strip. *Food Agric. Immunol.* **2018**, *29*, 498–510. [CrossRef]

42. Riberi, W.I.; Tarditto, L.V.; Zon, M.A.; Arévalo, F.J.; Fernández, H. Development of an electrochemical immunosensor to determine zearalenone in maize using carbon screen printed electrodes modified with multi-walled carbon nanotubes/polyethyleneimine dispersions. *Sens. Actuators B* **2018**, *254*, 1271–1277. [CrossRef]

43. Xu, L.; Zhang, Z.; Zhang, Q.; Zhang, W.; Yu, L.; Wang, D.; Li, H.; Li, P. An on-site simultaneous semi-quantification of aflatoxin b1, zearalenone, and T-2 toxin in maize- and cereal-based feed via multicolor immunochromatographic assay. *Toxins* **2018**, *10*, 87. [CrossRef] [PubMed]

44. Li, S.-J.; Sheng, W.; Wen, W.; Gu, Y.; Wang, J.-P.; Wang, S. Three kinds of lateral flow immunochromatographic assays based on the use of nanoparticle labels for fluorometric determination of zearalenone. *Microchim. Acta* **2018**, *185*, 238. [CrossRef] [PubMed]

45. Gadzała-Kopciuch, R.; Cendrowski, K.; Cesarz, A.; Kiełbasa, P.; Buszewski, B. Determination of zearalenone and its metabolites in endometrial cancer by coupled separation techniques. *Anal. Bioanal. Chem.* **2011**, *401*, 2069–2078. [CrossRef]

46. Lhotská, I.; Gajdošová, B.; Solich, P.; Šatínský, D. Molecularly imprinted vs. reversed-phase extraction for the determination of zearalenone: A method development and critical comparison of sample clean-up efficiency achieved in an on-line coupled SPE chromatography system. *Anal. Bioanal. Chem.* **2018**, *410*, 3265–3273.

47. Zhang, Y.; He, J.; Song, L.; Wang, H.; Huang, Z.; Sun, Q.; Ba, X.; Li, Y.; You, L.; Zhang, S. Application of surface-imprinted polymers supported by hydroxyapatite in the extraction of zearalenone in various cereals. *Anal. Bioanal. Chem.* **2020**, *412*, 4045–4055. [CrossRef]

48. Huang, X.; Aguilar, Z.P.; Xu, H.; Lai, W.; Xiong, Y. Membrane-based lateral flow immunochromatographic strip with nanoparticles as reporters for detection: A review. *Biosens. Bioelectron.* **2016**, *75*, 166–180. [CrossRef]

49. Wu, Y.; Zhou, Y.; Huang, H.; Chen, X.; Leng, Y.; Lai, W.; Huang, X.; Xiong, Y. Engineered gold nanoparticles as multicolor labels for simultaneous multi-mycotoxin detection on the immunochromatographic test strip nanosensor. *Sens. Actuators B* **2020**, *316*, 128107. [CrossRef]

50. Zhang, W.; Tang, S.; Jina, Y.; Yang, C.; He, L.; Wang, J.; Chen, Y. Multiplex SERS-based lateral flow immunosensor for the detection of major mycotoxins in maize utilizing dual Raman labels and triple test lines. *J. Hazard. Mater.* **2020**, *393*, 122348. [CrossRef]

51. McNamee, S.E.; Bravin, F.; Rosar, G.; Elliott, C.T.; Campbell, K. Development of a nanoarray capable of the rapid and simultaneous detection of zearalenone, T2-toxin and fumonisin. *Talanta* **2017**, *164*, 368–376. [CrossRef]

52. Chen, Y.; Meng, X.; Zhu, Y.; Shen, M.; Lu, Y.; Cheng, J.; Xu, Y. Rapid detection of four mycotoxins in corn using a microfluidics and microarray-based immunoassay system. *Talanta* **2018**, *186*, 299–305. [CrossRef]

53. Basova, E.Y.; Goryacheva, I.Y.; Rusanova, T.Y.; Burmistrova, N.A.; Dietrich, R.; Märtlbauer, E.; Detavernier, C.; Van Peteghem, C.; De Saeger, S. An immunochemical test for rapid screening of zearalenone and T-2 toxin. *Anal. Bioanal. Chem.* **2010**, *397*, 55–62. [CrossRef]

54. Chun, H.S.; Choi, E.H.; Chang, H.J.; Choi, S.W.; Eremin, S.A. A fluorescence polarization immunoassay for the detection of zearalenone in corn. *Anal. Chim. Acta* **2009**, *639*, 83–89. [CrossRef]

55. Mak, A.C.; Osterfeld, S.J.; Yu, H.; Wang, S.X.; Davis, R.W.; Jejelowo, O.A.; Pourmand, N. Sensitive giant magnetoresistive-based immunoassay for multiplex mycotoxin detection. *Biosens. Bioelectron.* **2010**, *25*, 1635–1639. [CrossRef]

56. Shao, Y.; Duan, H.; Guo, L.; Leng, Y.; Lai, W.; Xiong, Y. Quantum dot nanobead-based multiplexed immunochromatographic assay for simultaneous detection of aflatoxin B$_1$ and zearalenone. *Anal. Chim. Acta* **2018**, *1025*, 163–171. [CrossRef]

57. Urraca, J.L.; Benito-Peña, E.; Pérez-Conde, C.; Moreno-Bondi, M.C.; Pestka, J.J. Analysis of zearalenone in cereal and swine feed samples using an automated flow-through immunosensor. *J. Agric. Food Chem.* **2005**, *53*, 3338–3344. [CrossRef]

58. Xu, L.; Zhang, Z.; Zhang, Q.; Li, P. Mycotoxin determination in foods using advanced sensors based on antibodies or aptamers. *Toxins* **2016**, *8*, 239. [CrossRef]

59. Adányi, N.; Majer-Baranyi, K.; Székács, A. Evanescent field effect based nanobiosensors for agro-environmental and food safety. In *Nanobiosensors: Nanotechnology in the Agri-Food Industry*; Grumezescu, A.M., Ed.; Elsevier: Amsterdam, The Netherlands, 2017; Volume 8, pp. 429–474.

60. Goud, Y.K.; Kumar, S.V.; Hayat, K.; Gobi, V.K.; Song, H.; Kim, K.-H.; Marty, J.L. A highly sensitive electrochemical immunosensor for zearalenone using screen-printed disposable electrodes. *J. Electroanal. Chem.* **2019**, *832*, 336–342. [CrossRef]

61. Rhouati, A.; Catanante, G.; Nunes, G.; Hayat, A.; Marty, J.-L. Label-free aptasensors for the detection of mycotoxins. *Sensors* **2016**, *16*, 2178. [CrossRef]

62. Wu, S.; Liu, L.; Duan, N.; Li, Q.; Zhou, Y.; Wang, Z. Aptamer-Based Lateral Flow Test strip for rapid detection of zearalenone in corn samples. *J. Agric. Food Chem.* **2018**, *66*, 1949–1954. [CrossRef] [PubMed]

63. Taghdisi, S.M.; Danesh, N.M.; Ramezani, M.; Emrani, A.S.; Abnous, K. Novel colorimetric aptasensor for zearalenone detection based on nontarget-induced aptamer walker, gold nanoparticles, and exonuclease-assisted recycling amplification. *ACS Appl. Mater. Interfaces* **2018**, *10*, 12504–12509. [CrossRef] [PubMed]

64. He, D.; Wu, Z.; Cui, B.; Jin, Z.; Xu, E. A fluorometric method for aptamer-based simultaneous determination of two kinds of the fusarium mycotoxins zearalenone and fumonisin B-1 making use of gold nanorods and upconversion nanoparticles. *Microchim. Acta* **2020**, *187*, 254. [CrossRef] [PubMed]

65. Xu, J.; Chi, J.; Lin, C.; Lin, X.; Xie, Z. Towards high-efficient online specific discrimination of zearalenone by using gold nanoparticles@aptamer-based affinity monolithic column. *J. Chromatogr. A* **2020**, *1620*, 461026. [CrossRef] [PubMed]

66. Tan, H.; Guo, T.; Zhou, H.; Dai, H.; Yu, Y.; Zhu, H.; Wang, H.; Fu, Y.; Zhang, Y.; Ma, L. A simple mesoporous silica nanoparticle-based fluorescence aptasensor for the detection of zearalenone in grain and cereal products. *Anal. Bioanal. Chem.* **2020**, *412*, 5627–5635. [CrossRef] [PubMed]

67. Luo, L.; Liu, X.; Ma, S.; Li, L.; You, T. Quantification of zearalenone in mildewing cereal crops using an innovative photoelectro-chemical aptamer sensing strategy based on ZnO-NGQDs composites. *Food Chem.* **2020**, *322*, 126778. [CrossRef] [PubMed]

68. Navarro-Villoslada, F.; Urraca, J.L.; Moreno-Bondi, M.C.; Orellana, G. Zearalenone sensing with molecularly imprinted polymers and tailored fluorescent probes. *Sensors Actuators B* **2007**, *121*, 67–73. [CrossRef]

69. Choi, S.W.; Chang, H.J.; Lee, N.; Kim, J.H.; Chun, H.S. Detection of mycoestrogen zearalenone by a molecularly imprinted polypyrrole-based surface plasmon resonance (SPR) sensor. *J. Agric. Food Chem.* **2009**, *57*, 1113–1118. [CrossRef]

70. Radi, A.-E.; Eissa, A.; Wahdan, T. Molecularly imprinted impedimetric sensor for determination of mycotoxin zearalenone. *Electroanalysis* **2020**, *32*, 1788–1794. [CrossRef]

71. Hu, X.; Wang, C.; Zhang, M.; Zhao, F.; Zeng, B. Ionic liquid assisted molecular self-assemble and molecular imprinting on gold nanoparticles decorated boron-doped ordered mesoporous carbon for the detection of zearalenone. *Talanta* **2020**, *217*, 121032. [CrossRef]

72. Tittlemier, S.A.; Cramer, B.; Dall'Asta, C.; Iha, M.H.; Lattanzio, V.M.T.; Malone, R.J.; Maragos, C.; Solfrizzo, M.; Stranska-Zachariasova MStroka, J. Developments in mycotoxin analysis: An update for 2017–2018. *World Mycotoxin J.* **2019**, *12*, 3–29. [CrossRef]

73. Van der Gaag, B.; Spath, S.; Dietrich, H.; Stigter, E.; Boonzaaijer, G.; van Osenbruggen, T.; Koopal, K. Biosensors and multiple mycotoxin analysis. *Food Control* **2003**, *14*, 251–254. [CrossRef]

74. Hossain, M.Z.; Maragos, C.M. Gold nanoparticle-enhanced multiplexed imaging surface plasmon resonance (iSPR) detection of *Fusarium* mycotoxins in wheat. *Biosens. Bioelectron.* **2018**, *101*, 245–252. [CrossRef] [PubMed]

75. Nabok, A.; Tsargorodskaya, A.; Mustafa, M.K.; Székács, I.; Starodub, N.F.; Székács, A. Detection of low molecular weight toxins using an optical phase method of ellipsometry. *Sens. Actuators B* **2011**, *154*, 232–237. [CrossRef]

76. Adányi, N.; Levkovets, I.A.; Rodriguez-Gil, S.; Ronald, A.; Váradi, M.; Szendrő, I. Development of immunosensor based on OWLS technique for determining Aflatoxin B1 and Ochratoxin A. *Biosens. Bioelectron.* **2007**, *22*, 797–802. [CrossRef]

77. Majer-Baranyi, K.; Adányi, N.; Nagy, A.; Bukovkaya, O.; Szendrő, I.; Székács, A. Label-free immunosensor for monitoring vitellogenin as a biomarker for exogenous oestrogen compounds in amphibian species. *Int. J. Environ. Anal. Chem.* **2015**, *95*, 481–493. [CrossRef]

78. Székács, A.; Trummer, N.; Adányi, N.; Váradi, M.; Szendrő, I. Development of a non-labeled immunosensor for the herbicide trifluralin via OWLS detection. *Anal. Chim. Acta* **2003**, *487*, 31–42. [CrossRef]
79. EC (European Commission). Commission regulation (EC) No 519/2014. *Off. J. Eur. Union* **2014**, *L 147*, 29–43.
80. Tijssen, P. *Practice and Theory of Enzyme Immunoassays*; Elsevier: Amsterdam, The Netherlands, 1985; pp. 96–99.
81. Trummer, N.; Adányi, N.; Váradi, M.; Szendrő, I. Modification of the surface of integrated optical waveguide sensors for immunosensor applications. *Fresenius J. Anal. Chem.* **2001**, *371*, 21–24. [CrossRef]
82. Rodbard, D.; Hutt, D.M. Statistical analysis of radioimmunoassays and immunoradiometric (labeled antibody) assays: A generalized, weighted, iterative, least-squares method for logistic curve fitting. In *Radioimmunoassay and Related Procedures in Medicine*; International Atomic Energy Agency: Vienna, Austria, 1974; p. 165.

 *toxins*

Article

A Non-Enzyme and Non-Label Sensitive Fluorescent Aptasensor Based on Simulation-Assisted and Target-Triggered Hairpin Probe Self-Assembly for Ochratoxin a Detection

Mengyao Qian [1] , Wenxiao Hu [2], Luhui Wang [2], Yue Wang [1] and Yafei Dong [2,3,*]

[1] School of Computer Science, Shaanxi Normal University, Xi'an 710119, China; qmy@snnu.edu.cn (M.Q.); wy.wangyue@snnu.edu.cn (Y.W.)

[2] College of Life Sciences, Shaanxi Normal University, Xi'an 710119, China; huwenxiao@snnu.edu.cn (W.H.); wangluhui@snnu.edu.cn (L.W.)

[3] National Engineering Laboratory for Resource Developing of Endangered Chinese Crude Drugs in Northwest of China, Shaanxi Normal University, Xi'an 710119, China

* Correspondence: dongyf@snnu.edu.cn

Received: 10 May 2020; Accepted: 4 June 2020; Published: 6 June 2020

Abstract: The monitoring and control of mycotoxins has caused widespread concern due to their adverse effects on human health. In this research, a simple, sensitive and non-label fluorescent aptasensor has been reported for mycotoxin ochratoxin A (OTA) detection based on high selectivity of aptamers and amplification of non-enzyme hybridization chain reaction (HCR). After the introduction of OTA, the aptamer portion of hairpin probe H1 will combine with OTA to form OTA-aptamer complexes. Subsequently, the remainder of the opened H1 will act as an initiator for the HCR between the two hairpin probes, causing H1 and H2 to be sequentially opened and assembled into continuous DNA duplexes embedded with numerous G-quadruplexes, leading to a significant enhancement in fluorescence signal after binding with N-methyl-mesoporphyrin IX (NMM). The proposed sensing strategy can detect OTA with concentration as low as 4.9 pM. Besides, satisfactory results have also been obtained in the tests of actual samples. More importantly, the thermodynamic properties of nucleic acid chains in the monitoring platform were analyzed and the reaction processes and conditions were simulated before carrying out biological experiments, which theoretically proved the feasibility and simplified subsequent experimental operations. Therefore, the proposed method possess a certain application value in terms of monitoring mycotoxins in food samples and improving the quality control of food security.

Keywords: ochratoxin A; fluorescence; G-quadruplex; biosensor; computation; simulation

Key Contribution: A fast, sensitive and non-label fluorescence aptasensor for ochratoxin A detection based on computer simulation, G-quadruplex formation and signal amplification of hybrid chain reaction is reported.

---

1. Introduction

Ochratoxin A (OTA), another major contaminating mycotoxins in various food commodities such as grains, vegetables, nuts, spices, wine and animal feed [1,2], has attracted worldwide attention after aflatoxin [3]. The mycotoxin has highly kidney and liver toxicity, teratogenicity, carcinogenicity, mutagenicity and immunosuppressive effects on animals and humans [4–6]. As a consequence, many international agencies have specified the highest levels of OTA in different foodstuffs. For OTA in wine and grape-based beverages, the European Commission has set the maximum levels at 2 µg/kg. For unprocessed cereals, cereal-derived products and dried vine fruit, the levels were set at 5 µg/kg, 3 µg/kg and 10 µg/kg, respectively [7]. However, modern food processing technology cannot solve OTA contamination which can even survive in commercialized food systems such as bread, dried fruits, wine, and meat products, on account of its long half-life. Once ingested by human bodies, it will persist internally for more than 35 days [8,9]. Hence, it is essential to quantitatively detect OTA in foodstuffs for food control and human health.

The official methods for detecting OTA include thin-layer chromatography (TLC) [10], high performance liquid chromatography (HPLC) [11,12], high performance liquid chromatography–tandem mass spectrometry (HPLC–MS/MS) [13], gas chromatography (GC) [14], total internal reflection ellipsometry (TIRE) [15,16]. Although these methods show high sensitivity and low detection limits, they still need expensive and time-consuming pretreatments such as extraction, sample clean up and preconcentration. Such analytical processes are laborious and require advanced equipment and well-trained laboratory personnel. Immunoassays based on antigen-antibody interactions are also widely used in OTA detection in foodstuffs, such as enzyme linked immunosorbent assays (ELISA), surface plasmon resonance (SPR) and electrochemical immunosensors [17–20]. These assays show high sensitivity of detection, but often suffer from cross-reactively, matrix interference and poor shelf life of antibodies. Moreover, the production of antibodies are expensive and time-consuming which may take several weeks. Therefore, there is still a need to establish a simple, economical and reliable detection platform.

Aptamers, a sequence of oligonucleotides obtained by repeated screening from the library of random oligonucleotide sequences are synthesized artificially in vitro using the ligand index enrichment system evolution (SELEX) technique [21]. They can be used to identify different target elements, such as drugs, proteins, small molecules and cells [22] due to the characteristics of high affinity, miraculous selectivity, good thermal stability and easy to synthesize. Since the OTA aptamer has been reported [23], a great deal of aptamer-based biosensors for detecting OTA have been widely developed. Besides, many signal amplification techniques have also been reported to improve the sensitivity and dynamic detection range, including colorimetric and optical determination based on nanoparticles [24,25], nicking enzyme-assisted fluorescence signal enhancement [26,27], electrochemical impedance spectroscopy based on gold nanoparticles [28,29], fluorescence determination based on dye labeling [30], etc. These methods can greatly reduce the detection limit of the biosensors, but unfortunately, the use of nanomaterials can be disturbed by other electroactive substances coexisting in actual samples [31]. The protein enzyme's activity depends largely on the actual reaction conditions [32]. The preparation of gold nanoparticles and the fabrication of electrochemical biosensors are complicated and time-consuming [33]. In addition, the aptamers modified with chemical groups are expensive and may reduce the binding force between the aptamers and the molecular targets [34]. Therefore, it is necessary to develop a non-enzyme and label-free signal amplification method in order to detect OTA sensitively.

Hybridization chain reaction (HCR) is an isothermal, non-enzyme signal amplification technique originally proposed by Dirks and Pierce in 2004 [35]. In HCR, once the target DNA is introduced, two synthetic DNA hairpins coexisting in solution will hybridize into a continuous DNA nanowires. Due to the significant advantages of high amplification, controlled kinetics and non-enzyme natures, it has been widespread applied in the detection of proteins, nucleic acids and small molecules [36–38]. Nevertheless, most HCR-based detection methods require the modification of fluorophores and quenching groups, which increases the cost of detection and fluorescent background. Considering the above reasons, we introduced a unique high-order structure, G-quadruplex, instead of chemical labelling. Under the action of special cationic dyes, G-rich DNA sequences will pile up together to form G-quadruplex structure [39]. It can combine with NMM, which is a commercially available asymmetric anionic porphyrin that can specifically recognize the G-quadruplex structure rather than single-, duplex-, or triplex-stranded nucleic acid structure. After binding to the G-quadruplex, it shows a >20-fold increase in its fluorescence [40,41].

Actually, computer technology can solve the time-consuming, expensive and cumbersome shortcomings of biological experiments through simulation. Among them, by estimating the minimum free energy (MFE) of nucleic acids structures, the biological function of the relevant nucleotide sequence and the complete tertiary structure in the organism can be simulated and predicted [42]. Besides, the nucleotide sequences and reaction processes can be constructed by computer to simulate the corresponding nucleic acid models and perform related tasks dynamically. Therefore, we creatively combine computer simulation and biological experiments in our work for the sake of experiment time benefit and model execution efficiency.

To achieve non-enzyme and non-label sensitive detection of OTA in agricultural products, we use G-quadruplex/NMM as the fluorescent signal reporter gene, and use HCR to further amplify the fluorescent signal. More importantly, we performed a series of computer simulation analysis before the biological testing to simplify the subsequent experimental steps and eliminate some negative effects. In the presence of OTA, the two designed DNA hairpins will sequentially open up and self-assemble into continuous DNA duplexes embedded with numerous G-quadruplexes [43], thereby significantly enhancing the fluorescent signal. Our method displays high sensitivity, less time consumption, strong selectivity and has certain practical application towards OTA detection, which can open up new approaches for the use of aptasensor in the fields of food control and quality inspection.

2. Result and Discussion

2.1. Mechanism for OTA Detection

The designed biosensing platform for non-enzyme and non-label OTA fluorescence detection is shown in Figure 1. The biosensor model contains two hairpin probes (H1, H2), both of which have six nucleotide (nt) sticky ends. In particular, the 5′ end of hairpin probe H1 has OTA aptamer sequences (rose red and green, Figure 1) and the loop portion of hairpin probe H2 contains G-quadruplex sequence (blue, Figure 1). The same colored sections in H1 and H2 indicate that the sequence are complementary. After the introduction of OTA, due to the specificity and high affinity of aptamer and target, it will combine with the hairpin probe H1 to form OTA-aptamer complex, thereby opening the stem-loop structure of H1 and exposing the foothold region. Subsequently, the exposed part of H1 (3′ end) will hybridize to the longer end of H2 (3′ end). After opening the hairpin probe H2, the exposed 5′ end of H2 can open up hairpin H1 again. Therefore, the hairpin probes are sequentially opened and assembled into continuous DNA duplexes. Among them, G-rich sequences will shape numerous G-quadruplexes under the action of K$^+$. Finally, the fluorescent signals can be significantly enhanced by interact with NMM. In contrast, in the absence of OTA, the mixed solution of H1 and H2 merely showing a relatively low background fluorescence signal.

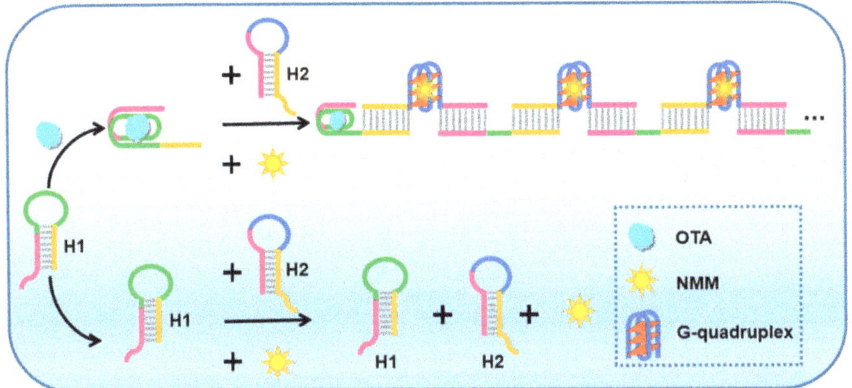

Figure 1. Schematic diagram for OTA detection based on HCR and G-quadruplex structures.

2.2. Verification of Feasibility by Computer Simulation and Biological Experiments

Before conducting actual experiments, it is essential to carry out computer simulations of the method proposed in order to economize time and cost of subsequent operations. Since nucleic acid secondary structure is critical to the function of the nucleic acid strands, we introduced an algorithm to predict and analyze the thermodynamic properties of the strands designed in this paper. In general, the combination of thermodynamic models and dynamic programming algorithms can estimate the minimum free energy (MFE) of nucleic acid structures with different loops and calculate the partition function [44]. MFE can be applied to predict the thermal stability of DNA and RNA strands [45] and partition function plays a major role in evaluating DNA and RNA sequences designed in the conformational ensemble.

Specifically, in the absence of a pseudoknot, thermodynamic model decompose the secondary structures of DNA and RNA molecules into different loops based on the base-pairing diagram. These loop configurations are associated with entropy and enthalpy values measured from loop sequence, type, and length [46]. Beginning with the study of Tinoco [47], numerous researchers have worked on the physical models of these structures. As shown in the base-pairing diagram of Figure 2, the recognized loop types include an interior loop, hairpin loops, a bulge loop, a multiloop and stacked bases. Meanwhile, the polymer main chain is represented by a straight line in the polymer graph, and the complementary paired bases are linked by arcs. All loop structures are nested with no crossing arcs. Furthermore, the free energy of a secondary structure S is vitally interrelated to the free energy F_L of each loop L it contains, so the total free energy F(S) can be calculated in Equation (1). The additivity of free energy means stronger impact on the partition function Q defined by Equation (2). Afterwards, the equilibrium probability of any nucleic acid secondary structure S can be calculated by weights (Equation (3)), where T and R represent temperature and universal gas constant.

$$F(S) = \sum_{L \in S} F_L \tag{1}$$

$$Q = \sum_{S} e^{-[F(S)/TR]} \tag{2}$$

$$P(S) = \frac{1}{Q} e^{-[F(S)/TR]} \tag{3}$$

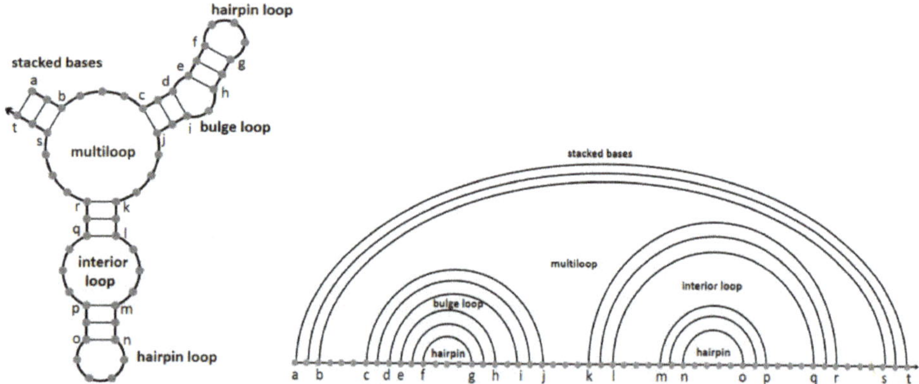

Figure 2. Canonical loop types of nucleic acid structure.

According to the above calculation methods, the MFE and secondary structure of the hairpin structures H1 and H2 designed in this model were estimated and simulated separately. It can be seen from Figure 3 that the two single strands were spontaneously folded into expected hairpin structures at

37 °C with relatively low free energy (F(H1) = −16.20 kcal/mol, F(H2) = −10.75 kcal/mol) by NUPACK simulation [48]. The lower free energy, the more stable structure. The results theoretically illustrate the feasibility of our design of the two hairpin probe sequences.

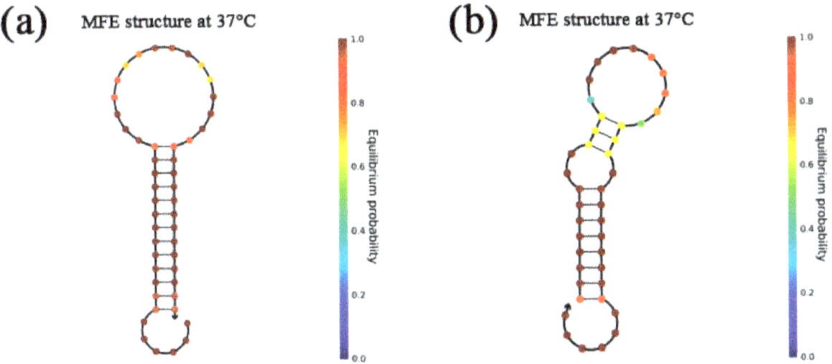

Figure 3. Simulation results of MFE structure at 37 °C of H1 (**a**) and H2 (**b**).

Moreover, in order to test whether the reaction meets the expectation and further simplify biological experiments, the experimental process, chain concentration and products were simulated and optimized by Visual DSD [49]. First of all, we processed the OTA into a single chain to facilitate computer input, which was complementary to its aptamer. It is worth noting that the HCR reaction products are long DNA nanowires, so we set reactants H1 and H2 forming the DNA duplex structures to a lower concentration than actual experiments for the convenience of computer output. In the presence of OTA (Figure 4a), due to the strong interaction between the aptamer in hairpin probe H1 and OTA, the concentration of OTA (Figure 4a, yellow curve) decreased rapidly to form OTA-H1 complexes, and eventually tends to 0. At the same time, the opened hairpin probe H1 can bind to H2 to shape Duplex 1 (Figure 4a, rose red curve). Afterwards, the exposed footholds of H2 could open the hairpin structure of H1 again, so that Duplex 1 (OTA-H1-H2) quickly disappeared and evolved into Duplex 2 (OTA-H1-H2-H1), which will immediately combined with new H1 into Duplex 3 (OTA-H1-H2-H1-H2). Therefore, the concentration of H1 and H2 gradually decreased during the reaction (Figure 4a, blue curve, red curve, respectively) and the duplexes were continuously produced and rapidly disappeared for evolving into longer DNA nanowires on account of constant hybridization between H1 and H2. In contrast, in the absence of OTA (Figure 4b), the concentration of hairpin probes H1 and H2 remained unchanged (Figure 4b, blue curve, red curve, respectively), indicating that no various duplexes were engendered. As we can see, the above results confirmed that our proposed strategy was theoretically feasible. At the same time, the experimental process can be simulated by changing the concentration and ratio of diverse reactive substances and the time and conditions of reactions, thereby greatly simplify subsequent actual experimental operations.

Subsequently, biological experiments were conducted to validate the practical feasibility of the proposed method for OTA detection. The fluorescence intensity changes of different solutions were recorded in Figure 5. The solution containing only OTA and H1 (curve c) or H2 (curve d) exhibited a relatively low fluorescence value. In the absence of OTA, H1 and H2 remained stable and the solution only showed a negligible change in fluorescence intensity (curve b). Then, adding OTA to test tubes containing H1 and H2 strands increased the fluorescence intensity significantly (curve a). These results are consistent with the demonstration in Figure 4, proving the feasibility of our model for OTA detection.

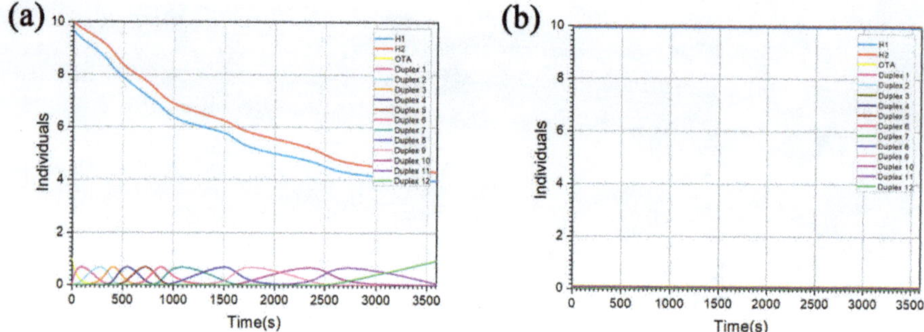

Figure 4. Visual DSD simulations. Changes in the concentration of various chains over time under different conditions: (**a**) with OTA; (**b**) without OTA.

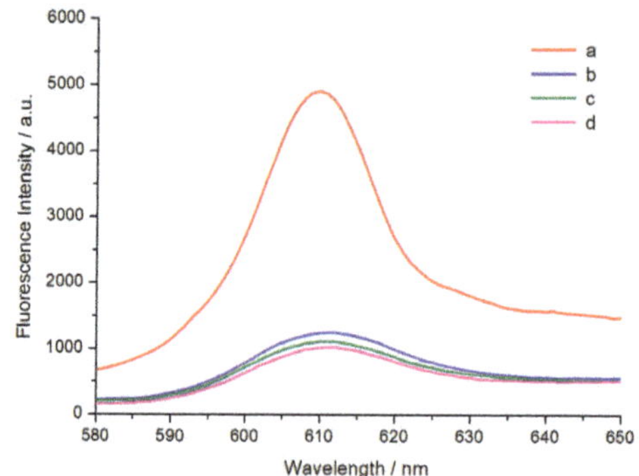

Figure 5. Fluorescence spectra of NMM with different substances: (**a**) H1, H2 and OTA; (**b**) H1 and H2; (**c**) H1 and OTA; (**d**) H2 and OTA. Experimental conditions: [H1] = [H2] = 300 nM, [OTA] = 10 nM.

2.3. Optimization of OTA Detection Conditions

For the sake of obtaining optimal performance of the proposed detection platform, several important parameters were optimized on the basis of single-factor experiments. Firstly, the effect of K^+ concentration on the changes of fluorescence intensity was studied because the sequences of G-rich oligonucleotide can form stable G-quadruplex structures under the action of K^+. Furthermore, the K^+ can mutual coordinate with carbonyl oxygen atoms of the G-residues and be embedded in the central of two stacked G-tetrads [36]. As shown in Figure 6a, F_0 and F represent the measured fluorescence value at 608 nm before and after adding OTA, respectively. With the increase of K^+ concentration, the amount of fluorescence intensity change ($F-F_0$) gradually increased and reached to the maximum at 20 mM, indicating that 20 mM of K^+ could completely accelerate the folding of G-quadruplex structures. Therefore, 20 mM K^+ concentration was chosen for the next experiment.

Subsequently, the HCR reaction time for H1 and H2 was also optimized. As the reaction time gradually increased in Figure 6b, the degree of hybridization between H1 and H2 deepened and the fluorescence change remained steady after 60 min, manifesting an equilibrium for HCR assembly between H1 and H2. Consequently, the time for HCR was set at 60 min in the subsequent experiments.

Moreover, the concentration of NMM and its reaction time with the G-quadruplexes produced in the experiment also directly affect the fluorescence intensity of the solution. According to Figure 6c, 1.5 μM of NMM displayed the highest fluorescence change in the detection because the lower concentration of NMM cannot provide sufficient fluorescence intensity for the G-quadruplexes produced in the reaction, and the higher concentration of NMM may engender increased background fluorescence signal. It was worth noting that the incubation time with NMM had little impact on the variation of fluorescence intensity (Figure 6d). Hence, in order to save the general time for the experiment, 10 min was selected as the combination time with NMM.

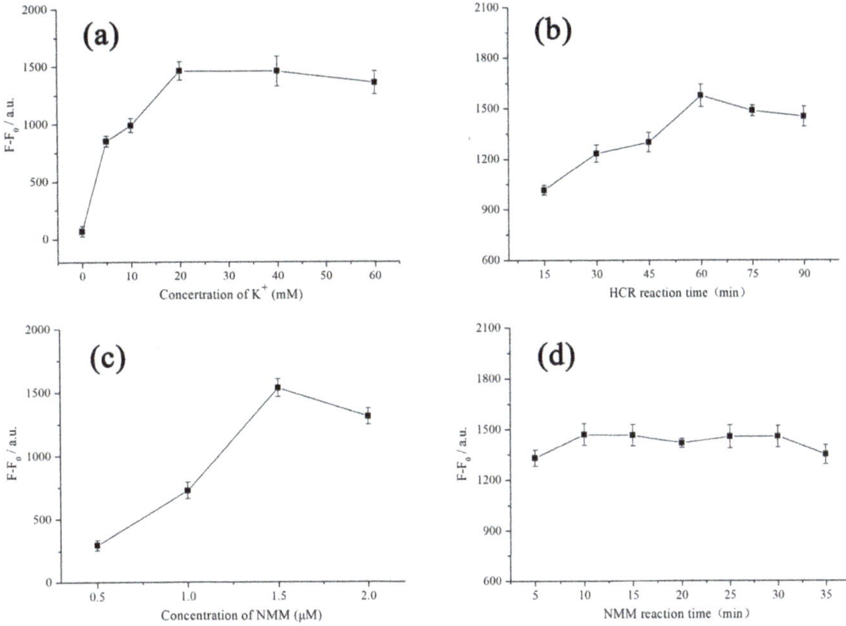

Figure 6. The fluorescence intensity change at 608 nm versus: (**a**) the concentration of K^+. [NMM] = 1.5 μM, and the reaction time of HCR and NMM are 60 min and 10 min, respectively; (**b**) the reaction time of HCR. [NMM] = 1.5 μM, [K^+] = 20 mM, and the combination time of NMM is 10 min; (**c**) the concentration of NMM. [K^+] = 20 mM, and the reaction time of HCR and NMM are 60 min and 10 min, respectively; (**d**) the reaction time of NMM. [NMM] = 1.5 μM, [K^+] = 20 mM, and the reaction time of HCR is 60 min. Other conditions: [H1] = [H2] = 300 nM. Error bars, SD, n = 3.

2.4. Sensitivity and Specificity

According to the optimal experimental conditions obtained from the above single-factor experiments, the sensitivity of the biosensor can be further analyzed by detecting the fluorescence intensity of different concentrations of OTA. As shown in Figure 7a, increasing OTA from 0.01 nM to 50 nM results in a gradual enhance in fluorescence signal. In addition, the fluorescence value of NMM at 608 nm is proportional to the logarithm of the OTA concentration from 0.01 nM to 0.5 nM (Figure 7b). The linear regression equation is y = 457.535lgx + 620.267 (x and y refer to OTA concentration and fluorescence intensity, respectively) with the correlation coefficient of 0.9948. Furthermore, the calculated detection limit (LOD) for OTA is 4.9 pM (according to 3σ/S rule). When the OTA concentration of the detected solution is above the linear range (i.e., 0.5 nM), by simply diluting the actual samples to a calculable concentration with buffer solution, the target concentration of OTA can be estimated quantitatively according to the multiple of dilution.

Compared with other proposed strategies for OTA detection (Table 1), although our method is not as sensitive as electrochemical and immunofluorescence assays, the platform is economical, convenient and fast, only takes one and a half hours from preparation to detection. In addition, this work uses non-enzyme and non-label strategies and has a lower detection limit compared with general colorimetry, fluorescence and chemiluminescence methods.

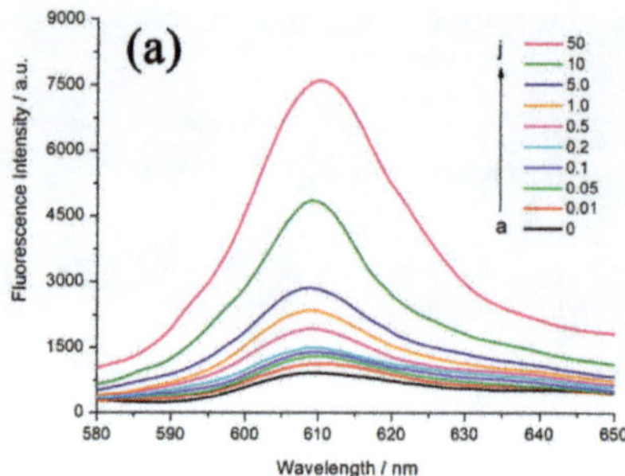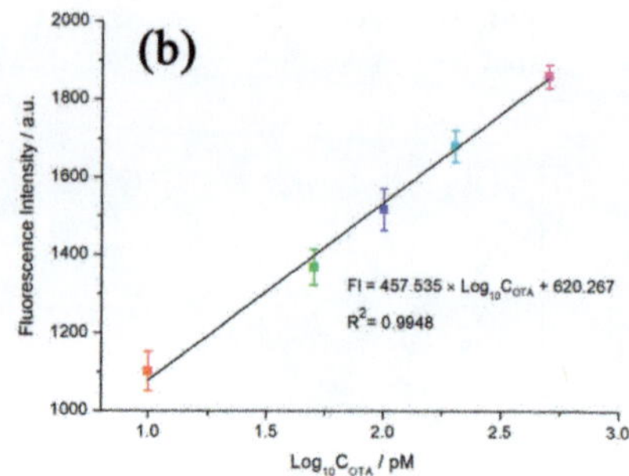

Figure 7. (**a**) Fluorescence spectra of different concentrations of OTA. From a to j, the concentrations of OTA is 0, 0.01, 0.05, 0.1, 0.2, 0.5, 1.0, 5.0, 10, 50 nM, respectively; (**b**) Linear relationship between the fluorescence value of (**a**) at 608 nm versus logarithmic concentration of OTA. Error bars, SD, *n* = 3.

Table 1. Comparison of proposed OTA detection strategies and this work.

Detection Method	Matrix	LOD	References
colorimetric	peanuts, corn	74.3 pM	[21]
fluorescence	bear, red wine	4.2 nM	[23]
fluorescence	red wine	198.1 pM	[24]
electrochemical	soybean	5.2 fg mL^{-1}	[26]
immunofluorescence	corn, rice, wheat	0.12 pM	[50]
fluorescence	corn flour	30 pM	[51]
chemiluminescence	wheat, rice, core	10.6 pM	[52]
chemiluminescence	coffee	0.5 nM	[53]
electrochemical	coffee	0.125 ng mL^{-1}	[54]
fluorescence	wheat flour, red wine	4.9 pM	this work

In order to verify the specificity of the developed sensor, other mycotoxins (OTB and AFB1) were also tested under the same conditions. According to Figure 8, even if there were other mycotoxins with concentrations ten times higher than OTA, negligible changes in fluorescence intensity could be observed. Nevertheless, the presence of OTA caused significant increase in fluorescence intensity. In addition, the fluorescence signal of the mixture of OTA and other control mycotoxins was similar to that of the OTA group. The results suggest that this method possesses an outstanding specificity of OTA detection.

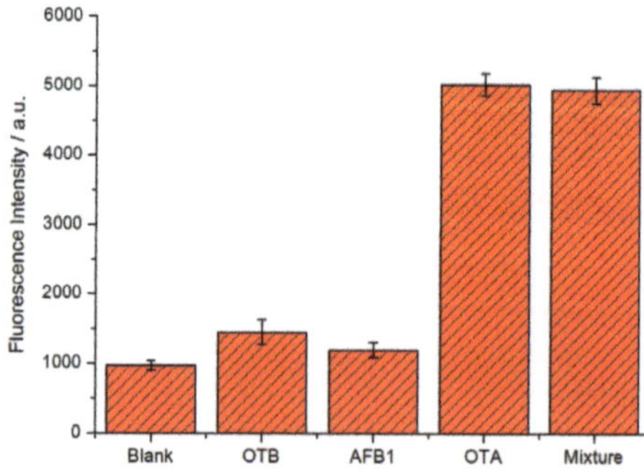

Figure 8. Specificity of the fluorescent biosensor to OTA other mycotoxins and the mixture of OTA, OTB and AFB1. The concentration of OTA is 10 nM and other mycotoxins are 100 nM. Error bars, SD, *n* = 3.

2.5. Application in Practical Samples

Subsequently, the practical application potential of the fluorescent aptasensor was evaluated by adding three different concentrations of OTA in actual wheat flour samples through standard addition methods to determine recovery rates. Table 2 shows that the recoveries determined in wheat flour samples were between 97.9% and 105%, and the relative standard deviation (RSD) was lower than ±4.8%. In addition, the detected recovery of three wine samples was higher than 94% and the RSD in the range from 3.7% to 5.1%. The experimental results suggest that our approach may be an effective and convenient method for OTA detection in actual agricultural commodities, and may provide promising strategies for improving food quality and safety.

Table 2. Application of fluorescent aptamer sensor for OTA determination in wheat flour and red wine samples.

Samples		Added (pM)	Found (pM)	Recovery (%) [a]	RSD (%) [b]
	1	10	10.5	105	±4.1
Wheat flour	2	50	48.9	97.9	±3.3
	3	100	104.8	104.8	±4.8
	1	10	9.4	94	±3.7
Red wine	2	50	48.4	96.8	±5.1
	3	100	95.7	95.7	±4.9

[a] The mean of three measurements. [b] RSD = The relative standard deviation.

3. Conclusions

In short, combining with the target-triggered structure-switching signaling aptamer and HCR technology, we have proposed a non-enzyme and non-label fluorescence biosensing system for OTA detection, proved by computer simulations and biological experiments. The approach we developed exhibits several advantages. First, the biosensor model has specific recognition and awesome detection ability for OTA with no modification of fluorescent groups and quenching groups, making the experiments more economical and unsophisticated. Second, by conducting computer simulations to verify and optimize the experimental process, making the subsequent biological experiments more facile and effective. Third, using HCR for signal amplification rather than other proteases and complex thermal cycling processes makes the operation more convenient and controllable with an OTA detection down to 4.9 pM. Furthermore, the detection and analysis of other interfering mycotoxins and actual samples showed that the proposed sensor system possesses high specificity and practical application potential. Finally, the strategy has good universal adaptability for detecting other small molecules and proteins by skillfully designing aptamer sequence of the hairpin probe. In summary, due to its simple design and operation, high sensitivity and specificity, and low cost, we anticipate that the platform could open up new opportunities for the detection of mycotoxins and contaminants in foodstuffs, and provide new ideas in future research of the interdisciplinary discipline of biology and computers.

4. Materials and Methods

4.1. Reagents

OTA, ochratoxin B (OTB), ochratoxin C (OTC) and aflatoxin B1 (AFB1) were purchased from Pribolab Co., Ltd. (Qingdao, China). N-methyl-mesoporphyrin IX (NMM) was bought from JKchemical Co., Ltd. (Beijing, China), stored at −20 °C in the dark before use. The ssDNA used in the experiments were synthesized and further purified by Sangon Biotech Co., Ltd. (Shanghai, China). The sequence of ssDNA (H1) is 5′-GAT CGG GTG TGG GTG GCG TAA AGG GAG CAT CGG ACA CGC CAC CCA CAC-3′, where the italic sequences are OTA aptamer. The sequence of ssDNA (H2) is 5′-CCA CAC CCG ATC CTG GGA GGG AGG GAG GGG TGT GGG TGG CG-3′, and some of which can form

G-quadruplex structure. By dissolving the DNA strands in 10 mM Tris buffer (200 mM NaCl, 20 mM MgCl$_2$, pH 7.4) to obtain the DNA stock solution, and then stored at −4 °C before use. The ultra pure water used in our experiment was purified by a Milli-Q system (18 MΩ cm). 96-Well microplates were purchased from Lingyi Biotech Co., Ltd. (Shanghai, China).

4.2. Instrumentation

The fluorescence spectrum of NMM was gauged by fluorescence scanning spectrometer under 399 nm excitation and 610 nm emission through EnSpire ELIASA from PerkinElmer USA company (Shanghai, China).

4.3. Fluorescent Detection of OTA

Before the experiments, the diluted solution with H1 (1 μM) and H2 (1 μM) were heated at 95 °C by PCR for 5 min, then cooled to room temperature slowly at a rate of 1 °C/min to form a hairpin structure. Subsequently, different concentrations of OTA, H1 (300 nM), KCl (20 mM) were mixed with working buffer and heated at 37 °C for 15 min. H2 (300 nM) was then put in with a pipette and the reaction was continued at 37 °C for 60 min. After self-assembly was completed, NMM (1.5 μM) was added and further incubated for 10 min at 25 °C. Finally, a part of the solution was removed to a 96-well plate, and the fluorescence intensity was detected and recorded by the instrument mentioned above.

4.4. OTA Detection in Wheat Flour and Red Wine Samples

Wheat flour and red wine samples were purchased from local supermarkets. Mix aliquots of wheat flour (1 g), different concentrations of OTA and 10 mL of extraction solvent (methanol: water, 6:4 (v/v)) in a vortex mixer for 5 min. After centrifuging for 10 min, the supernatant was removed and diluted with buffer, and then used for detection. At the same time, by adding different concentrations of OTA to the solution containing 1% red wine to prepare three samples with OTA concentration of 10 pM, 50 pM, and 200 pM respectively. All samples were then detected and analyzed as described above.

Author Contributions: M.Q. and W.H. designed biological experiments; M.Q. performed the experimental work, analyzed test data and drafted the manuscript; L.W. conducted the experiments and provided technical support; Y.W. contributed to the discussion of the algorithm and simulation software; Y.D. answered doubts during research and commented on the manuscript. All authors have read and agreed to the published version of the manuscript.

Funding: This work was funded by the Natural Science Foundation of Shaanxi Province (No. 2020JM-298).

References

1. Pfohl-Leszkowicz, A.; Manderville, R.A. Ochratoxin A: An overview on toxicity and carcinogenicity in animals and humans. *Mol. Nutr. Food Res.* **2007**, *51*, 61–99. [CrossRef] [PubMed]
2. Hussein, H.S.; Brasel, J.M. Toxicity, metabolism, and impact of mycotoxins on humans and animals. *Toxicology* **2001**, *167*, 101–134. [CrossRef]
3. Lei, L.; Cheng, B.C.; Cheng, Y.L.; Wu, R.Q.; Wang, S.H.; Guo, Z.J. Aptamer-based single-walled carbon nanohorn sensors for ochratoxin A detection. *Food Control* **2016**, *60*, 296–301.
4. Varga, J.; Kozakiewicz, Z. Ochratoxin A in grapes and grape-derived products. *Trends Food Sci. Technol.* **2006**, *17*, 72–81. [CrossRef]
5. Malir, F.; Ostry, V.; Novotna, E. Toxicity of the mycotoxin ochratoxin A in the light of recent data. *Toxin Rev.* **2013**, *32*, 19–33. [CrossRef]
6. Duarte, S.C.; Pena, A.; Lino, C.M. A review on ochratoxin A occurrence and effects of processing of cereal and cereal derived food products. *Food Microbiol.* **2010**, *27*, 187–198. [CrossRef]
7. European Commission. Commission regulation (EC) no 1881/2006 of 19 December 2006 setting maximum levels for certain contaminants in foodstuffs. *Off. J. Eur. Union* **2006**, *L364*, 5–24.

8.   Abrunhosa, L.; Paterson, R.R.M.; Venâncio, A. Biodegradation of ochratoxin A for food and feed decontamination. *Toxins* **2010**, *2*, 1078–1099. [CrossRef]

9.   O'Brien, E.; Dietrich, D.R. Ochratoxin A: The continuing enigma. *Crit. Rev. Toxicol.* **2005**, *35*, 33–60. [CrossRef]

10.   Pittet, A.; Royer, D. Rapid, low cost thin-layer chromatographic screening method for the detection of ochratoxin A in green coffee at a control level of 10 µg/kg. *J. Agric. Food Chem.* **2002**, *50*, 243–247. [CrossRef]

11.   Zhu, W.Y.; Ren, C.; Nie, Y.; Xu, Y. Quantification of ochratoxin A in Chinese liquors by a new solid-phase extraction clean-up combined with HPLC-FLD method. *Food Control* **2016**, *64*, 37–44. [CrossRef]

12.   Mariño-Repizo, L.; Goicoechea, H.; Raba, J.; Cerutti, S. A simple, rapid and novel method based on salting-out assisted liquid–liquid extraction for ochratoxin A determination in beer samples prior to ultra-high-performance liquid chromatography coupled to tandem mass spectrometry. *Food Addit. Contam. A* **2018**, *35*, 1622–1632. [CrossRef] [PubMed]

13.   Ali, M.; Yeun-Mun, C.; Guan, H.T.; Lukman, B.A. Determination of mycotoxins using hollow fiber dispersive liquid–liquid–microextraction (HF-DLLME) prior to high-performance liquid chromatography – tandem mass spectrometry (HPLC-MS/MS). *Anal. Lett.* **2019**, *52*, 1976–1990.

14.   Olsson, J.; Börjesson, T.; Lundstedt, T.; Schnürer, J. Detection and quantification of ochratoxin A and deoxynivalenol in barley grains by GC-MS and electronic nose. *Int. J. Food Microbiol.* **2002**, *72*, 203–214. [CrossRef]

15.   Al Rubaye, A.G.; Nabok, A.; Catanante, G.; Marty, J.L.; Takacs, E.; Szekacs, A. Detection of ochratoxin A in aptamer assay using total internal reflection ellipsometry. *Sens. Actuators B Chem.* **2018**, *263*, 248–251. [CrossRef]

16.   Al Rubaye, A.G.; Nabok, A.; Catanante, G.; Marty, J.L.; Takacs, E.; Szekacs, A. Label-free optical detection of mycotoxins using specific aptamers immobilized on gold nanostructures. *Toxins* **2018**, *10*, 291. [CrossRef]

17.   Radoi, A.; Dumitru, L.; Barthelmebs, L.; Marty, J.L. Ochratoxin A in some French wines: Application of a direct competitive ELISA based on an OTA-HRP conjugate. *Anal. Lett.* **2009**, *42*, 1187–1202. [CrossRef]

18.   Wei, T.; Ren, P.P.; Huang, L.L.; Ouyang, Z.C.; Wang, Z.Y. Simultaneous detection of aflatoxin B1, ochratoxin A, zearalenone and deoxynivalenol in corn and wheat using surface plasmon resonance. *Food Chem.* **2019**, *300*, 125176. [CrossRef]

19.   Zhu, Z.L.; Feng, M.X.; Zuo, L.M.; Zhu, Z.T.; Wang, F.W.; Chen, L. An aptamer based surface plasmon resonance biosensor for the detection of ochratoxin A in wine and peanut oil. *Biosens. Bioelectron.* **2015**, *65*, 320–326. [CrossRef]

20.   Prieto-Simon, B.; Campas, M.; Marty, J.L.; Noguer, T. Novel highly-performing immunosensor-based strategy for ochratoxin A detection in wine samples. *Biosens. Bioelectron.* **2008**, *23*, 995–1002. [CrossRef]

21.   Ellington, A.D.; Szostak, J.W. In Vitro selection of RNA molecules that bind specific ligands. *Nature* **1990**, *346*, 818–822. [CrossRef] [PubMed]

22.   Lorenz, C.; Renée, S. *The Aptamer Handbook: Functional Oligonucleotides and Their Applications*, 2nd ed.; Wiley–VCH: Weinheim, Germany, 2006; pp. 94–190.

23.   Cruz-Aguado, J.A.; Penner, G. Determination of ochratoxin A with a DNA aptamer. *J. Agric. Food Chem.* **2008**, *56*, 10456–10461. [CrossRef] [PubMed]

24.   Wang, C.Q.; Qian, J.; Wang, K.; Yang, X.W.; Liu, Q.; Hao, N.; Wang, C.K.; Dong, X.Y.; Huang, X.Y. Colorimetric aptasensing of ochratoxin A using Au@Fe3O4 nanoparticles as signal indicator and magnetic separator. *Biosens. Bioelectron.* **2015**, *77*, 1183–1191. [CrossRef] [PubMed]

25.   Goud, K.Y.; Reddy, K.K.; Satyanarayana, M.; Kummari, S.; Gobi, K.V. A review on recent developments in optical and electrochemical aptamer-based assays for mycotoxins using advanced nanomaterials. *Microchim. Acta* **2019**, *187*, 146–158. [CrossRef]

26.   Wu, H.L.; Kang, R.J.; Liang, X.J.; Cheng, Y.; Lv, L.; Guo, Z.J. Fluorometric aptamer assay for ochratoxin A based on the use of single walled carbon nanohorns and exonuclease III-aided amplification. *Mirochim. Acta* **2017**, *185*, 27–33. [CrossRef]

27.   Ma, C.B.; Wu, K.F.; Zhao, H.; Liu, H.S.; Wang, K.M.; Xia, K. Fluorometric aptamer-based determination of ochratoxin A based on the use of graphene oxide and RNase H-aided amplification. *Microchim. Acta* **2018**, *185*, 347–353. [CrossRef]

28.   Gu, C.; Yang, L.; Wang, M.; Zhou, N.; He, L.; Zhang, Z.; Du, M. A bimetallic (Cu-Co) Prussian Blue analogue loaded with gold nanoparticles for impedimetric aptasensing of ochratoxin A. *Microchim. Acta* **2019**, *186*, 1–10. [CrossRef]

29.   Wei, M.; Zhang, W.Y. A novel impedimetric aptasensor based on AuNPs–carboxylic porous carbon for the ultrasensitive detection of ochratoxin A. *RSC Adv.* **2017**, *7*, 28655–28660. [CrossRef]

30. Shao, X.L.; Zhu, L.J.; Feng, Y.X.; Zhang, Y.Z.; Luo, Y.B.; Huang, K.L.; Xu, W.T. Detachable nanoladders: A new method for signal identification and their application in the detection of ochratoxin A (OTA). *Anal. Chim. Acta* **2019**, *1087*, 113–120. [CrossRef]

31. Chen, J.H.; Fang, Z.Y.; Liu, J.; Zeng, L.W. A simple and rapid biosensor for ochratoxin A based on a structure-switching signaling aptamer. *Food Control* **2012**, *25*, 555–560. [CrossRef]

32. Fu, B.; Cao, J.C.; Jiang, W.; Wang, L. A novel enzyme-free and label-free fluorescence aptasensor for amplified detection of adenosine. *Biosens. Bioelectron.* **2013**, *44*, 52–56. [CrossRef] [PubMed]

33. Xu, S.P.; Chen, X.J.; Peng, G.; Ling, J.; He, H. An electrochemical biosensor for the detection of Pb^{2+} based on g-quadruplex dna and gold nanoparticles. *Anal. Bioanal. Chem.* **2018**, *410*, 5879–5887. [CrossRef] [PubMed]

34. Choi, M.S.; Yoon, M.; Baeg, J.O.; Kim, J. Label-free dual assay of DNA sequences and potassium ions using an aptamer probe and a molecular light switch complex. *Chem. Commun.* **2009**, *47*, 7419–7421. [CrossRef] [PubMed]

35. Dirks, R.M.; Pierce, N.A. From the cover: Triggered amplification by hybridization chain reaction. *Proc. Natl. Acad. Sci. USA* **2004**, *101*, 15275–15278. [CrossRef]

36. Ma, C.; Liu, H.Y.; Tian, T.; Son, X.R.; Yu, J.H.; Yan, M. A simple and rapid detection assay for peptides based on the specific recognition of aptamer and signal amplification of hybridization chain reaction. *Biosens. Bioelectron.* **2016**, *83*, 15–18. [CrossRef]

37. Shi, Z.l.; Zhang, X.F.; Cheng, R.; Li, B.X.; Jin, Y. Sensitive detection of intracellular RNA of human telomerase by using graphene oxide as a carrier to deliver the assembly element of hybridization chain reaction. *Analyst* **2016**, *141*, 2727–2732. [CrossRef]

38. Ren, D.X.; Sun, C.J.; Huang, Z.J.; Luo, Z.W.; Zhou, C.; Li, Y.X. A novel FRET biosensor based on four-way branch migration HCR for Vibrio parahaemolyticus detection. *Sens. Actuators B Chem.* **2019**, *296*, 126577. [CrossRef]

39. Sen, D.; Gilbert, W. A sodium-potassium switch in the formation of four-stranded G4-DNA. *Nature* **1990**, *344*, 410–414. [CrossRef]

40. Hu, D.; Pu, F.; Huang, Z.Z.; Ren, J.S.; Qu, X.G. A quadruplex-based, label-free, and real-time fluorescence assay for RNase H activity and inhibition. *Chem. Eur. J.* **2010**, *16*, 2605–2610. [CrossRef]

41. Guo, L.Q.; Nie, D.D.; Qiu, C.Y.; Zheng, Q.S.; Wu, H.Y.; Ye, P.R.; Hao, Y.L.; Fu, F.F. A G-quadruplex based label-free fluorescent biosensor for lead ion. *Biosens. Bioelectron.* **2012**, *35*, 123–127. [CrossRef]

42. Mccaskill, J.S. The equilibrium partition function and base pair binding probabilities for RNA secondary structure. *Biopolymers* **1990**, *29*, 1105–1119. [CrossRef] [PubMed]

43. Guo, Q.Q.; Chen, Y.; Song, Z.P.; Guo, L.Q.; Fu, F.F.; Chen, G.N. Label-free and enzyme-free sensitive fluorescent detection of human immunodeficiency virus deoxyribonucleic acid based on hybridization chain reaction. *Anal. Chim. Acta* **2014**, *852*, 244–249. [CrossRef] [PubMed]

44. Dirks, R.M.; Bois, J.S.; Schaeffer, J.M.; Winfree, E.; Pierce, N.A. Thermodynamic analysis of interacting nucleic acid strands. *SIAM Rev.* **2007**, *49*, 65–88. [CrossRef]

45. Waterman, M.S. Secondary Structure of Single-Stranded Nucleic Acids. In *Studies in Foundations and Combinatorics. Advances in Mathematics Supplementary Studies*; Rota, G.-C., Ed.; Academic Press: New York, NY, USA, 1978; Volume 1, pp. 167–212.

46. Mathews, D.H.; Sabina, J.; Zuker, M.; Turner, D.H. Expanded sequence dependence of thermodynamic parameters improves prediction of RNA secondary structure. *J. Mol. Biol.* **1999**, *288*, 911–940. [CrossRef] [PubMed]

47. Tinoco, I.; Uhlenbeck, O.C.; Levine, M.D. Estimation of secondary structure in ribonucleic acids. *Nature* **1971**, *230*, 362–367. [CrossRef]

48. Zadeh, J.N.; Steenberg, C.D.; Bois, J.S.; Wolfe, B.R.; Pierce, M.B.; Khan, A.R.; Dirks, R.M.; Pierce, N.A. Nupack: Analysis and design of nucleic acid systems. *J. Comput. Chem.* **2011**, *32*, 170–173. [CrossRef]

49. Lakin, M.R.; Youssef, S.; Polo, F.; Emmott, S.; Phillips, A. Visual DSD: A design and analysis tool for DNA strand displacement systems. *Bioinformatics* **2011**, *27*, 3211–3213. [CrossRef]

50. Huang, X.L.; Zhan, S.N.; Xu, H.Y.; Meng, X.W.; Xiong, Y.H.; Chen, X.Y. Ultrasensitive fluorescence immunoassay for detection of ochratoxin A using catalase-mediated fluorescence quenching of CdTe QDs. *Nanoscal* **2016**, *8*, 9390–9397. [CrossRef]

51. Liu, F.; Ding, A.; Zheng, J.S.; Chen, J.C.; Wang, B. A label-free aptasensor for ochratoxin A detection based on the structure switch of aptamer. *Sensors* **2018**, *18*, 1769. [CrossRef]

52. Shen, P.; Li, W.; Liu, Y.; Ding, Z.; Deng, Y.; Zhu, X.R.; Jin, Y.H.; Li, Y.C.; Li, J.L.; Zheng, T.S. High-throughput low background G-quadruplex aptamer chemiluminescence assay for ochratoxin a using a single photonic crystal microsphere. *Anal. Chem.* **2017**, *89*, 11862–11868. [CrossRef]

53. Jo, E.J.; Mun, H.; Kim, S.J.; Shim, W.B.; Kim, M.G. Detection of ochratoxin A (OTA) in coffee using chemiluminescence resonance energy transfer (CRET) aptasensor. *Food Chem.* **2016**, *194*, 1102–1107. [CrossRef] [PubMed]

54. Zejli, H.; Goud, K.Y.; Marty, J.L. Label free aptasensor for ochratoxin A detection using polythiophene-3-carboxylic acid. *Talanta* **2018**, *185*, 513–519. [CrossRef] [PubMed]

toxins

MDPI

Article

Improved Sample Selection and Preparation Methods for Sampling Plans Used to Facilitate Rapid and Reliable Estimation of Aflatoxin in Chicken Feed

James Kibugu [1,2,*], Raymond Mdachi [1], Leonard Munga [3], David Mburu [2], Thomas Whitaker [4], Thu P. Huynh [5], Delia Grace [6] and Johanna F. Lindahl [6,7,8]

1   Biotechnology Research Institute, Kenya Agricultural and Livestock Research Organization, P.O. Box 362, Kikuyu 00902, Kenya; rayelliemdachi@gmail.com
2   Department of Biochemistry, Microbiology and Biotechnology, School of Pure and Applied Sciences, Kenyatta University, P.O. Box 43844, Nairobi 00100, Kenya; nmburu01@gmail.com
3   Department of Animal Science, School of Agriculture and Enterprise Development, Kenyatta University, P.O. Box 43844, Nairobi 00100, Kenya; munga.leonard@ku.ac.ke
4   Department of Biological and Agricultural Engineering, North Carolina State University, Box 7625, Raleigh, NC 27695-7625, USA; whitaker@ncsu.edu
5   Hygiena LLC, Santa Ana, CA 92704-6804, USA; thuynh@hygiena.com
6   Department of Biosciences, International Livestock Research Institute, P.O. Box 30709, Nairobi 00100, Kenya; D.Randolph@cgiar.org (D.G.); J.Lindahl@cgiar.org (J.F.L.)
7   Department of Clinical Sciences, Swedish University of Agricultural Sciences, 75007 Uppsala, Sweden
8   Department of Medical Biochemistry and Microbiology, Uppsala University, 75123 Uppsala, Sweden
*   Correspondence: jkkibugu1@yahoo.com

Citation: Kibugu, J.; Mdachi, R.; Munga, L.; Mburu, D.; Whitaker, T.; Huynh, T.P.; Grace, D.; Lindahl, J.F. Improved Sample Selection and Preparation Methods for Sampling Plans Used to Facilitate Rapid and Reliable Estimation of Aflatoxin in Chicken Feed. *Toxins* **2021**, *13*, 216. https://doi.org/10.3390/toxins13030216

Received: 15 January 2021
Accepted: 10 March 2021
Published: 16 March 2021

Publisher's Note: MDPI stays neutral with regard to jurisdictional claims in published maps and institutional affiliations.

Abstract: Aflatoxin B1 (AFB1), a toxic fungal metabolite associated with human and animal diseases, is a natural contaminant encountered in agricultural commodities, food and feed. Heterogeneity of AFB1 makes risk estimation a challenge. To overcome this, novel sample selection, preparation and extraction steps were designed for representative sampling of chicken feed. Accuracy, precision, limits of detection and quantification, linearity, robustness and ruggedness were used as performance criteria to validate this modification and Horwitz function for evaluating precision. A modified sampling protocol that ensured representativeness is documented, including sample selection, sampling tools, random procedures, minimum size of field-collected aggregate samples (primary sampling), procedures for mass reduction to 2 kg laboratory (secondary sampling), 25 g test portion (tertiary sampling) and 1.3 g analytical samples (quaternary sampling). The improved coning and quartering procedure described herein (for secondary and tertiary sampling) has acceptable precision, with a Horwitz ratio (HorRat = 0.3) suitable for splitting of 25 g feed aliquots from laboratory samples (tertiary sampling). The water slurring innovation (quaternary sampling) increased aflatoxin extraction efficiency to 95.1% through reduction of both bias (−4.95) and variability of recovery (1.2–1.4) and improved both intra-laboratory precision (HorRat = 1.2–1.5) and within-laboratory reproducibility (HorRat = 0.9–1.3). Optimal extraction conditions are documented. The improved procedure showed satisfactory performance, good field applicability and reduced sample analysis turnaround time.

Keywords: aflatoxin; chicken feed; representative sampling; improved aflatoxin test procedure; validation

Key Contribution: This paper reports optimization and validation of processes employed to estimate aflatoxin levels in chicken feed. Modification of critical steps in sample collection and preparation segments of the aflatoxin test procedure enhanced the sampling representativeness necessary for rapid and accurate detection and estimation of dietary aflatoxin hazards.

1. Introduction

Aflatoxins are food-borne toxins produced by *Aspergillus* fungi sections *Flavi, Ochrace orosei* and *Nidulantes*. Some aflatoxin producing species are *A. flavus, A. parasiticus, A. nomius A. minisclerotigenes* and *A. arachidicola* whose aflatoxigenic strains are widespread in agricultural commodities, food and feed [1,2]. There are four types designated as aflatoxin B1, B2, G1 and G2 [3], found as natural dietary contaminants [4,5]. Metabolites such as aflatoxin M are found in edible animal products [5]. Aflatoxin B1 (AFB1), the most toxic and prevalent [2], is a potent human carcinogen [6]. Aflatoxins are moderately stable under normal cooking and industrial processing procedures [4,7,8]. There have been reports of acute human and animal aflatoxicosis outbreaks resulting in deaths [9,10] and widespread exposure to chronic dietary aflatoxins [2,11]. Prevalence data of aflatoxins contamination in poultry feeds are scanty particularly in low- and middle-income countries and characterized by wide variation. Levels of 36 ppb (mean) aflatoxin B1 was observed in Sudan [12], 100 ppb (mean) in India [13], 10–166 ppb aflatoxin in Pakistan [14], 74 ppb (mean) in Nigeria [15], 2.7 ppb (median) in Argentina [16] and 20–50 ppb in Kenya [17] Recently, aflatoxin levels of 7.5–393.5 ppb in feed processing plants samples and 19.0–188.5 ppb in samples collected from farmers in Uganda [18] and 0.2–318 ppb in 2020 in Kenya were reported [19]. Maximum allowable limits for aflatoxin content in human food and animal feed have been established in more than 100 countries [20]. For total aflatoxins, the United States set a maximum guidance level of 20–300 ppb in animal feed and 20 ppb in human food [21], while it is 4 ppb in human food as set by the European Union (EU) [22] Other than for dairy feed, AFB1 residues in animal feed are not usually regulated [20]. Uniquely, however, the EU has established a threshold for this mycotoxin in several animal feed matrices [23].

Chronic aflatoxicosis aggravates disease pathogenesis, impairs animal nutrition and productivity [6,24,25]. Aflatoxins are also teratogenic, carcinogenic, mutagenic, estrogenic, nephrotoxic, hepatotoxic and immunosuppressive [2,6,26–28]. Aflatoxins promote development of human primary hepatocellular carcinoma through synergy with the hepatitis B virus and has been associated with childhood stunting [29]. In chicken, dietary aflatoxins decrease feed intake and productivity and impair reproduction, causing economic losses, increased susceptibility to disease, poor vaccine response and toxin residues in poultry products [4,11,30]. In fact, dietary aflatoxin can reduce weight gain by 11% and increase mortality by 2.8% in chicken [31]. Aflatoxin contamination also causes food insecurity and economic impact through its adverse effect on international trade [8]. Dietary aflatoxin is therefore a public health concern of paramount importance that requires accurate estimation to enable employment of appropriate intervention strategies. Substantial efforts have been made to improve sensitivity and throughput of the analytical methods used for estimation of aflatoxin in food and feed. Thus, great achievements have been accomplished in the improvement of the analytical characteristics of the instrumental detection methods, e.g., liquid chromatography tandem mass spectrometry [4]. Nonetheless, for detection of trace levels of target analytes such as dietary aflatoxins, it is equally important that sample collection and preparation procedures are also optimized for accurate and rapid determination of the mycotoxin content.

Aflatoxin detection methods include immunoassays [16,32], fluorimetry [18] and chromatographic methods [14,32–34] such as LC-MS/MS for multi-mycotoxin analysis [15,19]. While these methods have different performance, the largest uncertainty associated with the measurement of aflatoxin content is due to lack of homogeneity of the contaminant in food and feed leading to variability [4,35]. It is indeed not easy to get a representative sample [11,36,37] that accurately estimates true aflatoxin content in a bulk consignment, as observed by Matumba et al. [38]. Another source of measurement uncertainty is bias, deviation from the true value due to sampling tools [39]. Aflatoxin analysis in food and feed is a three-step process: selection of the sample of a given size, sample preparation and quantification [4,40]. Development of chemical analysis often focuses on the last step, yet the sample selection step is the largest source of variability, followed by sample preparation, while quantification is the smallest contributor [35,37,39,41]. High variability necessitates increase in replicates to achieve required accuracy, thereby increasing the sample analysis turnaround time. There is need for test procedures with improved accuracy and precision for estimation of true aflatoxin exposure to ensure feed safety [36]. Recent data on aflatoxin contamination in figs [35] and maize [42] show that optimization of upstream procedures can considerably reduce the measurement uncertainty. In this study, sample selection, reduction and extraction steps were designed and validated to ensure representativeness of collected samples as well as appropriateness of the procedures and sampling tools used for estimation of aflatoxin residues in chicken feed. We first optimized sample selection procedures, and then incorporated a wet milling (water slurring) step in the feed sample preparation procedure, a critical modification that enhances sample homogenization more effectively than dry milling and a lesson learned from food analysis [35,39,41–43]. This reduces inter-assay variability and need for measurement replications thus decreasing sample analysis turnaround time. Because of lack of national, regional and international legal regulatory limits for AFB1 content in chicken feed, the EU legal framework was used as a reference in this study.

2. Results and Discussion

2.1. Highlights of Major Modification of the Improved Aflatoxin Test Procedure

The main modification in the five segments of the aflatoxin test procedure are as follows:

(a) Primary sampling or sample selection (number and size of incremental samples, type of sampling tools for open and closed sub-lots, random procedure, size of incremental and aggregate samples, Section 4.1.1)

(b) Secondary sampling (size of laboratory sample determined employing FAO Mycotoxin Sampling Tool, coning and quartering method improved by performing all coning and shoveling procedures under a steadfast funnel for mass-reduction of aggregate sample to 2 kg laboratory sample, Section 4.1.2)

(c) Tertiary sampling (size of test portion determined employing FAO Mycotoxin Sampling Tool, the improved coning and quartering method in Point (b) for mass-reduction of laboratory sample to 25 g test portion, Section 4.1.3)

(d) Quaternary sampling (homogenization and splitting of test portion by water slurring at matrix/water, 25:37.5, *w*/*w*; optimal matrix to organic solvent ratio for solid–liquid extraction, slurry/extraction solvent, 1.3:86.5, *w*/*v*, Section 4.1.4)

(e) Quantification of AFB1 (optimal organic solvent to aqueous buffer ratio for AFB1 extraction back to aqueous phase-modified extract to aqueous buffer mixture was modified to 80% acetonitrile extraction solvent: PBS-T mixture, 100:650, *v*/*v*, Section 4.1.5)

2.2. Enzyme-Linked Immunosorbent Assay of Prepared Standards for Determination of Aflatoxin Content in Chicken Feed Samples

Aflatoxin B1 cELISA, using AFB1-ELISA low matrix kit (Helica Biosystems Inc.®, Santa Ana, CA, USA) is the last segment of the improved aflatoxin test procedure (Section 4.1) and its details are in Section 4.1.5. Curve-fitting characteristics of this immunoassay are shown in Figure 1. The four-parameter logistic curve (4PLC) of spiked aflatoxin B1 (AFB1) concentrations was characterized by two plateau regions and an inflection point (Figure 1). This curve was used to study linearity of measurements of AFB1 spiked in the modified extract to aqueous buffer mixture, 80% acetonitrile extraction solvent: PBS-T mixture (100:650, *v*/*v*).

$$\text{Response (inhibition), } y = a - d/[1 + (x/c)b] + d \qquad (1)$$

where x = AFB1 concentration and a–d are described in Table 1.

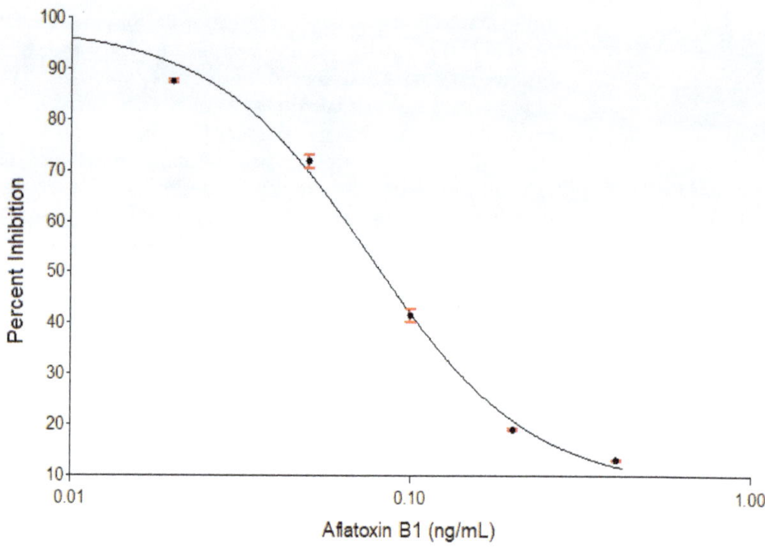

Figure 1. Four-parameter logistic standard curve of duplicate analysis of aflatoxin B1 standards in modified acetonitrile extract solvent: phosphate buffered saline-20 mixture (100:650, v/v). Extraction of aflatoxin B1 from organic into aqueous phase using the modified extraction solvent mixture did not affect the assay performance since the percent inhibition values were within the range specified by the manufacturer.

Table 1. Curve fitting characteristics for the AFB1-ELISA generated by AFB1 standards in modified acetonitrile extract solvent: phosphate buffered saline-20 mixture (100:650, v/v) showing both values of the parameters (a–d) and linearity-associated coefficients (r and r^2).

Parameters Values of 4-Parameter Logistic Curve					
Maximum signal intensity (a)	Slope at inflection point (b)	Concentration at inflection point ((50% B/B$_0$), IC$_{50}$ (c)	Minimum signal intensity (d)	Coefficients of correlation (r)	Coefficients of determination (r^2)
98.9	1.58	0.072 ng/mL	5.6	0.998	0.997

Numerous methods have been developed for analysis of aflatoxins in food and feed with high performance liquid chromatography (HPLC) as the gold standard. However, enzyme-linked immunosorbent assay (ELISA) has also been widely used given its many advantages over HPLC and other chromatographic techniques. ELISA is cost-effective because it does not need clean-up columns and expensive instrumentation, user-friendly, high-throughput, accurate and reproducible. One disadvantage of ELISAs is susceptibility to matrix effects, causing the reported analyte level to be falsely elevated or depressed. We used an ELISA method designed to be resistant to matrix interferences and which was previously validated to test disparate sample types, such as pet food [44], sorghum [45], maize [46], nuts, spices and many others [47]. Lack of requirement for sample clean-up and specialized equipment, skilled personnel for quantification by ELISA together with improved upstream sample handling procedures described herein, combined with high throughput (42 samples per run), guarantee rapid and reliable estimation of aflatoxin in complex and amorphous matrices such as animal feed. In addition, incorporation of a water slurring step in the extraction procedure reduces the number of test replications, considerably reducing sample analysis turnaround time.

2.3. Validation Results of the Improved Aflatoxin Test Procedure

Details of experimental design used for method validation are given in Section 4.2.2. Fourteen groups of native pseudo blank feed aliquots were used for method validation (Table 2). Briefly, Groups I–V were used to evaluate extraction efficiency and repeatability studies of surrogate aflatoxin (spiked analyte), Groups IV–XI were used for replication studies (within laboratory repeatability and reproducibility) of native aflatoxin (naturally occurring analyte) and Groups XII–XIV were used to evaluate limits of detection (LOD) and quantification (LOQ).

Table 2. Experimental design for recovery, replication and limits of detection (LOD) and quantification (LOQ) determination studies. Spike and recovery studies utilized surrogate aflatoxin (Groups I–V), whereas replication, LOD and LOQ studies used native aflatoxin (Groups VI–XIV) in feed aliquots.

Feed Aliquot Group	Extraction Conditions [ε]	Study
I (R = 6) *	(i) dry milling (ii) matrix/ACN [a] (25:133, w/v); (iii) ACN [a] extract/PBS-T [b] (10.6:989.4, v/v), (iv) final dilution factor = 500	Spike and recovery (comparison of dry and wet milling)
II (R = 6) *	(i) wet milling @ matrix/water (25:37.5, w/w); (ii) slurry/ACN [a], (2:133, w/v); (iii) ACN [a] extract/PBS-T [b] (100:650, v/v), (iv) final dilution factor = 1247	
III (R = 10) *	(i) wet milling @ matrix/water (25:37.5, w/w); (ii) slurry/ACN [a], (1.3:130, w/v); (iii) ACN [a] extract/PBS-T [b] (100:900, v/v), (iv) final dilution factor = 2500	Spike and recovery studies (comparison of three extraction conditions used in wet milling)
IV (R = 10) *	(i) wet milling @ matrix/water (25:37.5, w/w); (ii) slurry/ACN [a], (1.3:65, w/v); (iii) ACN [a] extract/PBS-T [b] (100:900, v/v), (iv) final dilution factor = 1250	
V (R = 10) *	(i) wet milling @ matrix/water (25:37.5, w/w); (ii) slurry/ACN [a], (1.3:86.5, w/v); (iii) ACN [a] extract/PBS-T [b] (100:650, v/v), (iv) final dilution factor = 1247	
VI (R = 10)	(i) dry milling (ii) matrix/ACN [a] (25:130, w/v); (iii) ACN [a] extract/PBS-T [b] (100:900, v/v), (iv) final dilution factor = 52	
VII (R = 10)	(i) wet milling @ matrix/water (25:37.5, w/w); (ii) slurry/ACN [a], (1.3:130, w/v); (iii) ACN [a] extract/PBS-T [b] (100:900, v/v), (iv) final dilution factor = 2500	Intra-laboratory variability (Repeatability)
VIII (R = 10)	(i) dry milling (ii) matrix/ACN [a] (1.3:130, w/v); (iii) ACN [a] extract/PBS-T [b] (100:900, v/v), (iv) final dilution factor = 1000	
IX (R = 10)	(i) wet milling @ matrix/water (25:37.5, w/w); (ii) slurry/ACN [a], (1.3:130, w/v); (iii) ACN [a] extract/PBS-T [b] (100:900, v/v) (iv) final dilution factor = 2500	
X (R = 15)	(i) wet milling @ matrix/water (25:37.5, w/w); (ii) slurry n/ACN [a], (1.3:130, w/v); (iii) ACN [a] extract: PBS-T [b] (100:900, v/v); (iv) final dilution factor = 2500	Intermediate variability (within-laboratory reproducibility)
XI (R = 15)	(i) wet milling @ matrix/water (25:37.5, w/w); (ii) slurry/ACN [a], (1.3:86.5, w/v); (iii) ACN [a] extract/PBS-T [b] (100:650, v/v) (iv) Final dilution factor = 1247	
XII (R = 4)	(i) dry milling (ii) ACN [a] extract/PBS-T [b] (1:1000, v/v) (iii) final dilution factor = 1000	
XIII (R = 5)	(i) dry milling ACN [a] extract/ACN [a]/PBS-T [b] (1:5:100, $v/v/v$) (ii) final dilution factor = 500	Determination of LOD and LOQ
XIV (R = 5)	(i) wet milling (ii) ACN [a] extract/PBS-T [b] (1:1247, v/v) (iii) final dilution factor = 1247	

[ε] Extraction conditions here mean sample homogenization, solid–liquid and liquid–liquid extraction solvent mixture components with Roman numbers (i–iii) standing for specific conditions whose details can be found in Sections 4.2.2–4.2.4; R is n = number of replicates; * number of "0" ppb feed aliquots not included in *n*.; [a] acetonitrile:water (80:20, v/v); [b] phosphate buffered saline tween 20.

We modified primary sampling procedures to improve representativeness and ensure sample integrity. Uncertainty associated with sample selection was minimized through careful calculation of incremental and laboratory samples' size using granulometry and particle size of matrix, together with increased number and size of test portions [39,40], correct design of sampling equipment to eliminate bias and random sampling procedures [48,49]. Specifically, the incremental samples were optimally spaced [48]. Sample size, sample selection and handling are usually dismissed as a "simple procedure", but they are major source of variation. Attention should be given to the sample selection process for a given sample size as described herein.

2.3.1. Method Accuracy and Precision

Extraction efficiency and variability associated with various aflatoxin extraction procedures for recovery studies are shown in Table 3 and raw data in Supplementary Materials (Dataset S1. Aflatoxin recovery data). Wet milling method (Group II) had mean aflatoxin recovery of 121%, which at $p = 0.05$ is not significantly different from mean recovery of 80% of conventional dry milling procedure (Group I). For estimation of spiked aflatoxin variability associated with dry milling was significantly higher compared to wet milling. Coefficient of variation (CV) also referred to as relative standard deviation (RSD) of dry milling procedure (Group I) was 2.3-fold compared to slurry, wet milling method (Group II). CV effect associated with Group II (wet milling procedure) was less (CV rank 2) compared to CV rank 4 of Group I (dry procedure) (Table 3). Multiple comparisons employing Welch ANOVA-associated Games–Howell post-hoc test showed that mean aflatoxin recovery associated with Group III procedure was significantly ($p < 0.05$) different from those of both Groups IV and V (Table 3). Of the three wet milling methods, variability (Observed CV) associated with Group V (CV rank 1) was remarkably low compared to Groups III (1.5-fold) and IV (3-fold) procedures. Additionally, Group V had the least bias and repeatability precision (HorRat value < 1). Wet milling procedure associated with Groups II and V had acceptable precision level (expressed as CV or RSD) as prescribed by modified Horwitz equation. Analyzed together, the extraction procedure associated with Groups II and V had the lowest percent bias of −4.95, acceptable variability and recovery (Table 3: HorRat value < 2; recovery = 95%).

Precision (variability) was evaluated using the Horwitz equation where the measure of variation, predicted relative standard deviation (RSD_p) or CV is a function of the analyte concentration [50–54]. This is given by modified and unmodified Horwitz equations:

$$\text{For modified Horwitz equation, } RSD_p < 2^{\,(1-0.5\log C)} \times 0.67 \tag{2}$$

$$\text{For unmodified Horwitz equation, } RSD_p < 2^{\,(1-0.5\log C)} \tag{3}$$

where C = AFB1 concentration.

The modified Horwitz equation was used to predict RSD under repeatability and routine inter-assay conditions while the unmodified form was used for within-laboratory reproducibility conditions (intermediate precision). The observed relative standard deviation (RSD_O) was compared with the RSD_p to give a Horwitz Ratio (HorRat) value, thus:

$$\text{HorRat value} = RSD_O / RSD_p \tag{4}$$

During method validation, the Commission of the European Communities (CEC) precision requirement for both repeatability and within-laboratory reproducibility conditions is a HorRat value < 2 [55]. However, for routine work (Sections 2.3.5 and 4.2.8), we adopted inter-assay precision level of HorRat ≤ 1 [56].

Variability associated with estimation of natural AFB1 in chicken feed employing two sample splitting techniques at two sampling stages are shown in Table 4 and raw data in supplementary material (Dataset S2. Precision data). HorRat values for repeatability ($HorRat_r$) and within-laboratory reproducibility ($HorRat_R$) ranged 0.3–5.8 and 0.9–1.3, respectively, with all groups but one having HorRat values below the maximum allowable limit of 2 set by European legislation [55]. The lowest intra-laboratory variation was observed in secondary sampling for splitting 25 g dry aliquots from the laboratory sample (Group VI; RSD_r = 6.5%, HorRat$_r$ value = 0.3; rank H1), and highest for preparing 1.3 g dry aliquots at tertiary sampling employing modified coning and quartering procedure (Group VIII; RSD_r = 91.5%, HorRat$_r$ value = 5.8; rank H6). For preparation of 1.3 g analytical samples, intra-laboratory variation associated with coning and quartering procedure (Group VIII) was 4.4-fold that for water slurry (wet milling) method (Group IX, RSD_r = 20.6%, rank H3). At tertiary sampling stage, intermediate precision (within-laboratory reproducibility) associated with Group X (RSD_R = 30.3%, HorRat$_R$ value = 1.3, rank H4) and Group XI (RSD_R = 26.9%, HorRat$_R$ value = 0.9, rank H2) of water slurry procedure were almost the same (HorRat value > 2), the latter having slightly lower variability. One-way ANOVA showed no significant effect ($p > 0.05$) of the analyst or the day of analysis on means of AFB1 levels for each condition of the two water slurry procedures.

The second most important source of variability after sample collection is the sample preparation segment of the aflatoxin test procedure [41,48]. Variability associated with splitting 25 g test portions from the comminuted laboratory sample employing modified coning and quartering procedure was much below the threshold level prescribed by European Union [52] and therefore suitable mass reduction method for this purpose. Density effects and matrix particle size influence performance of sample splitting methods [57]. We minimized this by efficient dry comminution of the aggregate sample prior to mass reduction [58]. Preparation of smaller size aliquots did not yield desirable precision under intra-laboratory conditions. Indeed, the FAO sampling tool [59] will not accept mass reduction in granular products beyond paired 25 g test portions for aflatoxin analysis because this will compromise representativeness. To enhance sampling precision, aggregate sample collected should not be less than 2 kg. Variability at this level can be reduced by increasing aggregate sample size before comminution through collection of 200 g incremental samples from all potential sampling locations. Removal of test portions larger than 25 g from a 2 kg (or larger) comminuted laboratory sample will reduce variability and can still be water slurried and a small (1.3 g) slurry aliquot selected for extraction. There are also commercial laboratory mills that incorporate a sample-splitting mechanism [4,41]. However, these are expensive and not readily available. Our novel aflatoxin test procedure is designed especially for laboratories that do not have automated sample splitting facilities.

Another critical innovation described here is inclusion of wet milling (water slurring), an additional comminuting step in the extraction procedure followed by processing of a smaller slurry aliquot. This allowed analysis of an adequately large test portion (25 g and larger), minimizing huge sampling uncertainty associated with aflatoxin estimation in animal feed and with reduced extraction cost. As reported in the literature, aflatoxin contamination is characterized by heterogeneous spatial distribution and nugget effect [37,60]. Wet milling is more efficient than dry milling in producing a more homogenous sample [35,39,43,61]. Indeed, water slurring was recently incorporated as a sample homogenizing procedure for aflatoxin analysis in maize [42] and dried figs [35]. This is the first report of using wet milling sample preparation method for aflatoxin analysis in animal feed. Through optimization of sample selection and mass reduction procedures and minimizing spatial heterogeneity of aflatoxin distribution in the test portion by wet milling, we were able to reasonably reduce measurement uncertainty and extraction cost. However, our modification does not completely eradicate inherent variability associated with sample selection, sample preparation and analytical steps of the aflatoxin test procedure, but minimizes this variability at each step, as well as reducing bias at both test portion selection and analytical segment. The aflatoxin diagnosis kit used in this study was not validated specifically for chicken feed. It is designed for various food matrices and animal feed grouped as one matrix. Because animal feed is an amorphous matrix with diverse physicochemical properties which can be a source of variation due to within-class matrix effects [62], we generated validation data associated with chicken feed. Single-laboratory validation of the modified aflatoxin test procedure was carried out through collection of data on spiking and recovery, replication, LOD and LOQ, robustness and ruggedness as the CEC prescribes [50–52,54,63].

Accuracy is a trade-off between bias and recovery data variability. We observed least bias (−4.95) and good aflatoxin recovery (95.1%) in wet milling procedure (Group. V). Recovery, precision and efficiency were compliant with CEC requirements of 75–125% for recovery and HorRat < 2 for precision [52,54,64]. AOAC and other authorities also recognize HorRat < 2 as a reliable precision criterion [64–66]. For replication studies using native aflatoxin, precision data collected under repeatability and reproducibility conditions met the EU guidelines. In absence of collaborative trial data, we estimated inter-laboratory precision using modified Horwitz equation against our within-laboratory reproducibility data, an internationally accepted practice [62]. Repeatability was both within the EU requirements and the same range for surrogate and native aflatoxin contents. Since native aflatoxin contamination is characterized by heterogeneity [4,35], we attribute the observed reduced variability to effective sample homogenization through the wet milling innovative procedure in our novel method described herein. By CEC requirements, the wet milling method described here has good repeatability and reproducibility. Reducing intrinsic variability associated with aflatoxin heterogeneity is cost effective in terms of time and resources [67].

Table 3. Extraction efficiency (percent recovery), bias and measurement precision of surrogate AFB1 content in chicken feed associated with dry and wet sample homogenization procedures at various extraction conditions (dilution factors) and ranked using HorRat* value effect.

Milling Method	Group	Expected Concentration	Size of Analytical Sample	DF of Water Slurry	DF in ACN *	DF in PBS-T [b]	Final DF	* Mean AFB1 % Recovery	% Bias	StDev.	CV	CV Effect	HorRat ** Value
Dry	I (R = 6)	20 ppb (R = 3)	25 g	-	5.3	94.3	500	57		±6	10.9		0.57
		100 ppb (R = 3)						103		±13	12.4		0.82
		All replicates						80 [a]	−20	±26	33.1	9.5 [R4]	0.57–0.82
Wet	II (R = 6)	20 ppb (R = 3)	2 g	2.5	66.5	7.5	1247	119		±25	21.2		0.95
		100 ppb (R = 3)						123		±9	7.4		0.8
		All replicates						121 [a]	21	±17	14.1 [h]	−9.5 [R2]	0.8–0.95
Wet	III (R = 10)	20 ppb (R = 5)	1.3 g	2.5	100	10	2500	145		±19	12.8		0.58
		100 ppb (R = 5)						112		±9	17.4		0.44
		All replicates						129 [b,c]	29	±22	17.2	−3.8 [R3]	0.44–0.58
	IV (R = 10)	20 ppb (R = 5)	1.3 g	2.5	50	10	1250	49		±12	23.2		1.04
		100 ppb (R = 5)						85		±15	18.1		1.04
		All replicates						67 [b]	−33	±23	34.2	13.3 [R5]	1.04
	V (R = 10)	20 ppb (R = 5)	1.3 g	2.5	66.5	7.5	1247	80		±13	15.8		0.71
		100 ppb (R = 5)						80		±6	6.9		0.39
		All replicates						80 [c]	−20	±9	11.5	−9.5 [R1]	0.39–0.71
Groups II and V replicates analyzed together (R = 16)		20 ppb (R = 8)	-	2.5	66.5	7.5	1247	94		±26	27.6		1.24
		100 ppb (R = 8)						96		±23	24.1		1.38
		All replicates						95	−5	±24	25.0		1.24–1.38

** Observed residue standard deviation/predicted residue standard deviation; R is n = number of replicates; * acetonitrile:water (80:20, v/v), [a] figures marked with this superscript were statistically compared by independent *t*-test; [b] phosphate buffered saline tween 20; DF, Dilution factor; StDev, Standard deviation; CV, Coefficient of variation (relative standard deviation); [b,c] figures marked with the same superscript were statistically compared by Welch's ANOVA (Games-Howel post-hoc test); [R1−5] method ranking based on CV effect with descending order of preference; * 75–125% mean recovery is the accepted recovery range for trace AFB1 levels [52].

Table 4. Within laboratory repeatability and reproducibility associated with measurement of native aflatoxin content in chicken feed at secondary and tertiary sampling stages after sample homogenization employing by either dry or wet (water slurring) milling techniques. Variability values were determined under different extraction conditions and ranked using HorRat * value effect.

| | Intra-Laboratory Assay Precision (Within-Laboratory Repeatability) | | | | | | Intermediate Precision (Within-Laboratory Reproducibility) | | | | | |
| Sampling Stage | Group Number/ Mass Reduction Method | Milling Method and Size of Analytical Sample | EC | AFB1 Level (ppb) Mean ± Sd. | Variability | | Group Number/ Mass Reduction Method | Milling Method and Size of Analytical Sample | EC | AFB1 Level (ppb) Mean ± Sd | Variability | |
					RSD$_r$	HorRat ValueEffect					RSD$_R$	HorRat Value Effect
Secondary	Group VI Coning and quartering (R = 10)	Dry 25 g	1	15 ± 1	6.5	0.3 [H1]				-		
Tertiary	Group VII Water slurry (R = 10)	Wet 1.3 g	2	68 ± 17	24.5	1.5 [H5]	Group X Water slurry (R = 15)	Wet 1.3 g	2	69 ± 21	30.3	1.3 [H4]
Tertiary	Group VIII Coning and quartering (R = 10)	Dry 1.3 g	3	73 ± 67	91.5	5.8 [H6]	Group XI Water slurry (R = 15)	Wet 1.3 g	4	19 ± 5	26.9	0.9 [H2]
Tertiary	Group IX Water slurry (R = 10)	Wet 1.3 g	2	50 ± 10	20.6	1.2 [H3]			-			

* Observed residue standard deviation/predicted residue standard deviation; R is n = number of replicates; EC (Extraction conditions) 1 enumerated as i–iv: (i) dry milling (ii) matrix/ 80% acetonitrile (25:130, *w/v*); (iii) 80% acetonitrile extract/ phosphate buffered saline tween 20 (100:900, *v/v*); (iv) final dilution factor = 52: EC 2 enumerated as i–iv: (i) wet milling with matrix/water (25:37.5, *w/w*); (ii) slurry/80% acetonitrile (1.3:130, *w/v*); (iii) 80% acetonitrile extract/ phosphate buffered saline tween 20 (100:900, *v/v*); (iv) final dilution factor = 2500: EC 3 enumerated as i-iv: (i) dry milling; (ii) matrix/ 80% acetonitrile (1.3:130, *w/v*); (iii) 80% acetonitrile extract/ phosphate buffered saline tween 20 (100:900, *v/v*); (iv) final dilution factor = 1000: EC 4 enumerated as i–iv: (i) wet milling with matrix/water (25:37.5, *w/w*); (ii) slurry/80% acetonitrile (1.3:86.5, *w/v*); (iii) 80% acetonitrile extract/ phosphate buffered saline tween 20 (100:650, *v/v*); (iv) final dilution factor = 1247: Sd, Standard deviation; RSD$_r$, within-laboratory repeatability relative standard deviation; RSD$_R$, within-laboratory reproducibility relative standard deviation; [H1-6] method ranking based on HorRat value effect with descending order of preference.

2.3.2. Limits of Detection (LOD) and Quantification (LOQ) Values

LOD and LOQ values associated with different sample preparation procedures are shown in Table 5 and raw data in Supplementary Materials (Dataset S3. LOD and LOQ data). The wet milling (slurry) procedure exhibited the strongest and most stable signals with the highest signal-to-noise ratio. These conditions yielded the lowest LOD (7.5 ppb) and the only value below the EU legal limit for AFB1 residues in animal feed of 10 ppb [23]. LOD and LOQ values associated with the slurry procedure were 7.5 and 16.0 ppb, respectively.

According to EU requirement, the maximum allowable AFB1 content in complete feed is 50 ppb (adult cattle, sheep and goats), 20 ppb (adult pigs and poultry) and 10 ppb (young animals) [23]. The LOD of our method is 7.5 ppb, well below the lowest regulatory limit. Other quantification methods can be incorporated with an obvious advantage of lowering LOD if so desired. Alternatively, levels below LOD can be estimated by extrapolation using a company program [68], substituting the values with the LOD divided by the square root of two [69] or replacing by half the LOD value if the below LOD results are less than 60% of the data [70,71], the latter being an approach used by several workers [72–75].

Table 5. Limits of detection (LOD) and quantification (LOQ) values associated with dry and wet (water slurring) sample homogenization procedures for chicken feed at different extraction conditions (dilution factors). The LOD and LOQ values were derived from signal and noise of baseline native aflatoxin content expressed as absorbance.

Baseline Response (OD) Statistics	Limits of Detection and Quantification in µg AFB1 Per kg of Chicken Feed		
	Dry Milling		Wet Milling
	Group XII (R = 4) (Final Dilution Factor = 1000)	Group XIII (R = 5) (Final Dilution Factor = 500)	Group XIV (R = 5) (Final Dilution Factor = 1247)
Mean-2StDev (Limit of detection)	16.3	10.6	7.5
Mean-5StDev (Limit of quantification)	31	22.3	16
Baseline response (OD) statistics	Magnitude of signal, noise, their ratio and precision (absorbance)		
	Group XII	Group XIII	Group XIV
B_0	1.773	1.752	1.871
StDev (blank)	0.0399	0.0485	0.0339
Signal/noise ratio	44.44	36.12	55.19
Coefficient of variance	2.34	2.99	1.97

R is n = number of replicates; OD, optical density; B_0, average OD of the 0 standard; Mean, average of B_0 OD values; StDev, standard deviation of OD values of blank feed material.

2.3.3. Linearity

Expected and observed AFB1 concentration values of the prepared standards are shown in Table 6. There was significant ($p < 0.01$) negative correlation between ODs and measured aflatoxin levels of the prepared standards (Pearson correlation coefficient, r = −0.932). The expected and observed aflatoxin levels of the standards had a significant ($p < 0.01$) positive correlation (Pearson correlation coefficient, r = 0.961). The linear response range of the assay is 0.02–0.4 ng/mL, beyond which linearity was lost (Table 6) These results and curve-fitting characteristics (Figure 1) indicate that the modified sample preparation conditions did not affect the assay range. Values of coefficients of correlation and determination (Section 2.2) showed good linearity of measured inhibition and AFB1 concentration in the range of 0.02–0.4 ng/mL.

2.3.4. Robustness and Ruggedness of the Aflatoxin Extraction Procedure

For robustness data, Groups II and V (slurry extraction method at final dilution factor of 1247) had HorRat values of less than 2 (Table 3), indicating stable extraction efficiency precision of the improved aflatoxin test procedure under intra-laboratory repeatability conditions suggesting more satisfactory robustness compared to the other extraction conditions (see raw data at Dataset S1. Aflatoxin recovery data of supplementary files). For ruggedness results, HorRat value for Group X (final dilution factor = 2500) was greater than 1 but less than 2, while, for Group XI (final dilution factor = 1247), it was less than 1 (Table 4), indicating more stable precision under reproducibility conditions for Group XI (see raw data at Dataset S2. Precision data of supplementary files). This suggests that the improved aflatoxin test procedure has more satisfactory ruggedness (Group XI) compared to the other extraction conditions. Efficiency of extraction solvent stored at ambient temperature were 79%, 80% and 74%, respectively (see raw data at Dataset S4. Robustness-ruggedness data of supplementary files), for freshly prepared, one- and three-month-old 80% acetonitrile:water (80:20, v/v). The HorRat value associated with these data was 0.22, suggesting minimal variability indicating stability of the solvent as an extraction solvent for AFB1 at ambient temperature for three months and satisfactory method ruggedness.

Table 6. Linearity of aflatoxin B1 standard solutions in modified acetonitrile extract solvent: phosphate buffered saline tween 20 mixture (100:650, *v/v*). The values are mean of duplicate analysis.

	Concentration and Optical Density of AFB1 Standard Solutions					
	Solution 1	Solution 2	Solution 3	Solution 4	Solution 5	Solution 6
Spiked AFB1 level (ng/mL)	0.02	0.05	0.1	0.2	0.4	0.5
Measured level (ng/mL)	0.03	0.06	0.12	0.28	0.36	>0.4
Measured optical density	1.429	1.106	0.663	0.297	0.226	0.199

These data, taken together, suggest that the wet milling procedure (slurry method associated with final dilution factor of 1247) had satisfactory robustness as demonstrated by suitable repeatability of extraction efficiency, while stability of extraction solvent in storage at ambient temperature for three months and good within-laboratory reproducibility indicated its acceptable ruggedness [52,54,76]. Good robustness and ruggedness indicate satisfactory stability of our improved aflatoxin test procedure.

2.3.5. Evaluation of the Improved Aflatoxin Test Procedure

The number of replicates of individual samples processed to achieve acceptable level of inter-assay precision (i.e., HorRat $\leq$ 1) during analysis of 251 field-collected feed samples (routine analysis) is shown in Table 7. A left-skewed one-tailed distribution was observed, with the majority of the samples (75%) adequately analyzed using paired test portions (recommended minimum number) with a further 20% requiring a third run, all these totaling to 95% of the samples. This indicates reduced cost due to sample re-testing in terms of resources and time, decreasing sample analysis turnaround time. Table 8 compares the characteristics of the novel aflatoxin testing procedure with those of other aflatoxin testing protocols recently used to collect aflatoxin residues data in chicken feed. Due to large uncertainty associated with estimation of dietary aflatoxin, FAO [59] has a sampling tool to guide workers on the size of laboratory and test portion samples. For granular products such as finished commercial animal feed, an aggregate sample of more than 2 kg, laboratory sample of not less than 2 kg and test portion sample of at least 50 g (2 × 25 g) have to be processed to reduce results variability. Our modified aflatoxin test adheres to these criteria, while the other published methods fail to meet the appropriate aggregate sample, laboratory sample and/or test portion size (Table 8). Additionally, all of the published methods employed dry milling method for sample homogenization. Because solid–liquid extraction procedures are expensive, workers are tempted to use small test portion samples (which increase inherent variation). Our improved test incorporates a wet milling step (water slurring) in the extraction procedure. This allows processing of the recommended size of the test portion sample of 50 g. Since wet milling is far more effective in sample homogenizing compared to dry milling, a small aliquot of homogenized slurry can be analyzed with minimal inherent variability [35,39,43,61]. For routine analysis, a minimum of two replicates was performed to achieve the required level of inter-assay precision of HorRat $\leq$ 1 and replicates were increased until this was accomplished (Section 4.2.8).

Table 7. Number of replicates (25 g test portions) required to attain acceptable level of precision during routine analysis of aflatoxin B1 in chicken feed employing the improved test procedure. Percentage of samples analyzed is included in brackets. Paired test portion (duplicate analysis) is the minimum number of replicates recommended by FAO Mycotoxin Sampling tool.

	Number of Replicates Required to Attain Acceptable Intra-Assay Precision				
	Two Replicates Required	**Three Replicates Required**	**Four Replicates Required**	**Five Replicates Required**	**Six Replicates Required**
Number and percentage of samples analyzed	188 (74.9%)	51 (20.3%)	8 (3.2%)	2 (0.8%)	2 (0.8%)

Table 8. Sample homogenization techniques, size of aggregate, laboratory and test portion samples of various aflatoxin analysis protocols compared to the modified (novel) test procedure. This highlights compliance of our modified to FAO criteria for aflatoxin analysis.

Study Reference	Characteristics of Aflatoxin Test Procedures				
	Homogenization Method	**Aggregate Sample (kg)**	**Laboratory Sample (kg)**	**Size of Test Portion**	**Analytical Method**
[15]	Dry milling	4	Not given	5 g	LC-MS/MS
[16]	Dry milling	1–2	1	5 g	ELISA
[19]	Dry milling	1	1	5 g	LC-MS/MS
[32]	Dry milling	Not given	Not given	Not given	HPLC/ELISA
[18]	Dry milling	Not given	Not given	50 g	VICAM Fluorimeter
[13]	Not given	Not given	Not given	Not given	Not given
[14]	Not given	1	Not given	Not given	TLC
Novel method	Wet milling (water slurring)	>2	2	at least 25 g × 2	ELISA or any other quantification method

3. Conclusions

The in-house validation data presented here show that the improved aflatoxin test procedure is suitable for estimation of aflatoxin contamination levels in chicken feed. The optimal aflatoxin B1 extraction conditions are wet milling (water slurring 25 g feed in 37.5 mL water), solid–liquid extraction 1.3 g slurry (0.52 matrix: 0.78 water, w/w) with acetonitrile:water (80:20, v/v) at dilution rate of slurry/acetonitrile solvent (1:66.5, w/v) followed by extraction back to aqueous phase at dilution rate of extract/phosphate buffered saline tween 20 (1:6.5, v/v). The improved aflatoxin test procedure is an accurate, precise, stable, reliable and cost-effective tool (in terms of time and resources) for surveillance of dietary aflatoxin. The improved aflatoxin test procedure described herein is especially suitable for laboratories that may not have access to automated sample-splitting equipment.

4. Materials and Methods

4.1. Description of Improved Aflatoxin Test Procedure

The improved aflatoxin test procedure for animal feed was designed using FAO mycotoxin sampling tool [40,77] with modification. Briefly, sample selection made from a stationary lot to get a representative aggregate sample (primary sampling) was representatively mass reduced to a 2 kg laboratory sample (secondary sampling) and then two 25 g test samples (tertiary sampling) selected from 2 kg laboratory sample. The test samples were then homogenized by slurring in water, a 1.3 g analytical portion was removed from the analytical sample slurry material (quaternary sampling) and aflatoxin extracted in an organic solvent. The analyte was then extracted back to aqueous phase, phosphate buffered saline tween 20 (PBST) prior to analysis for AFB1 by competitive ELISA.

4.1.1. Sample Selection (Primary Sampling)

The number of bags (sub-lots) sampled was determined as described earlier [49]. Incremental samples (200 g each) were collected by random sampling [49,78] from as many locations of the lot as possible [54], thoroughly mixed to make an aggregate sample (of at least 2 kg). If the number of containers (bags) was $\leq$10 bags, all were sampled and, if >10 bags, every 4th bag was sampled in every row. All units in the lot were made accessible. To collect incremental samples, a sampling cup was used for open containers (bags) after thorough mixing, while, for closed containers, a bag trier was used. Silica gel packs were added to the aggregate samples, double bagged in brown bags, transported to the laboratory and stored at 4 °C until required [54]. All aggregate samples should first be comminuted prior to mass reduction.

4.1.2. Comminution and Mass Reduction of Samples (Secondary Sampling)

Pellet and crumb aggregate feed samples (varying in size but at least 2 kg) were first comminuted in a laboratory grinder (Grindomix Retsch® Model Gm 200, Hann, Germany) at a rotary speed of 10 × 1000 RPM for 30 s before mass reduction to laboratory samples. This was carried out in 330–350 g aggregate sample portions to protect the grinder from malfunctioning due to overheating. The comminuted aggregate feed samples were then representatively mass-reduced to 2 kg laboratory samples employing coning and quartering technique recommended for sample-splitting of feed samples [49,60] and described earlier [57] but with modification. Briefly, the comminuted aggregate sample was mixed and shoveled into a cone, then flattened by pressing the top without further mixing and dividing the flat circular pile into equal quadrants. Two opposite portions were discarded while the remaining two opposite portions were mixed and shoveled into a cone and the procedure repeated until the material was reduced into four quadrants each of about 500 g. Three quadrants were randomly selected, pooled and the mass topped up to 2 kg laboratory sample on an electronic weighing balance (Mettler PM34, DoltaRange®, Zürich, Switzerland)by transferring many small portions randomly picked from the 4th quadrant and double bagged in fresh brown bags. The shoveling and coning process was carried out under a funnel fastened on a tripod stand to ensure uniform distribution of the material.

4.1.3. Preparation of Test Portions (Tertiary Sampling)

The 2 kg laboratory samples were split into four aliquots of about 31 g. Briefly, the 2-kg laboratory samples were representatively mass reduced through five runs of modified coning and quartering method described above until four quadrants each of between 30–35 g were formed. Two 25 g test portions were weighed out from each of opposite portions and the alternate quadrants used for adjusting weights of the selected corresponding test portions if they were below the desired size of 25 g.

4.1.4. Sample Extraction (Quaternary Sampling)

Before blending, dry-run of equipment was carried out to locate particulate contaminants and then cleaned with 70% ethanol. Two (2) 25 g test portions were homogenized by a wet milling step through 2.5-fold dilution proposed earlier for maize samples [59], followed by high speed slurring in water. Briefly, a 25 g test portion was transferred to a kitchen blender cup, 37.5 mL water added (matrix/water ratio of 1:1.5, w/v) and blended at high speed (Moulinex®, Model: Type LM240, Écully, France) for 5 min using a two-step milling protocol (3 min running; 1.5 min pulse; 2 min running) and 1.3 g slurry (0.52 g feed: 0.78 g water) immediately weighed out for extraction and quantification for AFB1. Regular tapping of the blending cup during blending was critical for successful sample homogenization. This reduced splashing of contents on sides of container away from the macerating rotor. Further, the slurry aliquot was weighed out immediately after slurring directly to the bottom avoiding the neck of the extraction bottle. In each slurry sample 86.5 mL of acetonitrile (HPLC Grade):water (Milli Q) (80:20, v/v) was added, contents agitated at 300 rpm for 15 min in orbital shaker-incubator (MRC®, Model: Tou-50, Holon Israel), allowed to settle for 40 min at ambient temperature, supernatant extract decanted and stored at 4 °C.

4.1.5. Quantification of AFB1 in Feed Samples

The extract was prepared and analyzed for AFB1 using AFB1-ELISA low matrix kit (Helica Biosystems Inc.®) according to manufacturer's instructions but with modification. Briefly, 100 μL of the extract, appropriately diluted in phosphate buffered saline containing tween 20 (PBS-T) was loaded on polystyrene ELISA micro-plate with immobilized mouse anti-AFB1 monoclonal antibodies, incubated at ambient temperature for 30 min, contents decanted and plate washed with PBS-T (this was prepared by dissolving contents of PBS-T packet in 1 L distilled water) to remove non-specific reactants, 100 μL of AFB1-horseradish peroxidase conjugate added and incubated again for 30 min, contents decanted and washed with PBS-T to remove non-specific reactants. Substrate-chromogen reagent (100 μL) was added and incubated at ambient temperature for 10 min before stopping the reaction with 100 μL of acid solution and the optical density (ODs), which is inversely proportional to aflatoxin concentration measured at 450 nm on a multi-channel micro-plate reader (BDSL Immunoskan Plus, Finland) inter-faced to a personal computer (Dell®, Precision 470 Cherrywood, Ireland) and the OD s recorded using Eiaquik program (© M.C Eisler, 1995). By setting the "zero" standard as 100% binding (B_0), percent binding for each standard and sample as a percent of the zero binding (% B/B_0) was calculated. Average absorbance values for each standard and sample extract dilution (B) and that of reagent blank (B_0) were used to calculate percentage inhibition (B/B_0%) for each standard and sample dilution, and for construction of 4-parameter calibration curve (log of standard concentration versus logit of B/B_0%) using programmed template spreadsheets (Microsoft Excel Office 1997–2003). The calibration curve was used to calculate AFB1 concentrations of unknown samples. For the conventional method (Groups VI and VIII), 80% acetonitrile (ACN) extract to PBS-T ratio was 100:900, v/v., while for the novel method (Groups II and V) this was modified to 100:650, v/v (Table 2).

4.2. In-House Validation of Improved Aflatoxin Test Procedure

4.2.1. Materials

Crystalline aflatoxin B1 (Fermentek Ltd., Jerusalem, Israel) was quantified on UV-Vis spectrophotometer (UV-1650 PC, Shimadzu®, Kyoto, Japan) by scanning using spectrum mode against methanol blank, absorbance peaks read between 200 and 500 nm as described earlier [79]. Chicken feed specimens (mash form) collected from animal feed retail shops were analyzed for AFB1 using the aforementioned commercial ELISA kit. Specimens with AFB1 levels below the LOD were selected and pooled to make native pseudo blank material [80] for spiking and recovery, and replication studies, and determination of LOD and LOQ. The blank feed material was representatively split to the desired size of test portions employing coning and quartering method recommended for mass reduction of feed samples [49,60] and described earlier [57] with modification described above. AFB1 maize reference (23 ppb) material was acquired from Biosciences eastern and central Africa-International Livestock Research Institute Hub (BecA-ILRI, Nairobi campus) for ruggedness studies.

4.2.2. Study Design

Completely randomized designs for spike and recovery and replication studies and determination of LOD and LOQ are shown in Table 2. There were 11 groups of native pseudo blank feed aliquots (Groupa I–XI) representatively split from the native pseudo blank feed material employing the modified coning and quartering method described above. In addition, 3 groups of acetonitrile extracts of feed aliquots (Groups XII–XIV) were used for LOD and LOQ studies. Five groups were used for spike and recovery studies. First, conventional dry milling (Group I) and our novel wet milling (Group II) procedure were compared (Table 2). Eighteen 25 g aliquots were randomly assigned to a 2 × 3 factorial arrangement i.e., 9 aliquots for each of the two milling methods and three aliquots randomly selected for spiking to achieve three levels of AFB1 (0, 20 or 100 ppb). The EU legal AFB1 residue limits in feed for adult poultry and young animals are 20 and 10 ppb respectively (EU, 2002), but in tropical countries it is common to find feed naturally contaminated with 100 ppb AFB1 and above [13,15,18,19]. Of the 9 aliquots assigned to each group, 3 sets each of 3 aliquots were randomly selected for either of the three AFB1 levels. Secondly, efficiency of various extraction conditions (sample: organic solvent ratio and organic solvent extract: PBST ratio) associated with wet milling procedure were compared. Forty-five 25 g aliquots were randomly assigned to a 3 × 1 factorial arrangement i.e., 15 aliquots for each of the three extraction conditions (Groups III–V) and five aliquots randomly selected for spiking to achieve either of the three levels of AFB1, 0, 20 or 100 ppb. The number of replicates for Groups I and II was therefore 6, and it was 10 for Groups III–V (Table 2). Dry milling procedure involving representative splitting of laboratory sample to obtain test portion, followed by solid–liquid extraction at dilution factor of approximately 5 (Group I) was considered here as the conventional method for recovery studies. The "0" ppb aliquots were used to determine baseline AFB1 levels and were therefore excluded in the sample size (Table 2).

The replication studies used 70 feed aliquots (10 25 g aliquots analyzed whole; 2 25 g aliquots spilt to 20 wet 1.3 g aliquots; 1 25 g aliquot spilt to 10 dry 1.3 g aliquots; and 30 25 g aliquots all spilt to 30 wet 1.3 g aliquots) divided into 6 groups (Table 2). For repeatability, 4 groups (Groups VI–IX) each of 10 feed aliquots were used in a 2 × 3 factorial arrangement to investigate which between dry and wet milling procedures had most suitable intra-laboratory precision for secondary and tertiary sampling stages of sample mass reduction. In addition, intra-laboratory precision for 3 extraction conditions were also investigated. For intermediate precision work, 2 groups (Groups X and XI) each of 15 feed aliquots were used in a 5 × 2 × 3 factorial design to investigate effect of analyst, extraction condition and time on measurement variability. The conventional method (dry milling conditions described in Table 2) for repeatability studies at secondary and tertiary sampling stages are represented by Groups VI and VIII, respectively. For LOD and LOQ determination, field collected feed specimens were screened for AFB1 and one with 8.9 ppb (the lowest level) used as native pseudo blank material. Three groups of extracts derived from this material were used under three extraction conditions. Group XII (dry milling, final dilution factor = 1000) had 4 extracts while Groups XIII (dry milling, final dilution factor = 500) and XIV (wet milling, final dilution factor = 1247) each had 5 extracts. Dry milling procedures (Groups XII and XIII) represent the conventional method (Table 2).

4.2.3. Aflatoxin Spike and Recovery Studies

The solution of AFB1 (Fermentek Ltd., Jerusalem, Israel) in methanol (HPLC Grade) was spiked to achieve 20 or 100 ppb AFB1 level, respectively, in 25 g native pseudo blank feed aliquots, while blank methanol was added to native pseudo blank feed aliquots designated as "0 ppb".

Comparison of Conventional (Dry) and Novel Water Slurry (Wet) Milling Procedure

To a 25 g feed aliquot, 1 mL blank methanol (HPLC Grade), 1 mL of 0.5 or 2.5 µg AFB1/mL solution in methanol (HPLC Grade) was spiked to achieve 0, 20 or 100 ppb AFB1 level, respectively. Group I aliquots were blended in a kitchen blender (Moulinex®, Model: Type LM240, Écully, France) at high speed for 1 min, while Group II aliquots were prepared by adding 37.5 mL of water (2.5-fold dilution) and blending as described above (Section 4.1.4), and 2 g slurry immediately weighed out. To each of the 25 g dry and 2 g slurry samples, 133 mL acetonitrile:water (80:20, *v/v*) was added and the mixture agitated as described above (Section 4.1.4). Extracts of dry- and wet-blended aliquots were diluted 94.3-fold (final dilution factor = 500) and 7.5-fold (final dilution factor = 1247), respectively, in PBS-T and analyzed for AFB1 by ELISA as described above.

Comparison of Different Extraction Conditions Associated with Wet Milling Procedure

The 25 g aliquots were water-slurried and paired 1.3 g slurry analytical samples weighed out as described above (Section 4.1.4). To each of the 1.3 g slurry analytical sample, 130 mL acetonitrile water (80:20, *v/v*) was added for Group III (dilution factor = 100), 65 mL acetonitrile:water (80:20, *v/v*) for Group IV (dilution factor = 50) and 86.5 mL acetonitrile:water (80:20) was added Group V (dilution factor = 66.5), and the mixture was agitated as described in Section 4.1.4. The extracts were diluted 10-fold for Groups III (total dilution factor = 2500) and IV (final dilution factor = 1250) and 7.5-fold for Group V (final dilution factor = 1247) in PBST. The extracts were analyzed for AFB1 by ELISA (Section 4.1.5) and the results of paired test portions for each slurry sample averaged. Mean background level was calculated, subtracted from each of the AFB1 levels of "20 and 100 ppb" replicates and these results expressed as percent recovery. For each group (Groups I–V), mean percent recovery, StDev and RSD of the replicates were calculated. Outlier values in the percent recovery data were identified on Excel 2013 using the following formula:

$$1\text{st quartile-}1.5\ IQR \leq x \geq 3\text{rd quartile} + 1.5\ IQR \tag{5}$$

where IQR is the interquartile range.

Outlier identification criterion also included recommended mean recovery range of 75–125% by European Commission and variance predicted by the modified Horwitz equation [50–53,63]. Where more than one out of five replicates were identified as outliers, the data were discarded and spiking and recovery procedure repeated [52,53].

4.2.4. Replication Studies

Representativeness of mass reducing methods was investigated as described earlier [48,51,52,63,81].

Estimation of Intra-Laboratory Variability (Repeatability)

Ten 25 g test portions were extracted whole while the 11th one (from the same laboratory sample), was water-slurried (Section 4.1.4) and ten 1.3 g slurry aliquots weighed out for extraction. To each of the 25 g of dry (Group VI) and 1.3 g slurry (Group VII) aliquots, 130 mL acetonitrile:water (80:20, *v/v*) were added and the mixture agitated (Section 4.1.4), extracts diluted 10-fold in PBST and analyzed for AFB1 by ELISA (Section 4.1.5) to compare variance associated with the dry and wet milling methods at secondary sampling level. Intra-laboratory precision of two procedures was further investigated at tertiary sampling stage. Feed material (50 g) was representatively split into four equal portions by coning and quartering method (Section 4.1.2) and two opposite quadrants pooled to make two equal portions each weighing 25 g. One portion was further split by two runs of 16 aliquots each weighing approximately 1.56 g from which 10 were randomly selected, dry 1.3 g analytical samples weighed (Group VIII) while the other 25 g portion was water slurried and ten 1.3 g slurry aliquots weighed (Group IX). Both dry and slurry samples were extracted and analyzed for AFB1 as described above.

Estimation of Intermediate Variability Associated with Wet Milling Procedure

Intermediate precision of the wet milling (slurry) method was also investigated. Five analysts were first trained, each randomly assigned three 25 g native pseudo blank feed aliquots, and each prepared a water slurry daily from one aliquot and weighed out one 1.3 g analytical sample as described above on three consecutive days. Each analytical sample was diluted 100-fold in acetonitrile:water (80:20, *v/v*), mixture agitated (Section 4.1.4), extracts diluted 10-fold in PBST and analyzed for AFB1 by ELISA as described above (Group X). This was repeated with another set of fifteen 25 g test portions, but each slurry aliquot was diluted 66.5-fold in extraction solvent, agitated, diluted 7.5-fold in PBST and analyzed for AFB1 (Group XI).

4.2.5. Determination of Limits of Detection (LOD) and Quantification (LOQ)

To counteract matrix effect of chicken feed, we used blank measurement approach to determine LOD and LOQ [81,82]. Briefly a blank specimen was quantified for AFB1 in five measurements and mean OD value and standard deviation obtained. From this, an OD value of mean OD minus 2 standard deviations and mean OD minus 5 standard deviations were calculated to get two OD values which were extrapolated from the standard curve to get LOD and LOQ respectively. Details are given below. Two acetonitrile extracts of the blank specimen prepared by conventional dry milling procedure were pooled, then diluted 100-fold in PBST to attain baseline level of analyte signal and used for LOD and LOQ determination at two assay conditions. Six 100 µL extract aliquots were diluted 100-fold in PBST (Group XII), while the other five 100 µL aliquots were first diluted 5-fold in acetonitrile:water (80:20, v/v) and then 10-fold in PBS-T (for Group XIII). All the aliquots were further diluted 10-fold in PBST to attain final dilution factor of 1000 for Group XII and 500 for Group XIII for dry milling procedure, before analysis by ELISA. For wet milling procedure, five 25 g fresh feed aliquots were separately water slurried and blended, two 1.3 g analytical portions from each water slurry, weighed and diluted 66.5-fold in acetonitrile:water (80:20, v/v), contents agitated and extract harvested (Section 4.1.4). The extract was further diluted 66.5-fold in PBS-T to attain baseline level of analyte signal, then diluted 7.5-fold in PBS-T and assayed by ELISA to give a final dilution factor of 1247 (Group XIV) for wet milling procedure. The limit values were calculated from mean absorbance of the 0-standard, B_0 [63] minus 2-fold (for LOD) and 5-fold (for LOQ) the StDev of absorbance [81] of replicate wells of the blank samples analyzed by ELISA. The means of the concentrations corresponding to the %B/B_0 value of B_0 minus 2-fold, and 5-fold the standard deviation generated from the calibration curve was taken as the lowest detectable (LOD) and quantifiable (LOQ) concentration of AFB1.

4.2.6. Linearity Studies

Using pure crystalline AFB1 (Fermentek Ltd., Jerusalem, Israel), 0.5, 0.4, 0.2, 0.1, 0.05 and 0.02 ng/mL AFB1 standards in 80% acetonitrile (HPLC Grade):water (Milli Q) (80:20):PBS-T mixture (10:65) were prepared, analyzed by AFB1-ELISA and the optical density (ODs) of duplicate wells read, as described in Section 4.1.5.

4.2.7. Robustness and Ruggedness Studies

We tested stability of our modified aflatoxin test procedure by examining influence of external factors on its most vulnerable performance parameters [51], precision and accuracy. Intra-laboratory repeatability was used to measure robustness while relatively long-term parameters such as intra-laboratory reproducibility and effect of storage time on extraction efficiency of extraction solvent measured of ruggedness. Determination of robustness was carried out by calculation of repeatability of surrogate recovery (Section 4.2.3). For ruggedness, intra-laboratory reproducibility (effects of day of analysis and analyst) was calculated (Section 4.2.4). Further, the effect of storage on extraction efficiency of organic solvent was investigated. Fresh acetonitrile:water (80:20, v/v) extraction solvent was prepared and divided into three batches; 1st batch was used immediately, while 2nd and 3rd batches were stored at ambient temperature (minimum = 9.4 $\pm$ 1.7 °C; maximum = 25.1 $\pm$ 2.9 °C) for one and three months, respectively, prior to use. One gram of 23 ppb maize reference specimen was extracted with 5 mL solvent by agitating and extract harvested (Section 4.1.4), analyzed for AFB1 (Section 4.1.5) and percent recoveries calculated.

4.2.8. Evaluation of the Improved Aflatoxin Test Procedure

In total, 251 chicken feed samples were collected in the field (one sample per lot) as described in Section 4.1.1, prepared and analyzed. Using a programmed spreadsheet

(Microsoft Excel Office), minimum requirement criterion for variability associated with sample preparation and analysis was incorporated as a quality control measure. Inter-assay variance associated with sample analysis was evaluated as described earlier [56]. Observed inter-sample variance, observed relative standard deviation (RSD) associated with measurement of paired test portions of each laboratory sample was compared with the predicted relative standard deviation (PRSD) to give a Horwitz Ratio (HorRat) value thus, HorRat value = RSD_O/RSD_p, where RSD_O and RSD_p are the observed RSD and PRSD respectively. PRSD was calculated from the modified Horwitz equation which gives RSD as a function of analyte concentration [50–54]: $RSD < 2^{(1-0.5\log C)} \times 0.67$, where C is the AFB1 concentration. For samples with HorRat values above 1 and AFB1 levels exceeding the regulatory limit for feed for animal feed (20 ppb), fresh test portions were re-tested until HorRat value of ≤ 1 was achieved [56]. However, all results from the same sample were averaged to give final mean result and none were discarded. Replicates per sample were recorded and the percent of samples analyzed using 2–6 replicates computed.

4.3. Statistical Analysis

Descriptive and inferential data analyses were done on statistical computer program (IBM SPSS Statistics 20). For spiking and recovery data, bias was calculated accordingly: %bias = mean% recovery-100. AFB1 recovery of the conventional dry milling procedure and the novel slurry method were analyzed by comparing means of Groups I and II employing independent *t*-test. Variability expressed as variance, standard deviation and coefficient of variation (relative standard deviation) for Groups I and II were computed and compared. Means of all wet milling procedures recovery data (Groups III–V) were compared employing Welch ANOVA to determine between and within the group variation. Games–Howell test (a post-hoc test) incorporated for multiple comparison of the groups. For spiking and recovery studies, RSD (or CV) effect were determined accordingly,

$$E_{CV} = G_{CV} - M_{CV} \tag{6}$$

where E_{CV} = CV effect, G_{CV} = Group CV and M_{CV} = Grand mean CV

The extraction methods ranked according to CV effect values. HorRat values and ORSD/PRSD were used in replication studies to rank sample splitting procedures. For repeatability data, PRSD values were calculated from modified Horwitz equation of $PRSD < 2^{(1-0.5\log C)} \times 0.67$, while, for intermediate precision (within-laboratory reproducibility), PRSD was calculated from the unmodified Horwitz equation: $PRSD < 2^{(1-0.5\log C)}$, where C is the concentration of the analyte. The intermediate precision data (from replication studies) were further subjected to one-way ANOVA to determine whether the day of analysis and the analyst affected the data. For LOD and LOQ data, outlier OD values were identified using the unmodified Horwitz equation: $RSD < 2^{(1-0.5\log C)}$ where, C is the aflatoxin concentration. OD values above B_0 were also treated as outliers. The HorRat value associated with efficiency of extraction solvent of different shelf ages was used to evaluate variability due to solvent storage time.

Supplementary Materials: The following are available online at https://www.mdpi.com/2072-6651/13/3/216/s1, Dataset S1. Aflatoxin recovery data, Dataset S2. Precision data, Dataset S3. LOD and LOQ data, Dataset S4. Robustness-ruggedness data.

Author Contributions: Conceptualization, J.K., T.W., R.M., J.F.L. and D.G.; Data curation, J.K.; Formal Analysis, J.K., J.F.L. and R.M.; Funding acquisition, D.G., J.F.L., J.K. and L.M.; Investigation, J.K., R.M., T.W., D.M. and L.M.; Methodology, J.K., R.M., T.W. and T.P.H.; Project administration, J.K.; Resources, J.K., T.W. and T.P.H., Supervision, J.K. and R.M.; Visualization, J.K.; Writing—original draft, J.K.; and Writing—review and editing, T.W., J.F.L., L.M., T.P.H., R.M., D.G., D.M. and J.K. All authors have read and agreed to the published version of the manuscript.

Funding: This research was funded by Consultative Group on International Agricultural Research (CGIAR)-Research Program (https://www.cgiar.org/research; accessed on 12 March 2021), Agriculture for Nutrition and Health (CRA between ILRI and KALRO Ref. Contract No. 1/2014)

and Government of Kenya through Kenya Agricultural & Livestock Research Organization (http://www.kalro.org; accessed on 12 March 2021). The APC was funded by CGIAR.

Institutional Review Board Statement: Not applicable.

Informed Consent Statement: Not applicable.

Data Availability Statement: The data presented in this study are available as supplementary material (Dataset S1. Aflatoxin recovery data; Dataset S2. Precision data; Dataset S3. LOD and LOQ data; Dataset S4. Robustness-ruggedness data).

Acknowledgments: This study was funded by CGIAR Research Program, Agriculture for Nutrition and Health (CRA between ILRI and KALRO Ref. Contract No. 1/2014) and the Government of Kenya (through KALRO) whom we sincerely thank. Former Director, Biotechnology Research Institute (BioRI) Sylvance Okoth is appreciated for his support and permission to carry out this work at KALRO-BioRI, Muguga. We acknowledge Andrew Slate (posthumously) of North Carolina State University, and Aloo of Public Health Laboratory(Kenyatta Hospital Complex, Nairobi) for their technical assistance. KALRO staff (Ben Wanyonyi, George Maina, Andrew Mageto, Gilbert Ouma, Sarah Kairuthi, Jackline Kagendo, Jacqueline Arusei, Agnes Nekesa, Mercyline Ong'ale and Clarah Jebet) also provided technical expertise while David Kinoti and M/s Joanna Auma (KALRO) provided statistical and bibliography softwares respectively. Phyllis Alusi of illustration unit KALRO, formatted the graphical abstract of this article. David Gikungu, the Deputy Director, Climate Services of the Kenya Metrological Department provided room temperature data for laboratory 5 (BioRI) for the period when this work was carried out. A former science teacher, Robert S. I. Karuku is highly acknowledged posthumously for inspiring the first author to the world of food poisoning, the main drive in this communication.

Conflicts of Interest: The authors declare no conflict of interest. TH is an employee of the manufacturer of aflatoxin diagnostic kit used in this study.

References

1.   Oloo, R.G.; Okoth, S.; Wachira, P.; Mutiga, S.; Ochieng, P.; Kago, L.; Nganga, F.; Entfellner, J.B.D.; Ghimire, S. Genetic profiling of aspergillus isolates with varying aflatoxin production potential from different maize-growing regions of Kenya. *Toxins* **2019**, *11*, 467. [CrossRef]

2.   Benkerroum, N. Aflatoxins: Producing-molds, structure, health issues and incidence in Southeast Asian and Sub-Saharan African Countries. *Int. J. Environ. Res. Public Health* **2020**, *17*, 1215. [CrossRef] [PubMed]

3.   Kumar, P.; Mahato, D.K.; Kamle, M.; Mohanta, T.K.; Kang, S.G. Aflatoxins: A global concern for food safety, human health and their management. *Front. Microbiol.* **2017**, *7*, 2170. [CrossRef] [PubMed]

4.   Pereira, C.S.; Sara, C.C.; Fernandes, J.O. Prevalent mycotoxins in animal feed: Occurrence and analytical methods. *Toxins* **2019**, *11*, 290. [CrossRef] [PubMed]

5.   World Health Organization. *Aflatoxins*; REF. No.: WHO/NHM/FOS/RAM/18.1; World Health Organization: Geneva, Switzerland, 2018.

6.   World Health Organization; International Agency for Research on Cancer. IARC monographs on the evaluation of carcinogenic risks to humans. In *Some Traditional Herbal Medicines, Some Mycotoxins, Naphthalene and Styrene*; World Health Organization/International Agency for Research on Cancer: Lyon, France, 2002; Volume 82, pp. 1–590.

7.   Carballo, D.; Font, G.; Ferrer, E.; Berrada, H. Evaluation of mycotoxin residues on ready-to-eat food by chromatographic methods coupled to mass spectrometry in tandem. *Toxins* **2018**, *10*, 243. [CrossRef] [PubMed]

8.   Alshannaq, A.; Yu, J.-H. Occurrence, toxicity, and analysis of major mycotoxins in food. *Int. J. Environ. Res. Public Health* **2017**, *14*, 632. [CrossRef]

9.   Peraica, M.; Radic, B.; Lucic, A.; Pavlovic, M. Toxic effects of mycotoxins in humans. *Bull. World Health Organ.* **1999**, *77*, 754–766. [PubMed]

10. Probst, C.; Njapau, H.; Cotty, P.J. Outbreak of an acute aflatoxicosis in Kenya in 2004: Identification of the causal agent. *Appl. Environ. Microbiol.* **2007**, *73*, 2762–2764. [CrossRef] [PubMed]

11. Grace, D.; Kang'ethe, E.; Lindahl, J.; Atherstone, C.; Wesonga, T. *Aflatoxin: Impact on Animal Health and Productivity. Building an Aflatoxin Safe East African Community—Technical Policy Paper 4*; IITA: Dar es Salam, Tanzania, 2015.

12. Suleiman, E.A.; Elgabbar, M.A.; Khaliefa, K.A.; Omer, F.A. Aflatoxins in broiler chicks feed. *Sudan J. Vet. Res.* **2010**, *25*, 5–8.

13. Banday, M.T.; Darzi, M.M.; Khan, A.A. Clinico-pathological and haemo-biochemical changes in broiler chicken following an outbreak of aflatoxicosis. *Appl. Biol. Res.* **2006**, *8*, 40–43.

14. Rashid, N.; Bajwa, M.A.; Rafeeq, M.; Khan, M.A.; Ahmad, Z.; Tariq, M.M.; Wadood, A.; Abbas, F. Prevalence of aflatoxin B1 in finished commercial broiler feed from west central Pakistan. *J. Anim. Plant. Sci.* **2012**, *22*, 6–10.

15. Akinmusire, O.O.; El-Yuguda, A.-D.; Musa, J.A.; Oyedele, O.A.; Sulyok, M.; Somorin, Y.M.; Ezekiel, C.N.; Krska, R. Mycotoxins in poultry feed and feed ingredients in Nigeria. *Mycotoxin Res.* **2019**, *35*, 149–155. [CrossRef]

16. Greco, M.V.; Franchi, M.L.; Golba, S.L.R.; Pardo, A.G.; Pose, G.N. Mycotoxins and mycotoxigenic fungi in poultry feed for food-producing animals. *Sci. World J.* **2014**, *2014*, 1–9. [CrossRef] [PubMed]

17. Gathumbi, J.K.; Bebora, L.C.; Muchiri, D.J.; Ngatia, T.A. A survey of mycotoxins in poultry feeds used in Nairobi, Kenya. *Bull. Anim. Health Prod. Afr.* **1995**, *43*, 243–245.

18. Nakavuma, J.; Kirabo, A.; Bogere, P.; Nabulime, M.M.; Kaaya, A.N.; Gnonlonfin, B. Awareness of mycotoxins and occurrence of aflatoxins in poultry feeds and feed ingredients in selected regions of Uganda. *Int. J. Food Contam.* **2020**, *7*, 1–10. [CrossRef]

19. Kemboi, D.C.; Ochieng, P.E.; Antonissen, G.; Croubels, S.; Scippo, M.-L.; Okoth, S.; Kangethe, E.K.; Faas, J.; Doupovec, B.; Lindahl, J.F.; et al. Multi-mycotoxin occurrence in dairy cattle and poultry feeds and feed ingredients from Machakos town, Kenya. *Toxins* **2020**, *12*, 762. [CrossRef] [PubMed]

20. FAO. Worldwide Regulations for Mycotoxins in Food and Feeds in 2003. FAO Food and Nutrition Paper 81. Available online: http://www.fao.org/3/y5499e/y5499e02.htm (accessed on 14 February 2021).

21. FDA. *Guidance for Industry: Action Levels for Poisonous or Deleterious Substances in Human Food and Animal Feed*; US Department of Health and Human Services, US Food and Drugs Administration: Washington, DC, USA, 2017; Volume 2018.

22. EU. Commission regulation (EU) No 165/2010 of 26 February 2010. *Off. J. Eur. Union* **2010**, *50*, 8–12.

23. EC. Directive 2002/32/EC of the European parliament and of the council of 7 may 2002 on undesirable substances in animal feed. *Off. J. Eur. Communities* **2002**, *140*, 110–121.

24. Tang, L.; Xu, L.; Afriyie-Gyawu, E.; Liu, W.; Wang, P.; Tang, Y.; Wang, Z.; Huebner, H.J.; Ankrah, N.A.; Ofori-Adjei, D.; et al. Aflatoxin-albumin adducts and correlation with decreased serum levels of vitamins A and E in an adult Ghanaian population. *Food Addit. Contam. Part. A* **2009**, *26*, 108–118. [CrossRef] [PubMed]

25. Williams, J.H.; Phillips, T.D.; Jolly, P.E.; Stiles, J.K.; Jolly, C.M.; Aggarwal, D. Human aflatoxicosis in developing countries: A review of toxicology, exposure, potential health consequences and interventions. *Am. J. Clin. Nutr.* **2004**, *80*, 1106–1122. [CrossRef] [PubMed]

26. Amin, Y.A.; Mohamed, R.H.; Zakaria, A.M.; Wehrend, A.; Hussein, H.A. Effects of aflatoxins on some reproductive hormones and composition of buffalo's milk. *Comp. Clin. Pathol.* **2019**, *28*, 1191–1196. [CrossRef]

27. El Mahady, M.M.; Ahmed, K.A.; Badawy, S.A.; Ahmed, Y.F.; Aly, M.A. Pathological and hormonal effects of aflatoxins on reproduction of female albino rats. *Middle East. J. Appl. Sci.* **2015**, *5*, 998–1006.

28. Umar, S.; Younus, M.; Rehman, M.U.; Aslam, A.; Shah, M.A.A.; Munir, T.; Hussain, S.; Iqbal, F.; Fiaz, M.; Ullah, S. Role of aflatoxin toxicity on transmissibility and pathogenicity of H9N2 avian influenza virus in turkeys. *Avian Pathol.* **2015**, *44*, 305–310. [CrossRef] [PubMed]

29. Trench, P.C.; Narrod, C.; Roy, D.; Tiongco, M. Responding to health risks along the value chain. 2020 Conference Paper 5. In Proceedings of the Leveraging Agriculture for Improving Nutrition and Health, New Delhi, India, 10–12 February 2011; pp. 1–54.

30. Fouad, A.M.; Ruan, D.; El-Senousey, H.K.; Chen, W.; Jiang, S.; Zheng, C. Harmful effects and control strategies of aflatoxin B1 produced by Aspergillus flavus and Aspergillus parasiticus strains on Poultry: Review. *Toxins* **2019**, *11*, 176. [CrossRef] [PubMed]

31. Andretta, I.; Kipper, M.; Lehnen, C.R.; Hauschild, L.; Vale, M.M.; Lovatto, P.A. Meta-analytical study of productive and nutritional interactions of mycotoxins in broilers. *Poult. Sci.* **2011**, *90*, 1934–1940. [CrossRef] [PubMed]

32. Morrison, D.M.; Ledoux, D.R.; Chester, L.F.B.; Samuels, C.A.N. A limited survey of aflatoxins in poultry feed and feed ingredients in Guyana. *Vet. Sci.* **2017**, *2017*, 60. [CrossRef]

33. Bata-Vidács, I.; Kosztik, J.; Mörtl, M.; Székács, A.; Kukolya, J. Aflatoxin B1 and sterigmatocystin binding potential of non-Lactobacillus LAB Strains. *Toxins* **2020**, *2020*, 799. [CrossRef]

34. Kosztik, J.; Mörtl, M.; Székács, A.; Kukolya, J.; Bata-Vidács, I. Aflatoxin B1 and sterigmatocystin binding potential of lactobacilli. *Toxins* **2020**, *2020*, 756. [CrossRef]

35. Ozer, H.; Basegmez, H.O.; Whitaker, T.B.; Slate, A.B.; Giesbrecht, F.G. Sampling dried figs for aflatoxin—Part 1: Variability associated with sampling, sample preparation, and analysis. *World Mycotoxin J.* **2017**, *10*, 31–40. [CrossRef]

36. Grace, D.; Lindahl, J.; Atherstone, C.; Kang'ethe, E.; Nelson, F.; Wesonga, T.; Manyong, V. Aflatoxin standards for feed. In *Building an Aflatoxin Safe East African Community—Technical Policy Paper 7*; IITA: Dar es Salam, Tanzania, 2015.

37. Whitaker, T.B. Sampling foods for mycotoxins. *Food Addit. Contam.* **2006**, *23*, 50–61. [CrossRef]

38. Matumba, L.; Whitaker, T.; Slate, A.; De Saeger, S. Current trends in sample size in mycotoxin in grains. Are we measuring accurately? *Toxins* **2017**, *9*, 276.

39. Cheli, F.; Campagnoli, A.; Pinotti, P.; Fusi, E.; Dell'Orto, V. Sampling feed for mycotoxins: Acquiring knowledge from food. *Ital. J. Anim. Sci.* **2009**, *8*, 5–22. [CrossRef]

40. Food and Agriculture Organization of the United Nations. *Mycotoxin Sampling Tool. User Guide. Version 1.0 (December 2013)*, Version 1.1 ed.; Food and Agriculture Organization of the United Nations: Rome, Italy, 2013; p. 62.

41. Whitaker, T.B.; Slate, A.B. Comparing the USDA/AMS subsampling mill to a vertical cutter mixer type mill used to comminute shelled peanut samples for aflatoxin analysis. *Peanut Sci.* **2012**, *39*, 69–81. [CrossRef]

42. Kumphanda, J.; Matumba, L.; Whitaker, T.B.; Kasapila, W.; Sandahl, J. Maize meal slurry mixing: An economical recipe for precise aflatoxin quantitation. *World Mycotoxin J.* **2019**, *12*, 203–212. [CrossRef]

43. Oulkar, D.; Goon, A.; Dhanshetty, M.; Khan, Z.; Satav, S.; Banerjee, K. High-sensitivity direct analysis of aflatoxins in peanuts and cereal matrices by ultra-performance liquid chromatography with fluorescence detection involving a large volume flow cell. *J. Environ. Sci. Health Part B* **2018**, *53*, 255–260. [CrossRef] [PubMed]

44. Okuma, T.A.; Huynh, T.P.; Hellberg, R. Use of Enzyme-linked immunosorbent assay to screen for aflatoxins, ochratoxin A, and deoxynivalenol in dry pet foods. *Mycotoxin Res.* **2018**, *34*, 69–75. [CrossRef] [PubMed]

45. Taye, W.; Ayalew, A.; Chala, A.; Dejene, M. Aflatoxin B1 and total fumonisin contamination and their producing fungi in fresh and stored sorghum grain in East Hararghe, Ethiopia. *Food Addit. Contam. Part B* **2016**, *9*, 237–245. [CrossRef] [PubMed]

46. Nguyen, X.T.T.; Nguyen, T.T.T.; Nguyen-Viet, H.; Tran, K.N.; Lindahl, J.; Randolph, D.G.; Ha, T.M.; Lee, H.S. Assessment of aflatoxin B1 in maize and awareness of aflatoxins in Son La, Vietnam. *Infect. Ecol. Epidemiol.* **2018**, *8*. [CrossRef]

47. Huynh, T.P.; Ly, C. Recovery of aflatoxin B1 in a range of food commodities utilizing a matrix resistant ELISA. In Proceedings of the IAFP 2015 Annual Meeting, Portland, OR, USA, 25–28 July 2015.

48. Wagner, C. Critical practicalities in sampling for mycotoxins in feed. *J. Aoac Int.* **2015**, *98*, 301–308. [CrossRef]

49. Herrman, T. Sampling: Procedures for feed. MF-2036. Feed Manufacturing. Kansas State University Agricultural Experiment Station and Cooperative Extension Service: 2001. Department of Grain Science. Available online: http://www.oznet.ksu.edu/library/grsci2/MF2036 (accessed on 23 January 2014).

50. Rao, T.N. *Validation of Analytical Methods*; IntechOpen: London, UK, 2018; Chapter 7. [CrossRef]

51. BPU. Guidelines for Validation of Analytical Methods for Non-Agricultural Pesticide Active Ingredients and Products. Available online: www.hse.gov.uk/biocides/copr/pdfs/validation. (accessed on 14 October 2016).

52. EC. Technical Material and Preparations: Guidance for Generating and Reporting Methods of Analysis in Support of Pre- and Post-Registration Data Requirements for Annex II (Part A, Section 4) and Annex III (Part A, Section 5) of Directive 91/414. Available online: https://ec.europa.eu/food/sites/food/files/plant/docs/pesticides_ppp_app-proc_guide_phys-chem-ana_tech-mat-preps.pdf (accessed on 14 October 2016).

53. OECD. Guidance document for single laboratory validation of quantitative analytical methods-guidance used in support of pre-and-post-registration data requirements for plant protection and biocidal products. In *Environment, Health and Safety Publications Series on Testing and Assessment*; No. 204 and Series on Biocides No. 9. ENV/JM/MONO (2014) 20; Organisation for Economic Co-operation and Development: Paris, France, 2014; Volume JT03360406.

54. EC. Laying down the methods of sampling and analysis for the official control of the levels of mycotoxins in foodstuffs. *Off. J. Eur. Union* **2006**, *70*, 12–33.

55. EU. Commission recommendation of 17 August 2006 on the presence of deoxynivalenol, zearalenone, ochratoxin A, T-2 and HT-2 and fumonisins in products intended for animal feeding. *Off. J. Eur. Union* **2006**, *229*, 227–229.

56. Dragacci, S.; Grosso, F. *Validation of Analytical Methods to Determine the Content of Aflatoxin, Ochratoxin and Patulin in Foodstuffs of Vegetable Origin*; European Commission BCR Information Chemical Analysis: Paris, France, 1999; pp. 1–40.

57. Gerlach, R.W.; Dobb, D.E.; Raab, G.A.; Nocerino, J.M. Gy sampling theories in environmental studies. 1. Assessing soil splitting protocols. *J. Chemom.* **2002**, *16*, 321–328. [CrossRef]

58. Walker, M.; Colwell, P.; Cowen, S.; Ellison, S.L.R.; Gray, K.; Elahi, S.; Farnell, P.; Slack, P.; Burns, D.T. Aflatoxins in Groundnuts–Assessment of the effectiveness of EU sampling and UK enforcement sample preparation procedures. *J. Assoc. Public Anal.* **2017**, *45*, 1–22.

59. Food and Agriculture Organization of the United Nations. *Sampling Plans for Aflatoxin Analysis in Peanuts and Corn. FAO Food and Nutrition Paper 55. Report of an FAO Technical Consultation Rome*; Food and Agriculture Organization of the United Nations: Rome, Italy, 1993.

60. Reiter, E.V.; Dutton, M.F.; Agus, A.; Nordkvist, E.; Mwanza, M.F.; Njobeh, P.B.; Prawano, D.; Haggblom, P.; Razzazi-Fazeli, E.; Zentek, J.; et al. Uncertainty from sampling in measurement of aflatoxins in animal feedstuffs: Application of the Eurachem/CITAC guidelines. *Analyst* **2011**, *136*, 4059–4069. [CrossRef] [PubMed]

61. Spanjer, M.C.; Scholten, J.M.; Kastrup, S.; Jorissen, U.; Schatzki, T.F.; Toyofuku, N. Sample comminution for mycotoxin analysis: Dry milling or slurry mixing? *Food Addit. Contam.* **2006**, *23*, 73–83. [CrossRef]

62. IUPAC. Harmonized guidelines for single laboratory validation of methods of analysis. IUPAC Technical Report. *Pure Appl. Chem.* **2002**, *74*, 835–855. [CrossRef]

63. APVMA. Guidelines for the Validation of Analytical Methods for Active Constituent, Agricultural and Veterinary Chemical Products. Available online: http://www.apvma.gov.au (accessed on 14 October 2016).

64. Shao, D.; Imerman, P.M.; Schrunk, D.E.; Ensley, S.M.; Rumbeiha, W.K. Intralaboratory development and evaluation of a high-performance liquid chromatography-fluorescence method for detection and quantitation of aflatoxins M1, B1, B2, G1, and G2 in animal liver. *J. Vet. Diagn. Investig.* **2016**, *28*, 646–655. [CrossRef]

65. Trucksess, M.W.; Weaver, C.M.; Oles, C.J.; Fry, F.S. Determination of aflatoxins B1, B2, G1, and G2 and ochratoxin A in ginseng and ginger by multitoxin immunoaffinity column cleanup and liquid chromatographic quantitation: Collaborative study. *J. AOAC Int.* **2008**, *91*, 511–522. [CrossRef]

66. Brera, C.; Debegnach, F.; Minardi, V.; Pannunzi, C.; Santis, B.D. Immunoaffinity column cleanup with liquid chromatography for determination of aflatoxin b1 in corn samples: Interlaboratory study. *J. AOAC Int.* **2007**, *90*, 765–772. [CrossRef]

67. Davis, J.P.; Jackson, M.D.; Leek, J.M.; Samadpour, M. Sample preparation and analytical considerations for the US aflatoxin sampling program for shelled peanuts. *Peanut Sci.* **2018**, *45*, 19–31. [CrossRef]

68. Kang'ethe, E.K.; Gatwiri, M.; Sirma, A.J.; Ouko, E.O.; Mburugu-Mosoti, C.K.; Kitala, P.M.; Nduhiu, G.J.; Nderitu, J.G.; Mungatu, J.K.; Hietaniemi, V.; et al. Exposure of Kenyan population to aflatoxins in foods with special reference to Nandi and Makueni counties. *Food Qual. Saf.* **2017**, *1*, 131–137. [CrossRef]

69. Daniel, J.H.; Lewis, L.W.; Redwood, Y.A.; Kieszak, S.; Breiman, R.F.; Flanders, W.D.; Bell, C.; Mwihia, J.; Ogana, G.; Likimani, S. et al. Comprehensive Assessment of Maize Aflatoxin Levels in Eastern Kenya, 2005–2007. *Environ. Health Perspect.* **2011**, *119*, 1794–1799. [CrossRef] [PubMed]

70. Food and Agriculture Organization/World Health Organization. Dietary exposure assessment of chemicals in food. In *Environmental Health Criteria 240: Principles and Methods for the Risk Assessment of Chemicals in Food. A Joint Publication of the Food and Agriculture Organization of the United Nations and the World Health Organization. Chapter 6*; Food and Agriculture Organization/World Health Organization: Geneva, Switzerland, 2009; pp. 1–95.

71. EURO. GEMS/Food-EURO Second Workshop on Reliable evaluation of low-level contamination of food. Report on a workshop in the Frame of GEMS/Food-EURO. In Proceedings of the Workshop in the Frame of GEMS/Food-EURO, Kulmbach, Germany, 26–27 May 1995.

72. Huong, B.T.M.; Tuyen, L.D.; Madsen, H.; Brimer, L.; Friis, H.; Dalsgaard, A. Total dietary intake and health risks associated with exposure to aflatoxin B1, ochratoxin A and fuminisins of children in Lao Cai Province, Vietnam. *Toxins* **2019**, *11*, 638. [CrossRef] [PubMed]

73. Li, P.; Ding, X.; Bai, Y.; Wu, L.; Yue, X.; Zhang, L. *Risk Assessment and Prediction of Aflatoxin in Agro-Products*; IntechOpen: London, UK, 2018; Chapter 18; pp. 319–334.

74. Senerwa, D.M.; Sirma, A.J.; Mtimet, N.; Kang'ethe, E.K.; Grace, D.; Lindahl, J.F. Prevalence of aflatoxin in feeds and cow milk from five counties in Kenya. *Afr. J. Food, Agric. Nutr. Dev.* **2016**, *16*, 11004–11021. [CrossRef]

75. Raad, F.; Nasreddine, L.; Hilan, C.; Bartosik, M.; Parent-Massin, D. Dietary exposure to aflatoxins, ochratoxin A and deoxynivalenol from a total diet study in an adult urban Lebanese population. *Food Chem. Toxicol.* **2014**, *73*, 35–43. [CrossRef]

76. Bellio, A.; Daniela Manila Bianchi, D.M.; Gramaglia, M.; Loria, A.; Nucera, D.; Gallina, S.; Gili, M.; Decastelli, L. Aflatoxin M1 in Cow's Milk: Method validation for milk sampled in Northern Italy. *Toxins* **2016**, *8*, 57. [CrossRef]

77. Krska, R.; Richard, J.L.; Schuhmacher, R.; Slate, A.B.; Whitaker, T.B. *Romer Labs Guide to Mycotoxins*, 4th ed.; Anytime Publishing Services: Leicestershire, UK, 2012.

78. Rodrigues, I.; Handl, J.; Binder, E.M. Mycotoxin occurrence in commodities, feeds and feed ingredients sourced in the Middle East and Africa. *Food Addit. Contam. Part B* **2011**, *4*, 168–179. [CrossRef]

79. AOAC. Preparation of Standards for aflatoxins: Thin-layer chromatographic-Spectrometric methods (AOAC Official Method 970.44). In *Official Methods of Analysis*; Cunniff, P., Ed.; Health Protection Branch, AOAC International: Ottawa, ON, Canada, 2005; pp. 43–49.

80. European Union Reference Laboratory. *Guidance Document on the Estimation of LOD and LOQ for Measurements in the Field of Contaminants in Feed and Food*; European Union Reference Laboratory: Lyngby, Denmark, 2016; pp. 1–52.

81. Rossi, C.N.; Takabavashi, C.R.; Ono, M.A.; Saito, G.H.; Itano, E.N.; Kawamura, O.; Hirooka, E.Y.; Ono, E.Y.S. Immunoassay based on monoclonal antibody for aflatoxin detection in poultry feed. *Food Chem.* **2012**, *132*, 2211–2216. [CrossRef]

82. Huynh, T.; Ly, C.; Knight, P.; Wolde-Mariam, W. Quantitation of aflatoxin B1 by ELISA in commodities that pose a matrix effect. In Proceedings of the AOAC 126th Annual Meeting, Las Vegas, Nevada, 30 September–3 October 2012.

Article

A Liquid Chromatographic Method for Rapid and Sensitive Analysis of Aflatoxins in Laboratory Fungal Cultures

Ahmad F. Alshannaq and Jae-Hyuk Yu * 

Department of Bacteriology, University of Wisconsin-Madison, 1550 Linden Drive, Madison, WI 53706, USA
* Correspondence: jyu1@wisc.edu

Received: 16 December 2019; Accepted: 28 January 2020; Published: 30 January 2020

Abstract: Culture methods supplemented with high-performance liquid chromatography (HPLC) technique provide a rapid and simple tool for detecting levels of aflatoxins (AFs) produced by fungi. This study presents a robust method for simultaneous quantification of aflatoxin (AF) B1, B2, G1, and G2 levels in several fungal cultivation states: submerged shake culture, liquid slant culture, and solid-state culture. The recovery of the method was evaluated by spiking a mixture of AFs at several concentrations to the test medium. The applicability of the method was evaluated by using aflatoxigenic and non-aflatoxigenic *Aspergilli*. A HPLC coupled with the diode array (DAD) and fluorescence (FLD) detectors was used to determine the presence and amounts of AFs. Both detectors showed high sensitivity in detecting spiked AFs or AFs produced in situ by toxigenic fungi. Our methods showed 76%–88% recovery from medium spiked with 2.5, 10, 50, 100, and 500 ng/mL AFs. The limit of quantification (LOQ) for AFs were 2.5 to 5.0 ng/mL with DAD and 0.025 to 2.5 ng/mL with FLD. In this work, we described in detail a protocol, which can be considered the foremost and only verified method, to extract, detect, and quantify AFs employing both aflatoxigenic and non-toxigenic *Aspergilli*.

Keywords: aflatoxins; laboratory culture; extraction; HPLC; recovery; detection limits

Key Contribution: Development a verified, rapid, and sensitive HPLC method for extraction and quantification of aflatoxins from different fungal cultures.

---

1. Introduction

Aflatoxins (AFs) are a group of mycotoxins that are toxic, carcinogenic, and mutagenic. Amongst them, aflatoxin B1 (AFB1) is the most potent carcinogen found in nature and thus is classified as a group 1 carcinogen to humans by the International Agency for Research on Cancer (IARC) [1,2]. AFs are produced mainly by a common fungus *Aspergillus flavus* in fields, transportation, and storage conditions. AFs consistently and increasingly contaminate both human food and animal feed, and thus have been strictly regulated by the government authorities in over 100 countries in the world [3–5]. Trace levels of AFs, 4–20 parts per billion (ppb), can be considered hazardous, and foods with higher amounts are not fit for human consumption [6]. As global warming progresses, AF-producing molds will expand their growing regions, leading to an increased burden of AF contamination in the world [7–9].

AFs are fluorescent heterocyclic secondary metabolites with molecular weights of 286 to 346 Da. Although more than 13 types of AFs have been discovered, AFB1, AFB2, AFG1, AFG2, and AFM1 (in milk) are particularly hazardous to humans and animals, as they have been commonly present in food and feed. The "B" and "G" refer to the blue and green-blue fluorescent colors emitted under ultraviolet (UV) light (Figure 1A), and the numbers represent the travelled position from the front line on the thin layer chromatography (TLC); moreover, AFB2 and AFG2 are the dihydroxy derivatives of AFB1 and AFG1, respectively [10,11]. Due to their oxygenated pentaheterocyclic structure, which is known as coumarinic nucleus, AFs have natural fluorescence properties (Figure 1A,B). This ability to fluoresce has paved the way for most analytical methods for the detection and quantification of these toxins [12]. Because of the absence of a double bond in the furan ring, AFB2 and AFG2 have a higher fluorescence quantum yield of fluorophore than the unsaturated compounds AFB1 and AFG1 [13].

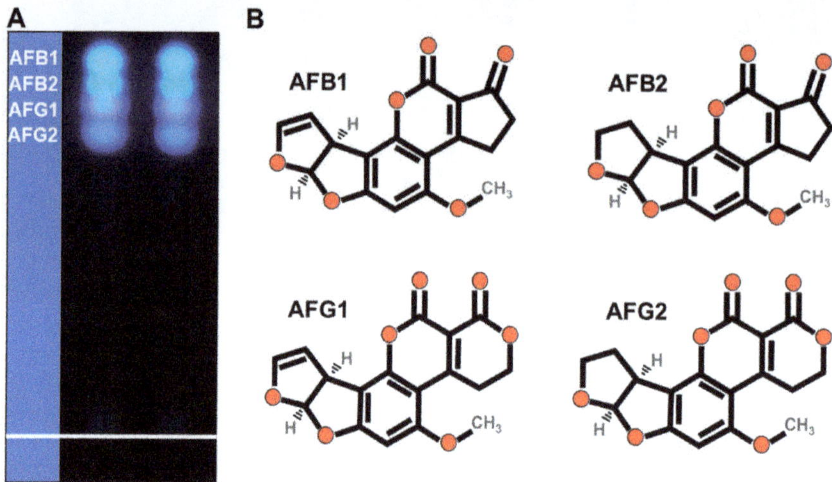

Figure 1. Aflatoxin (AF) thin layer chromatography (TLC) and structures. (**A**) A thin layer chromatograph of standard aflatoxin mixture containing AFB1, AFB2, AFG1, and AFG2. Note the color and the separation order. The photo was taken in a UV chamber at 365 nm. (**B**) Chemical structure of AFB1, AFB2, AFG1, and AFG2.

Aspergilli residing in field soil of *A. flavus* specifically, is considered as the main source of AF contamination of agricultural products; however, not all strains of *A. flavus* produce AFs [14]. Communities of AF-producing fungal residents in varying agricultural environments are complex groups of diverse individuals. Thus, knowing the AF-producing potential of *A. flavus* populations is an important factor for the predicting the incidence and severity of AF contamination. On the other hand, although it was thought that *A. flavus* only produced B type AFs, recent reports have demonstrated that several *A. flavus* strains can also produce the G type AFs [15–18].

To detect and differentiate aflatoxigenic and non-toxigenic *Aspergilli*, several methods have been developed including molecular marker-based methods and fungal culture methods [19,20]. Currently, in most cases, aflatoxigenic fungi are being identified by culture methods coupled with thin layer chromatography (TLC) or high-performance liquid chromatography (HPLC). However, to the best of our knowledge, no methods have been optimized and validated for simultaneous quantifications of aflatoxin cocktail (AFB1, AFB2, AFG1, AFG2) in the fungal cultures. Gell and Carbone have used HPLC-FLD (fluorescence) for quantification of AFB1 from fungal mycelium culture after sample purification by solid phase extraction tubes (SPE), and they were able to achieve a limit of detection (LOD) and limit of quantification (LOQ) of 2 and 3.9 ng/mL, respectively [21]. Culture method has a number of advantages over others, including it being inexpensive, rapid, available in most labs, and requiring minimal technical skills. However, due to the lack of verifications of these methods, they are generally regarded as being less precise than the other methods. Here, we report a new method for a rapid, sensitive, and simultaneous detection of AFB1, AFB2, AFG1, and AFG2 produced in laboratory culture conditions using HPLC equipped with a conventional diode array (DAD) and a fluorescence detector (FLD) without using any pre- or post-column derivatization reagents, SPE and immune affinity column (IAC), or fluorescent enhancers. Moreover, we have optimized and validated the method through a series of experiments to meet the research laboratory needs for a robust, fast, easy to use, cheap, and environmentally friendly protocol with minimum organic solvents waste.

2. Results and Discussion

Sample preparation plays a key role for the quality of chromatographic results. The selection of extraction solvent and condition are very important for achieving the true value of the assigned analyte. Prior to validation of the method, we optimized the AF extraction efficiency conditions by testing the effect of different extraction solvents, the effect of extraction solvent amount, and the effect of shaking time (unpublished data). We tested five solvents: chloroform, ethyl acetate, acetone, petroleum ether, and methanol in six different sample to solvent ratios (1:1, 1:1.5, 1:2, 1:2.5, 1:3, and 1:5). We also assessed the effect of the shaking time by vortexing the samples for 30, 60, 90, 120, 150, and 300 s. Through it all, we found that chloroform and ethyl acetate were the best extraction solvents with the highest recovery values. The data revealed that the extraction yield with chloroform was a little higher, with no significant differences, when compared with ethyl acetate. We chose chloroform as the AF extraction solvent in this study because (1) higher recover values were achieved and (2) AFs are more stable and soluble in chloroform than in ethyl acetate [22]. Technically, complete obtaining of the lower organic layer (chloroform) was achievable and easier than those on the top (ethyl acetate). We also found that a minimum 1.5-fold volume of chloroform and 30 s shaking time produced the maximum AF recovery values. Results indicated that total transfer of AFs can be accomplished by two extractions with chloroform. Although there was no detectable AFs in the third chloroform extract, three extractions are recommended to preclude loss of toxin.

Method validation is a crucial prerequisite to performing an analysis [23]. Several methods are available for analysis of AFs in food and feed that have been validated and accepted by official authorities, such as the European Committee for Standardization, the Association of Official Analytical Chemists (AOAC), and the International Organization for Standardization (ISO). Here, we employed a reverse-phase chromatography for the analysis of AFs by using a nonpolar bonded silica surface column and a polar mobile phase. With this reversed phase mode, AFs were eluted in the order of AFG2, AFG1, AFB2, and AFB1 (Figure 2A,B). This order was confirmed by comparing the obtained retention times in an AF mixture with the retention times of the individual AFs. All separated AFs were then detected by DAD and FLD detectors, connected in series, at parts per billion (ppb; ng/mL) concentrations (Figure 2A,B). It needs to be noted that, in using the FLD detector, AFG2 and AFB2 could be detected even at lower levels, as they fluoresce 40-fold more than AFB1 and AFG1 (Figure 2B). The LOQ is defined as the minimum concentration or mass of analyte in a given matrix that can be reported as a quantitative result with a certain level of precision [24]. On the contrary, the LOD is defined as the lowest concentration of the analyte that can be detected, but not necessarily quantitated, under the stated experimental conditions [25]. The LOD and LOQ for all AFs as detected by the UV detector was 1.0 ng/mL and 2.5–5.0 ng/mL, respectively. Using an FLD detector, the LOD and LOQ for AFB1 and AFG1 were 1.0 ng/mL and 2.5 ng/mL, respectively. Importantly, the LOD and LOQ for AFB2 and AFG2 using our method were 0.01 and 0.025 ng/mL, respectively. This method was designed for detection and quantification of aflatoxins mixture in laboratory cultures medium of growing fungi and it is not intended to use for food or feed for regulatory purpose. We found that DAD could at most detect as low as 1.0 ng/mL and quantify as low as 2.5–5.0 ng/mL for all aflatoxins. However, we injected several concentrations below 2.5 ng/mL to check the sensitivity of the FLD for detection of aflatoxins, specifically B2 and G2. We found that AFB2 and AFG2 could be easily detected, as expected and previously proved, at parts per trillion (ppt) level by FLD because of the absence of a double bond in the furan ring. To the best of our knowledge with fungal and fungal genetic studies, 1.0–5.0 ng/mL as LOD is sufficient to help researchers to distinguish between aflatoxigenic and non-aflatoxigenic strains, as well as the relative amounts of aflatoxins B and G produced between aflatoxigenic strains.

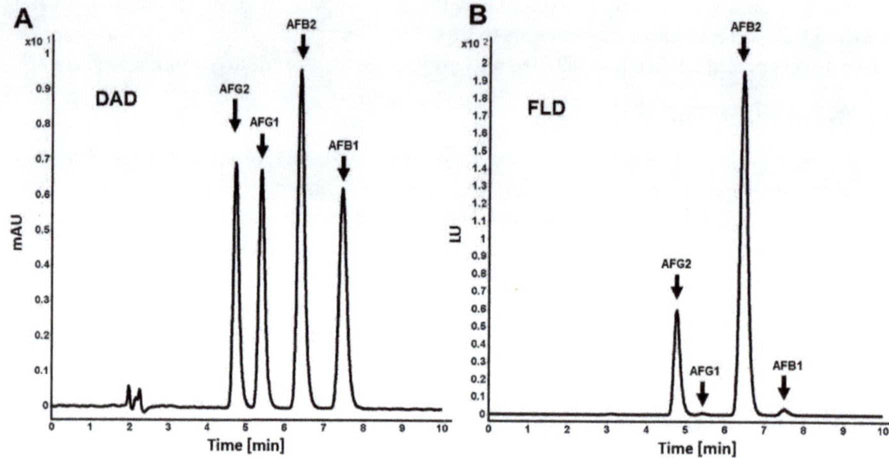

Figure 2. High-performance liquid chromatography (HPLC) chromatograms of the standard solution containing four aflatoxins (100 ng/mL each of AFB1, AFB2, AFG1, and AFG2) detected by diode array (DAD) (**A**) and fluorescence (FLD) (**B**).

Selectivity is defined as the ability to separate the analyte from other components (including impurities) that may be present in the sample [26]. Our method demonstrated a good separation ability and selectivity that allowed simultaneous quantification of four different AFs in the culture medium without interference between the AFs. Both detection methods (DAD and FLD) were able to differentiate the AF peaks in the same HPLC run with minimal background interference. In order to demonstrate a proportional relationship of response versus AF concentrations over the working range, the linearity of the method was tested from the calibration curves using seven points over the range of 5.0–1000 ng/mL for each AF and defined using the correlation coefficient (coefficients of determination, R^2) and the slope. Calibration curves were constructed by plotting the peak area (y) versus the concentrations of the AFs (x) (Figure 3A,B). Calibration curves fitted by linear regression showed R^2 ranging from 0.9987 to 1.0 for both detectors, indicating an excellent linearity for all four AFs (Figure 3C).

The fraction or percentage of the analyte that is recovered when the test sample is analyzed using the entire method is referred to as the method recovery [27]. Table 1 shows the percentage of AF recovery at a low, three-point intermediate, and high concentration levels spiked in three culture conditions. Recovery of AFs in solid, submerged, and slant culture states showed similar retention times with an overall average recovery of 76%–88%, 77%–88.4%, and 77%–86%, respectively. All spiked samples were detected by both DAD and FLD in a series manner, and the mean of both was calculated. This recovery range was within the guideline of acceptable recovery limits of AOAC and the Codex Alimentarius. The AOAC guideline for the acceptable recovery at the 10 ng/mL level is 70%–125%. The Codex Alimentarius acceptable recovery range is 70%–110% for a level of 10–100 ng/mL and 60%–120% for a level of 1–10 ng/mL. The repeatability of the method for AF analysis, as evaluated by the percentage of the RSD, ranged from 0.8% to 8.9%. These values agree with the AOAC guideline for a validated analytical method. The AOAC guidelines for acceptable repeatability (RSD) at 10 ng/mL are less than 15% and less than 8% at 1000 ng/mL.

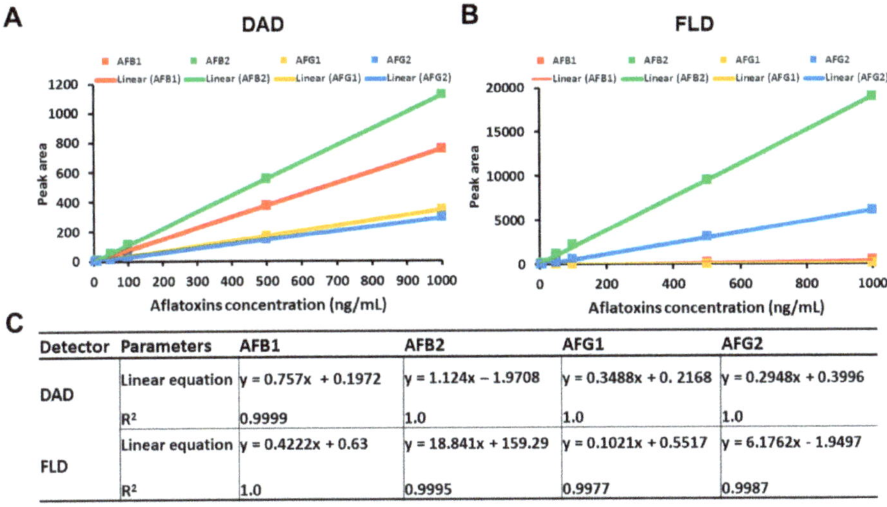

Figure 3. Calibration of aflatoxins. (**A**) Calibration curves of standard aflatoxin solutions (AFB1, AFB1, AFG1, and AFG2) over the concentrations of 5, 10, 50, 100, 500, and 1000 ng/mL as detected by DAD. (**B**) Those detected by FLD. (**C**) Linear relationship between aflatoxin concentrations and peak areas in the range of 5 to 1000 ng/mL. Correlation coefficient (R^2) and regression equation values were determined by plotting area values (y-axis) against aflatoxin concentration (x-axis).

Table 1. Recovery (%) of spiked aflatoxins from three culture methods (solid, submerged, and slant cultures); mean with (RSD) in percentages.

Spiked Levels (ng/mL)	Recovery of Aflatoxins (%)			
	AFB1	AFB2	AFG1	AFG2
Solid Culture				
500	86.1 (3.6)	88.2 (4.2)	85.6 (5.4)	86.4 (5.4)
100	79.9 (1.2)	87.0 (0.8)	81.7 (4.1)	85.2 (1.9)
50	79.7 (1.5)	79.6 (3.4)	79.8 (5.9)	82.8 (5.2)
10	78.1 (3.2)	81.6 (6.4)	78.8 (7.6)	79.0 (1.5)
2.5	79.9 (1.2)	78.1 (5.2)	77.2 (7.7)	76.1 (4.6)
Submerged Culture				
500	87.5 (3.1)	88.4 (4.2)	84.4 (4.7)	86.4 (5.0)
100	81.4 (3.8)	83.3 (4.4)	78.6 (5.1)	81.1 (4.1)
50	82.2 (3.2)	81.2 (4.1)	80.7 (2.3)	78.2 (4.2)
10	83.2 (6.1)	83.4 (5.4)	84.1 (6.7)	82.4 (4.9)
2.5	77.6 (2.3)	83.4 (3.4)	78.2 (1.7)	78.3 (1.5)
Slant Culture				
500	85.4 (1.4)	85.2 (3.4)	86.3 (2.6)	83.4 (0.9)
100	85.2 (3.6)	84.3 (4.2)	81.7 (3.8)	81.2 (1.3)
50	85.7 (8.9)	78.2 (3.2)	81.1 (2.7)	79.4 (1.1)
10	84.2 (7.1)	83.2 (5.7)	82.2 (5.0)	83.4 (7.3)
2.5	79.2 (3.1)	82.2 (3.1)	78.5 (0.9)	77.3 (7.2)

To validate our method, AFs were extracted from known aflatoxigenic and non-aflatoxigenic *Aspergillus* strains grown in three different culture conditions. *A. flavus* NRRL 3357 was able to produce 879 and 7.8 ng/mL of AFB1 and AFB2 (Figure 4A,B), respectively, when the fungus was cultured in solid agar with a total amount of 13,302 ng per plate. This strain produced 2041.9 and 221.1 ng/mL of

AFB1 and AFB2, respectively, when grown in liquid culture medium. On a slant cultivation, NRRL 3357 yielded 1100 and 11.49 ng/mL of AFB1 and AFB2, respectively. *A. parasiticus* NRRL 2999 was able to produce 398.27, 2.98, 207.8, and 10.19 ng/mL of AFB1, AFB2, AFG1, and AFG2, respectively, when it was inoculated onto an agar plate with a total amount of 9288.6 ng per plate. In liquid cultivation, this fungus was able to produce 508.2, 24.21, 339.3, and 42.6 ng/mL for AFB1, AFB2, AFG1, and AFG2 (Figure 4C,D), respectively. It was able to produce 310, 10.1, 437.86, and 37.66 ng/mL of AFB1, AFB2, AFG1, and AFG2, respectively, when the fungus was cultured in a slant tube. No peaks were detected within the expected retention times for *Aspergillus oryzae* NRRL 3483 grown on any of three cultivation mediums. Representative chromatograms of *A. oryzae* NRRL 3483 grown in slant cultivation medium are shown in Figure 4E,F. In addition, *A. oryzae* M2040 and *A. oryzae* NRRL RIB40 were unable to produce any types of AFs when grown in different culture medium, as shown in previous studies [28,29].

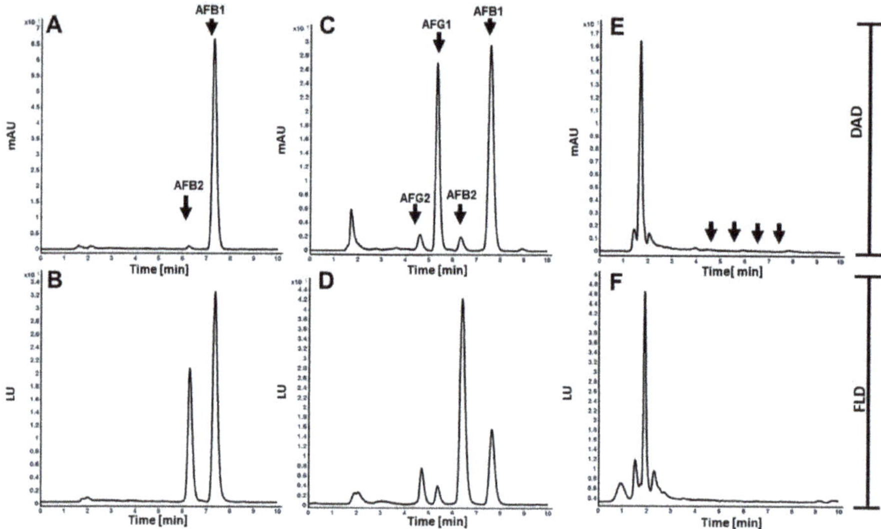

Figure 4. Analyses of aflatoxins from three different *Aspergillus* species. Representative HPLC chromatograms of aflatoxins in five-day potato dextrose agar (PDA) solid culture of *Aspergillus flavus* NRRL 3357 detected by DAD (**A**) and FLD (**B**), in five-day potato dextrose broth (PDB) submerged culture of *A. parasiticus* NRRL 2999 detected by DAD (**C**) and FLD (**D**), and in five-day PDB slant culture of *Aspergillus oryzae* M2040 detected by DAD (**E**) and FLD (**F**) are shown. Note that no aflatoxins are detectable from the culture of this food-grade strain.

In summary, in this work, we report a HPLC method coupled with DAD and FLD detectors that would be the first and only validated tool for simultaneous quantitation of AFB1, AFB2, AFG1, and AFG2 in three different laboratory culture conditions. This was an effective tool for quantitative screening of AFs in diverse *Aspergillus* strains. Chloroform was used as the extraction solvent to avoid emulsion formation—the mixture separates into two layers with AFs in the chloroform layer, thus reducing toxin loss and leaving other compounds in the aqueous layer. The extraction and cleanup procedures can be performed in less than 10 min and do not require the use of large amount of solvent or immune-affinity columns (IAC). The HPLC analysis is to be performed without any pre- or post-column derivatization reagents or any fluorescent enhancers. Peaks of the four AFs are separated in less than 10 min with high selectivity, linearity, and recovery. Finally, our method provides sufficient sensitivity to enable AF detection within mixtures at ppb levels for AFB1 and AFG1, and at parts per trillion (ppt) levels for AFB2 and AFG2 via FLD detection. In addition, our method can by readily available and easily applied in most mycology laboratories.

3. Materials and Methods

3.1. Chemicals and Materials

Individual AF standards of AFB1, AFB2, AFG1, and AFG2 were purchased from Sigma Chemical Co. (St. Louis, MO, USA). HPLC-grade acetonitrile and methanol (Merck, Darmstadt, Germany) were used for the preparation of the mobile phase. Analytical grade chloroform (Fischer Scientific, Leicestershire, United Kingdom) was used for extraction of aflatoxins. Reverse osmosis (RO) water was used for the preparation of the mobile phase, culture medium, and 0.1% Tween solution. Membrane filters (0.45 μm with 47 mm diameter), polytetrafluoroethylene (PTFE) syringe filters (0.45 μm with 17 mm diameter) and Polysorbate 80 (Tween-80) were purchased from Thermo Fisher Scientific (Rockwood, TN, USA). Both potato dextrose agar (PDA) and potato dextrose broth (PDB) were purchased from BD Difco Laboratories (Sparks, MD, USA).

3.2. Preparation of Aflatoxin Standards

Standard solutions of each of the four representative AFs AFB1, AFB2, AFG1, and AFG2 (Sigma) were prepared in acetonitrile at a final concentration of 10 μg/mL (part per million; ppm) according to the Association of Official Analytical Chemists (AOAC) method [30]. To prepare 10 μg/mL individual AF stock standard solutions, 10 mg of each AF was weighed into a separate 100 mL volumetric flask. Acetonitrile (50 mL) was added to each flask, mixed, and further added to the mark and mixed again. Then, 10 mL of this solution was transferred into another 100 mL volumetric flask and diluted to the mark with acetonitrile. Working solutions (individual or mixture) were prepared in acetonitrile and stored at −20 °C in amber glass vials for the study period (up to 3 months). The AF standard solutions used for the HPLC calibration curve were prepared by further dilution of the working solutions with the mobile phase.

3.3. Fungal Strains

Aspergillus flavus NRRL 21,882 (Afla-Guard), in which the entire AF biosynthetic gene cluster was deleted, was used as a non-aflatoxigenic strain [28,31] for the recovery experiments. The food-grade *Aspergillus oryzae* RIB40 [29], *A. oryzae* M2040 [28], and *A. oryzae* NRRL 3483 were used as non-aflatoxigenic strains. The aflatoxigenic strain *A. flavus* NRRL 3357 was used as a positive control, as it is a well-known AFB1 and AFB2 producer in lab and fields [3]. In addition, *A. parasiticus* NRRL 2999 was used as a positive control for AFB and AFG production [32]. All fungal strains were maintained on potato dextrose agar (PDA) medium (containing 4 g potato starch, 20 g glucose, and 15 g agar in 1 L of distilled water) at 4 °C. This medium, which has a high carbohydrate content and an acidic pH (5.1), was selected because it enhanced mold growth and aflatoxin production [28]. To prepare inoculum for these fungi, all were grown on PDA for 7 days at 30 ± 2 °C. *A. parasiticus* NRRL 2999 was grown on PDA for 7 d at 25 ± 2b °C. Spores were harvested from individual cultures using 0.1% Tween-80 solution. Asexual spores (conidia) were counted with a hemocytometer, and numbers were adjusted to 1 × 10^8 conidia/mL with sterile RO water. Fungal spore suspensions were stored at 4 °C and used within one week of preparation.

3.4. Culture Conditions

Fungal spore suspensions were used as a source of the inoculum for all cultivation states. *A. flavus* NRRL 21,882 was inoculated into three cultivation states: submerged cultivation, solid state, and semi-solid (slant). The incubation temperature for all strains except *A. parasiticus* (25 ± 2 °C) was 30 ± 2 °C. For submerged cultivation, 250 mL Erlenmeyer flasks were filled with 100 mL PDB and inoculated with fungal strains at a 5 × 10^5 conidia/flask. All flasks were incubated at 30 ± 2 °C with shaking at 220 rpm for 5 days. For solid state cultivation, Petri dishes (100 × 15 mm) containing about 25 mL of PDA were inoculated with fungal strains (5×10^5 conidia/plate) using a micropipette, and the spores were spread onto the plate surface by streaking. The plates were inverted and incubated at $30 \pm 2\,^{\circ}$C for 5 days. For slant cultivation, screwcap 25 mL glass test tubes were filled with 10 mL of PDB and inoculated with fungal strains (5×10^5 conidia/tube) using a micropipette, and then the spores were streaked back and forth from the bottom to the top of the slant using an inoculating loop. The tubes were placed in a rack and positioned at a 45° angle in the incubator at $30 \pm 2\,^{\circ}$C for 5 days.

3.5. Extraction of Aflatoxins from Cultures

A flow diagram for the extraction of aflatoxins from three different culture conditions is shown in Figure 5. Aflatoxins (B1, B2, G1, and G2) were extracted from the submerged culture by liquid–liquid extraction. Briefly, 1.0 mL aliquot of the fungal culture broth was mixed with 1.5 mL of chloroform in a 15 mL centrifuge tube and vigorously shaken by hand for 10 s followed by vortexing for 30 s. The mixture was then centrifuged for 2.5 m at $5000\times g$. The organic phase in the lower layer was transferred to a new glass vial. The sample residue was re-extracted with another 1.5 mL of chloroform to recover traces of AFs, which might have been present after the first extraction. The two chloroform extracts were combined and evaporated to complete dryness under a gentle stream of air. The dried extract was reconstituted with 1.0 mL of mobile phase. AFs from solid culture were extracted by adding 15.0 mL of 0.1% Tween-80 solution; the conidia were then harvested by gently scraping the top layer and collected into a 50 mL centrifuge tube. Spore suspension was homogenized by vortexing for 30 s. One mL of this suspension was transferred to a new centrifuge tubes (15 mL), and 1.5 mL of chloroform was added. The extractions were performed as described above for the liquid culture. For the liquid slant culture, the cultivated tubes were vortexed for 30 s, and 1.0 mL of the suspension was transferred to a new centrifuge tube (15 mL). Then, 1.5 mL of chloroform was added to the tubes and treated as described above. All samples were filtered into HPLC vials through a PTFE 0.45 μm syringe filter prior to HPLC analysis.

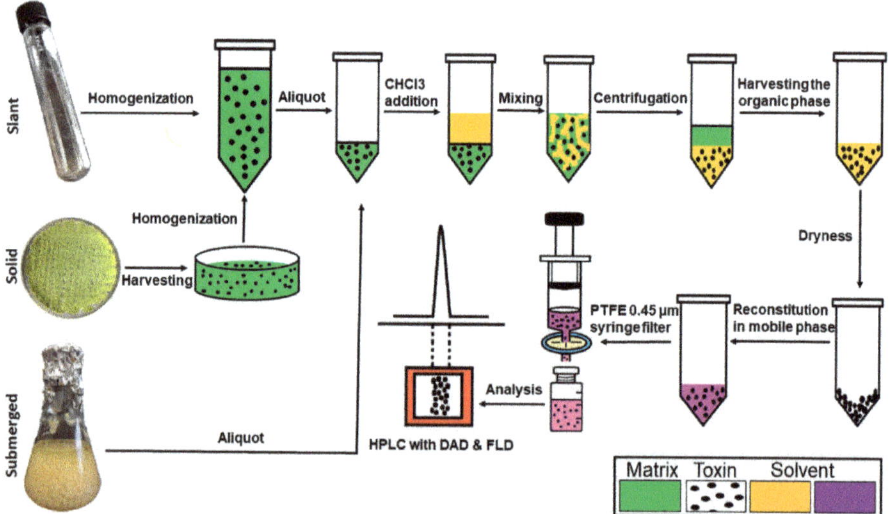

Figure 5. A process flow diagram for the extraction of aflatoxins from three different culture conditions.

3.6. HPLC Analysis of Aflatoxins

Samples were analyzed for AFB1, AFB2, AFG1, and AFG2 using a model 1100 HPLC system consisting of a degasser, an autosampler, and a quaternary pump, and equipped with a diode array

1260 Infinity (DAD) and fluorescence 1260 Infinity II (FLD) detectors connected in series (Agilent Technologies, Santa Clara, CA, USA). The separation was performed via a Zorbax Eclipse XDB-C18 4.6 × 150 mm, 3.5 µm column with a temperature set at 30 °C. Samples were monitored at a wavelength of 365 nm for UV detection and at 365 nm excitation and 450 nm emission for FLD detection. The samples were eluted with the mobile phase of water (H_2O)/methanol (CH_3OH)/acetonitrile (CH_3CN) (50:40:10) at a flow rate of 0.8 mL/m. The mobile phase was degassed and filtered through a membrane filter (47 mm × 0.45 µm) prior to use. The injection volume was 100 µL. Peak areas of AFs were recorded and integrated using the ChemStation software (Agilent).

3.7. Method Validation

The method employed for the extraction and simultaneous analysis of AFB1, AFB2, AFG1, and AFG2 in the laboratory culture conditions was validated according to the AOAC Guidelines Appendix F [33], with slight modifications, by determining the recovery, precision, selectivity, linearity, and the sensitivity. A mixture of known concentrations of four AFs (500, 100, 50, 10, and 2.5 of each) were spiked into blank culture samples (submerged, slant, and solid state) for the recovery validation. Each concentration was spiked in triplicate, and the experiments were repeated three times within a day and repeated in 3 consecutive days by the same operator. Precision was demonstrated as repeatability, which was evaluated by calculating the relative standard deviation (% RSD) of the spiked toxins repeated within 1 day and over 3 consecutive days. Blank samples were prepared by inoculating non-aflatoxin-producing *A. flavus* NRRL 21,881 in submerged, solid, and slant cultures. The samples were then harvested, and AFs were extracted and analyzed by HPLC coupled with DAD and FLD. The selectivity of this method was assured as there were no interfering chromatographic peaks corresponding to the retention time of the four AFs. The linearity was demonstrated for the AFs in the range of 2.5 to 1000 ppb in three replicates. The calibration standard of each concentration was constructed using the peak-area ratio of the AFs versus the concentration of the analytes. The linear relationship was evaluated by the correlation coefficient, y-intercept, and slope of the regression line. The sensitivity of the method was determined by measuring the LOD and the LOQ on the basis of a signal-to-noise ratio (S/N) of 3:1 and 6:1, respectively.

3.8. Aflatoxin Analyses from Cultures of Aspergillus Species

To verify the protocol of the AF quantification from the cultures, two common strains of aflatoxin-producing fungi, *A. flavus* NRRL 3357 and *A. parasiticus* NRRL 2999, were used and tested for aflatoxin production. In addition, three of the non-aflatoxin producing *Aspergillus* strains were used as a negative control: *A. oryzae* M2040, *A. oryzae* NRRL RIB40, and *A. oryzae* NRRL 3483. Fungal strains were grown in three culture conditions in triplicate, as mentioned previously; samples were then harvested after 5 days of incubation, and AFs were extracted and analyzed by HPLC coupled with DAD and FLD.

3.9. Statistical Analyses

The method was optimized and validated with a statistical treatment to increase the AF recovery and to save time and reagent waste. AFs peaks were simultaneously separated with no interfering. Significance ($p < 0.05$) of the data was analyzed by using a Student's *t*-test. The relative standard deviation (RSD) was calculated using Equation (1).

$$RSD = Si \times 100/x \tag{1}$$

S = standard deviation, and x = mean of the data.

Author Contributions: Designed the experiments, A.F.A. and J.-H.Y.; Performed the whole experiments and drafted the paper, A.F.A.; Supervised the experiments and revised the draft, J.-H.Y. All authors have read and agreed to the published version of the manuscript.

Funding: We thank Wendy Bedale of the Food Research Institute at the University of Wisconsin-Madison for critically reviewing the manuscript. Support for this research was provided by Food Research Institute and the University of Wisconsin-Madison Office of the Vice Chancellor for Research and Graduate Education (OVCRGE) with funding from the Wisconsin Alumni Research Foundation.

Conflicts of Interest: The authors declare that the research was conducted in the absence of any commercial or financial relationships that could be construed as a potential conflict of interest.

References

1. Ostry, V.; Malir, F.; Toman, J.; Grosse, Y. Mycotoxins as human carcinogens-the IARC Monographs classification. *Mycotoxin Res.* **2017**, *33*, 65–73. [CrossRef]
2. Loomis, D.; Guha, N.; Hall, A.L.; Straif, K. Identifying occupational carcinogens: an update from the IARC Monographs. *Occup. Environ. Med.* **2018**, *75*, 593. [CrossRef]
3. Klich, M.A. Aspergillus flavus: the major producer of aflatoxin. *Mol. Plant Pathol.* **2007**, *8*, 713–722. [CrossRef]
4. Alshannaq, A.; Yu, J.-H. Occurrence, Toxicity, and Analysis of Major Mycotoxins in Food. *Int. J. Env. Res. Public Health* **2017**, *14*, 632. [CrossRef]
5. Wu, F.; Guclu, H. Aflatoxin regulations in a network of global maize trade. *Plos One* **2012**, *7*, e45151. [CrossRef]
6. Weaver, M.A.; Mack, B.M.; Gilbert, M.K. Genome Sequences of 20 Georeferenced Aspergillus flavus Isolates. *Microbiol. Resour. Announc.* **2019**, *8*, e01718-18. [CrossRef]
7. Van der Fels-Klerx, H.J.; Vermeulen, L.C.; Gavai, A.K.; Liu, C. Climate change impacts on aflatoxin B1 in maize and aflatoxin M1 in milk: A case study of maize grown in Eastern Europe and imported to the Netherlands. *Plos One* **2019**, *14*, e0218956. [CrossRef]
8. Medina, A.; Rodriguez, A.; Magan, N. Effect of climate change on Aspergillus flavus and aflatoxin B1 production. *Front. Microbiol.* **2014**, *5*, 348. [CrossRef]
9. Assuncao, R.; Martins, C.; Viegas, S.; Viegas, C.; Jakobsen, L.S.; Pires, S.; Alvito, P. Climate change and the health impact of aflatoxins exposure in Portugal—An overview. *Food Addit. Contam. Part A* **2018**, *35*, 1610–1621. [CrossRef]
10. Hruska, Z.; Yao, H.; Kincaid, R.; Brown, R.; Cleveland, T.; Bhatnagar, D. Fluorescence Excitation–Emission Features of Aflatoxin and Related Secondary Metabolites and Their Application for Rapid Detection of Mycotoxins. *Food Bioprocess Technol.* **2014**, *7*, 1195–1201. [CrossRef]
11. Fente, C.A.; Ordaz, J.J.; Vázquez, B.I.; Franco, C.M.; Cepeda, A. New additive for culture media for rapid identification of aflatoxin-producing Aspergillus strains. *Appl. Env. Microbiol.* **2001**, *67*, 4858–4862. [CrossRef]
12. Maragos, C.M.; Appell, M.; Lippolis, V.; Visconti, A.; Catucci, L.; Pascale, M. Use of cyclodextrins as modifiers of fluorescence in the detection of mycotoxins. *Food Addit. Contam. Part A* **2008**, *25*, 164–171. [CrossRef]
13. Sharma, A.; Khan, R.; Catanante, G.; Sherazi, A.T.; Bhand, S.; Hayat, A.; Marty, L.J. Designed Strategies for Fluorescence-Based Biosensors for the Detection of Mycotoxins. *Toxins* **2018**, *10*, 197. [CrossRef]
14. Drott, M.T.; Fessler, L.M.; Milgroom, M.G. Population Subdivision and the Frequency of Aflatoxigenic Isolates in Aspergillus flavus in the United States. *Phytopathology* **2019**, *109*, 878–886. [CrossRef]
15. Camiletti, B.X.; Torrico, A.K.; Maurino, M.F.; Cristos, D.; Magnoli, C.; Lucini, E.I.; Pecci, M.D.L.P.G. Fungal screening and aflatoxin production by Aspergillus section Flavi isolated from pre-harvest maize ears grown in two Argentine regions. *Crop Prot.* **2017**, *92*, 41–48. [CrossRef]
16. Baranyi, N.; Despot, D.J.; Palagyi, A.; Kiss, N.; Kocsube, S.; Szekeres, A.; Kecskemeti, A.; Bencsik, O.; Vagvolgyi, C.; Klaric, M.S.; et al. Identification of Aspergillus species in Central Europe able to produce G-type aflatoxins. *Acta Biol. Hung.* **2015**, *66*, 339–347. [CrossRef]
17. Okoth, S.; De Boevre, M.; Vidal, A.; Diana Di Mavungu, J.; Landschoot, S.; Kyallo, M.; Njuguna, J.; Harvey, J.; De Saeger, S. Genetic and Toxigenic Variability within Aspergillus flavus Population Isolated from Maize in Two Diverse Environments in Kenya. *Front. Microbiol.* **2018**, *9*, 57. [CrossRef]
18. Saldan, N.C.; Almeida, R.T.R.; Avincola, A.; Porto, C.; Galuch, M.B.; Magon, T.F.S.; Pilau, E.J.; Svidzinski, T.I.E.; Oliveira, C.C. Development of an analytical method for identification of Aspergillus flavus based on chemical markers using HPLC-MS. *Food Chem.* **2018**, *241*, 113–121. [CrossRef]
19. Abbas, H.K.; Zablotowicz, R.M.; Weaver, M.A.; Horn, B.W.; Xie, W.; Shier, W.T. Comparison of cultural and analytical methods for determination of aflatoxin production by Mississippi Delta Aspergillus isolates. *Can. J. Microbiol.* **2004**, *50*, 193–199. [CrossRef]

20. Sadhasivam, S.; Britzi, M.; Zakin, V.; Kostyukovsky, M.; Trostanetsky, A.; Quinn, E.; Sionov, E. Rapid Detection and Identification of Mycotoxigenic Fungi and Mycotoxins in Stored Wheat Grain. *Toxins* **2017**, *9*, 302. [CrossRef]
21. Gell, R.M.; Carbone, I. HPLC quantitation of aflatoxin B1 from fungal mycelium culture. *J. Microbiol. Methods* **2019**, *158*, 14–17. [CrossRef]
22. Garcia, M.; Blanco, J.L.; Suarez, G. Aflatoxins B1 and G1 solubility in standard solutions and stability during cold storage. *Mycotoxin Res.* **1994**, *10*, 97–100. [CrossRef]
23. Rogers, H.A. How Composition Methods Are Developed and Validated. *J. Agric. Food Chem.* **2013**, *61*, 8312–8316. [CrossRef]
24. Şengül, Ü. Comparing determination methods of detection and quantification limits for aflatoxin analysis in hazelnut. *J. Food Drug Anal.* **2016**, *24*, 56–62. [CrossRef]
25. Currie, L.A. Detection and quantification limits: origins and historical overview1Adapted from the Proceedings of the 1996 Joint Statistical Meetings (American Statistical Association, 1997). Original title: "Foundations and future of detection and quantification limits". Contribution of the National Institute of Standards and Technology; not subject to copyright.1. *Anal. Chim. Acta* **1999**, *391*, 127–134. [CrossRef]
26. Danzer, K. Selectivity and specificity in analytical chemistry. General considerations and attempt of a definition and quantification. *Fresenius' J. Anal. Chem.* **2001**, *369*, 397–402. [CrossRef]
27. Trucksess, M.W.; Weaver, C.M.; Oles, C.J.; Fry, F.S.; Noonan, G.O.; Betz, J.M.; Rader, J.I. Determination of aflatoxins B1, B2, G1, and G2 and ochratoxin A in ginseng and ginger by multitoxin immunoaffinity column cleanup and liquid chromatographic quantitation: collaborative study. *J. Aoac Int.* **2008**, *91*, 511–523. [CrossRef]
28. Alshannaq, A.F.; Gibbons, J.G.; Lee, M.K.; Han, K.H.; Hong, S.B.; Yu, J.H. Controlling aflatoxin contamination and propagation of Aspergillus flavus by a soy-fermenting Aspergillus oryzae strain. *Sci. Rep.* **2018**, *8*, 16871. [CrossRef]
29. Rank, C.; Klejnstrup, M.L.; Petersen, L.M.; Kildgaard, S.; Frisvad, J.C.; Held Gotfredsen, C.; Ostenfeld Larsen, T. Comparative Chemistry of Aspergillus oryzae (RIB40) and A. flavus (NRRL 3357). *Metabolites* **2012**, *2*, 39–56. [CrossRef]
30. AOAC Official Methods of Analysis. *Preparation of Standards for Aflatoxins*; AOAC International: Gaithersburg, MD, USA, 2005; Chapter 49.
31. Dorner, J.W.; Lamb, M.C. Development and commercial use of afla-Guard((R)), an aflatoxin biocontrol agent. *Mycotoxin Res.* **2006**, *22*, 33–38. [CrossRef]
32. Wilson, D.M.; King, J.K. Production of aflatoxins B1, B2, G1, and G2 in pure and mixed cultures of Aspergillus parasiticus and Aspergillus flavus. *Food Addit. Contam.* **1995**, *12*, 521–525. [CrossRef] [PubMed]
33. AOAC Official Methods of Analysis. *Guidelines for Standard Method Performance Requirements*; Appendix F; AOAC International: Gaithersburg, MD, USA, 2016.

Article

Contamination of Zearalenone from China in 2019 by a Visual and Digitized Immunochromatographic Assay

Xia Hong [†], Yuhao Mao [†], Chuqin Yang, Zhenjiang Liu, Ming Li * and Daolin Du *

Institute of Environmental Health and Ecological Security, School of the Environment and Safety Engineering, Jiangsu University, Xuefu Road 301, Zhenjiang 212013, China; hongxiaujs@hotmail.com (X.H.); maoyuhao7@gmail.com (Y.M.); yangchuqin18@gmail.com (C.Y.); lzj1984@ujs.edu.cn (Z.L.)
* Correspondence: liming@ujs.edu.cn (M.L.); ddl@ujs.edu.cn (D.D.); Tel.: +86-511-8879-0955 (M.L. & D.D.)
† These authors contributed equally to this work.

Received: 10 July 2020; Accepted: 12 August 2020; Published: 14 August 2020

Abstract: Zearalenone (ZEN) is a prevalent mycotoxin that needs intensive monitoring. A semi-quantitative and quantitative immunochromatographic assay (ICA) was assembled for investigating ZEN contamination in 187 samples of cereal and their products from China in 2019. The semi-quantitative detection model had a limit of detection (LOD) of 0.50 ng/mL with visual judgment and could be completely inhibited within 5 min at 3.0 ng/mL ZEN. The quantitative detection model had a lower LOD of 0.25 ng/mL, and ZEN could be accurately and digitally detected from 0.25–4.0 ng/mL. The ICA method had a high sensitivity, specificity, and accuracy for on-site ZEN detection. For investigation of the authentic samples, the ZEN-positive rate was 62.6%, and the ZEN-positive levels ranged from 2.7 to 867.0 ng/g, with an average ZEN-positive level being 85.0 ng/g. Of the ZEN-positive samples, 6.0% exceeded the values of the limit levels. The ZEN-positive samples were confirmed to be highly correlated using LC-MS/MS (R^2 = 0.9794). This study could provide an efficiency and accuracy approach for ZEN in order to achieve visual and digitized on-site investigation. This significant information about the ZEN contamination levels might contribute to monitoring mycotoxin occurrence and for ensuring food safety.

Keywords: zearalenone; immunochromatographic assay; semi-quantification; quantification

Key Contribution: A semi-quantitative and quantitative test strip had been developed and applied for investigating zearalenone contamination in 187 authentic samples in China in 2019.

---

1. Introduction

Zearalenone (ZEN) is a common secondary metabolite from the *Fusarium* species, which has become one of the most widespread mycotoxins and has caused substantial economic losses to grains around the world [1,2]. ZEN exposure could cause the genotoxic, hepatoxic, immunotoxic, and even estrogenic effects [3–6]. Thus, ZEN is categorized as a class III carcinogen [7]. Recent investigations into mycotoxin contamination from all over the world found that a high percentage of ZEN contamination exists in cereals and animal feed [8–10]. To better monitor ZEN contamination and maintain human health, the maximum limits (MLs) of ZEN in unprocessed cereals and unprocessed maize have been regulated by the European Commission (EC) to be no more than 100 ng/g and 350 ng/g, respectively [11]. The Expert Committee of both the Food and Agriculture Organization (FAO) and the World Health Organization (WHO) had set a provisional maximal tolerable daily intake of 0.5 ng/g of body weight for ZEN [12]. The China National Standard regulates that ZEN should be no more than 60 ng/g in wheat, wheat flour, corn, and corn flour [13].

High performance liquid chromatography (HPLC) [14], HPLC-tandem mass spectrometry (HPLC-MS) [15], and immunoassays [16,17] are used for detecting ZEN. Compared with the referenced HPLC and HPLC-MS methods, immunoassays have the characteristics of rapidity, being low cost, and being able to achieve high throughput screening on-site for a large number of samples [18–21]. Of the variety of immunoassays, the immunochromatographic assay (ICA) has attracted more attention. It is widely used for detecting contamination, because of its outstanding characteristics of simplicity, readability, and portability [22,23].

ICA methods have been developed and applied for ZEN detection. An ICA strip has been used for the rapid detection of ZEN in wheat from Jiangsu, China, with a limit of detection (LOD) of 50 ng/mL, which was applied in 202 real wheat samples [24]. The ICA method based on the quantum dot nanobead and biotin-streptavidin system for the determination of aflatoxin B_1 (AFB_1) and ZEN was developed, which improved the sensitivity of ZEN detection to 59.15 pg/mL [25]. A quantum dot microbead based fluorometric lateral flow ICA was developed for the simultaneous detection of AFB_1, deoxynivalenol (DON), and ZEN, and the LOD reached 1.92 ng/g for ZEN in the real cereal samples [26]. Furthermore, amorphous carbon nanoparticles, aptamer, dyed latex microspheres, and other novel materials have been used to improve the performance of ICA for detecting ZEN [27–29]. The above-mentioned ICA methods for ZEN promoted detection sensitivity, for which it also showed the characteristics of convenience, rapidity, economy, visual detection on-site, and could even get accurate levels of ZEN contamination using the quantitative approach.

Given the high ZEN contamination rate and the large number of samples to be examined, a higher performance detection method for ZEN was developed and improved in this study. For this purpose, a practical ICA based on two judgment models for the semi-quantitative detection and quantitative detection of ZEN was developed and applied in authentic cereals and feeds (Figure 1). Combining the naked eye and strip reader, the ICA could quickly achieve the visualization and digitalization for ZEN detection, while improving the detection sensitivity and providing an alternative detection method for ZEN. The proposed ICA method was applied to detect the ZEN contamination levels in 187 samples of cereal and their products from China in 2019. Then, the referenced LC-MS/MS was used to verify the accuracy of the ICA method.

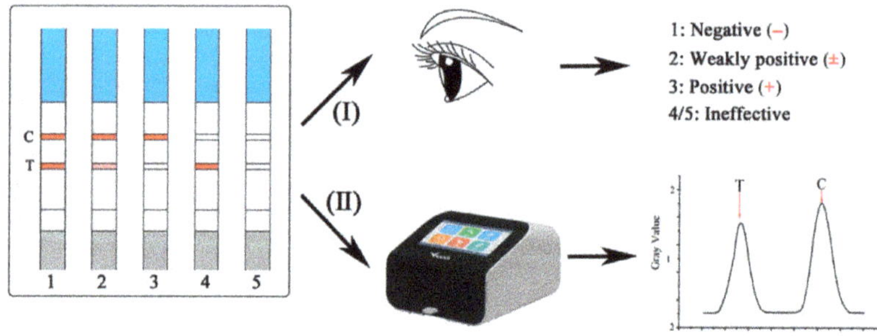

Figure 1. Two models of result judgment for the immunochromatographic assay (ICA).

2. Results and Discussion

2.1. Identification of the Gold Nanoparticles

The image of the transmission electron micrograph showed that the prepared colloidal gold nanoparticles (GNPs) were about 17 nm and the particles exhibited a uniform distribution (Figure 2A). The ultraviolet–visible spectra (UV/vis) for the GNPs solution had a smooth protruding peak at 518 nm. Qualitative differences were found between the UV/vis spectra of the ZEN-monoclonal antibody (McAb)-GNP probes and GNPs (Figure 2B). The ZEN-McAb-GNP probes showed a smooth protruding peak at 527 nm, indicating that the GNPs had been successfully coupled in ZEN-McAb to produce probes. Further studies of the ICA evaluation and application also demonstrated the availability and excellent performance of the prepared GNPs.

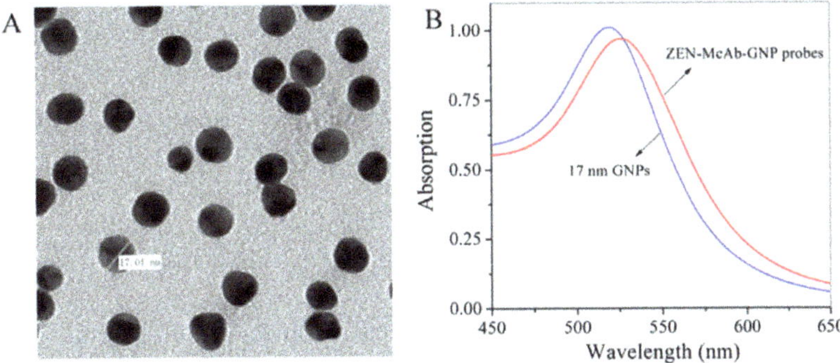

Figure 2. The transmission electron micrograph image for the gold nanoparticles (GNPs) (**A**) and the UV/vis spectra for the GNPs and zearalenone (ZEN)-monoclonal antibody (McAb)-GNP probes (**B**).

2.2. Key Parameters for the ICA

It was a key procedure to optimize the parameters for the preparation of the ZEN-McAb-GNP probes and the ICA strips, which could be beneficial for improving the sensitivity of the ICA method. The evaluation criterion was based on using fewer ZEN in order to reduce the red color intensity in the T line and for maintaining the effectiveness of ICA. The optimal parameters for the ICA are shown in Table 1. For preparing the ZEN-McAb-GNP probes, 2.3 µL of 0.2 mol/L K_2CO_3 in an aqueous solution, 3.0 µL of a 1.0 mg/mL ZEN-McAb in a 10 mmol/L PB buffer, and 100 µL of 10% bovine serum albumin (BSA) in purified water were used in 1 mL of GNPs. For spraying a 30 cm of nitrocellulose (NC) membrane, the 0.5 µL of 13.1 mg/mL goat anti-mouse IgG (GAM-IgG) and 1.0 µL of 8.8 mg/mL ZEN-antigen in 30 µL of 10 mmol/L of phosphate buffered saline (PBS buffer) were immobilized on the C line and T line, respectively.

Table 1. The optimization of key parameters for the preparation of ICA for ZEN.

Parameter	Concentration	Volume	Solvent
GAM-IgG	13.1 mg/mL	0.5 µL	10 mmol/L PBS (containing 0.15 mol/L NaCl, pH 7.4)
ZEN-antigen	8.8 mg/mL	1.0 µL	10 mmol/L PBS (containing 0.15 mol/L NaCl, pH 7.4)
ZEN-McAb	1.0 mg/mL	3.0 µL	10 mmol/L PB
BSA	10%	100 µL	Purified water
K_2CO_3	0.2 mol/L	2.3 µL	Purified water

Note: The concentrations and volumes of ZEN-McAb and K_2CO_3 in this table are used in 1 mL GNPs. Other parameters were used for preparing a 30 cm of NC membrane.

2.3. Semi-Quantitative Detection of the ICA

A series of ZEN standards (0, 0.25, 0.5, 0.7, 1.0, 1.5, 2.0, 3.0, 4.0, and 5.0 ng/mL in 20% methanol of 10 mmol/L of PBS buffer) were detected by the developed ICA. With the increasing in ZEN standards, the red color intensities were in a gradually decreasing process, and eventually disappeared for the T lines (Figure 3). The red color intensities for the C lines always existed to prove the effectivity of the ICA. It was observed that a significant reduction in the red color in the T line was shown in 0.50 ng/mL of ZEN. Thus, the semi-quantitative LOD of the ICA could be established as 0.50 ng/mL. The red color intensities eventually disappeared in the T line when the concentration of ZEN exceeded 3.0 ng/mL. These results of the ICA judged by the naked eye indicated that the visual detection of the ZEN levels could be in three intervals: <0.50 ng/mL (−, negative), 0.50 ng/mL ≤ZEN concentration <3.0 ng/mL (±, weakly positive), and ≥3.0 ng/mL (+, positive).

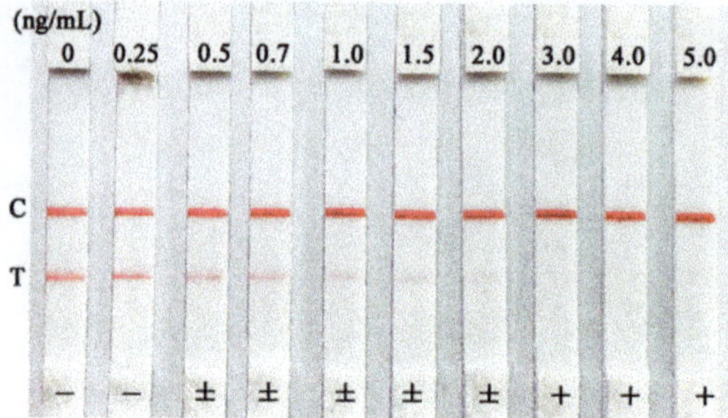

Figure 3. The semi-quantitative detection of ICA for the series of ZEN standards.

2.4. Quantitative Detection of the ICA

The strips of ICA were digitally detected by the strip reader, and the gray values of the T lines were plotted against the concentrations of ZEN in order to get a reduction curve (Figure 4A). This showed a good linear relationship between the inhibition ratios of the gray values in the T lines and the logarithm of the ZEN concentrations (Figure 4B; Y = 81.88X + 37.86, R^2 = 0.9977). In this case, ZEN could achieve linear detection in the range of 0.25 to 4.0 ng/mL, and the quantitative LOD could be defined as 0.25 ng/mL (Table 1). With the semi-quantitative and quantitative LOD of the ICA all below the MLs for ZEN, the sensitivity of the developed ICA was satisfied for detecting ZEN.

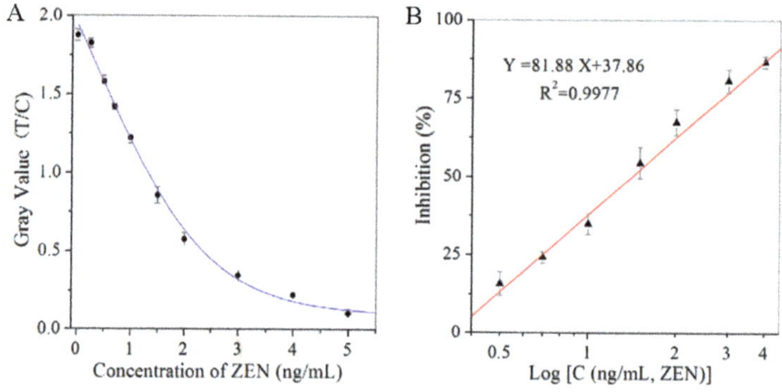

Figure 4. The quantitative detection of ICA for the series of ZEN standards. (**A**) The corresponding relationship between the gray value (T/C) and the concentration of ZEN (ng/mL); (**B**) The linear relationship between the inhibition (%) and the logarithm of the ZEN concentrations (ng/mL).

Comparing the developed ICA with the enzyme-linked immunosorbent assay (ELISA) for ZEN, ICA showed more convenient and rapid operation steps for on-site detection, which could achieve ZEN detection within 5 min using one step. Two models of judgment using the naked eye and strip reader could allow for the ICA detection to be more flexible. The digitized detection of ICA showed a lower LOD and a wider detection range than the visual detection. It was close to the sensitivity of the enzyme-linked immunosorbent assay (ELISA; Table S1 and Figure S1). Moreover, the digitized detection of ICA could obtain accurate ZEN levels in order to realize quantitative detection. It was suggested that the visual and digitized detection of ICA for ZEN contamination could be selected as needed or used simultaneously in order to achieve semi-quantitative and quantitative detection for on-site testing.

2.5. Specificity of the ICA

The cross-reactivity tests (CR; %) of the ICA for the analogs of ZEN and the common mycotoxins were carried out in order to assess the specificity of the developed ICA method. It could be observed that ZEN made the red color completely disappear, and α-Zearalenol or β-Zearalenol left the red color markedly diminished, while other compounds caused the T lines to be obviously weaker than that the C line when the compounds were 5 ng/mL in the sample solutions (Figure 5). Further study showed that the CRs of the ICA method for α-Zearalenol, β-Zearalenol, α-Zearalanol, β-Zearalanol, and Zeranol were 13.0%, 10.5%, 0.5%, 0.48%, and 0.35%, respectively, which indicated that the ICA method had a slight specificity for the ZEN analogs. At the same time, the CRs of the ICA method for common mycotoxins were less than 0.1%, which indicated that the ICA method could not identify other mycotoxins. Thus, the developed ICA method had a high specificity for ZEN detection.

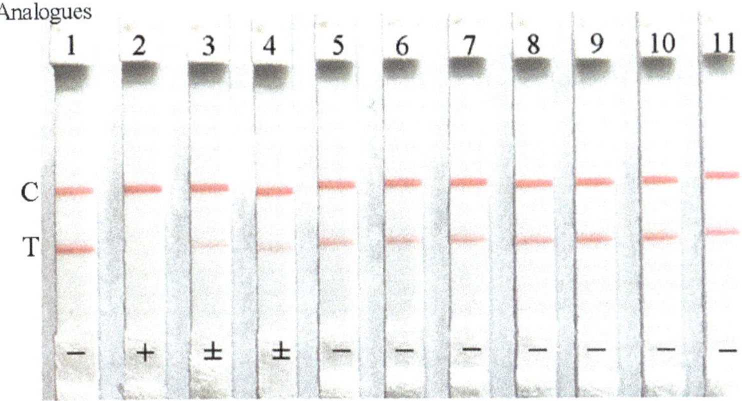

Figure 5. The cross-reactivity tests of the ICA for ZEN detection. The analogs were 5 ng/mL. (1) negative; (2) ZEN; (3) α-Zearalenol; (4) β-Zearalenol; (5) α-Zearalanol; (6) β-Zearalanol; (7) Zeranol; (8) Aflatoxin B$_1$; (9) Deoxynivalenol; (10) Fumonisin B1; (11) Ochratoxin A.

2.6. Stability of the ICA

The effective test and the sensitivity test of the ICA for ZEN over 3 months is shown in Figure S2. During the 3 months of storage life and interval sampling detection, the red color intensity of the T lines and C lines were kept clear and had distinct effects. Meanwhile, the T lines completely disappeared at 5 ng/mL of ZEN standard, which meant that the sensitivity of the ICA was well represented in this storage life. This indicates that the developed ICA for ZEN had a desirable stability, and the useful life could be at least 3 months.

2.7. Accuracy and Precision of the ICA

After the steps of pretreatment and being diluted at 10:1, the matrix effect for the spiked samples effectively decreased. The free of ZEN and 3 ng/g of ZEN in the spiked samples were judged to be negative (−), and 20 ng/g of ZEN was weakly positive (±) and 50 ng/g of ZEN was positive (+) using the ICA detection using naked eye, which we achieved by visualization and semi-quantitative detection. Then, the spiked sample extracts were quantitatively detected by the ICA coupling strip reader in order to achieve digitization with ZEN that ranged from 2.6 to 59.3 ng/g. The quantitative recoveries of the spiked samples were between 86.7 to 118.6%, with the standard deviation (SD) being from 3.1% to 6.2%. These results indicate that the accuracy and precision of the ICA were desirable for the semi-quantitative and quantitative detection for ZEN contamination in the cereal and feed samples (Table 2).

Table 2. The ZEN recoveries from using the ICA method in the spiked samples.

Sample	Spiked (ng/g)	Dilution Times	Visualization [a]	Digitization (ng/g)	Quantitative Recovery ± SD (%, $n = 5$) [b]
Corn	0	10	−	ND [c]	ND
	3		−	3.4	113.3 ± 4.2
	20		±	18.6	93.0 ± 3.9
	50		+	59.3	118.6 ± 5.5
Wheat	0	10	−	ND	ND
	3		−	2.6	86.7 ± 4.6
	20		±	22.4	112.0 ± 3.7
	50		+	45.2	90.4 ± 5.8
Feed	0	10	−	ND	ND
	3		−	3.1	103.3 ± 4.9
	20		±	23.3	116.5 ± 6.2
	50		+	56.4	112.8 ± 3.1

[a] The visual detection was based on the ZEN levels: <5 ng/g (−), 5 ng/g ≤ ZEN concentration < 30 ng/g (±); >30 ng/g (+). [b] Each value is the mean of three replicates. [c] Not detected.

2.8. Investigation of Authentic Samples by the ICA and LC-MS/MS

Using the ICA method, the ZEN contamination in the authentic samples was visually judged following the digitized detection. The results of the semi-quantitative and quantitative detection by the two models of visualization and digitization for ICA were in good agreement. Furthermore, the digitized detection of ICA could obtain accurate ZEN levels in the authentic samples, and some negative samples might have also detected ZEN contamination, which could further expand the detection ability and sensitivity of the developed ICA method. A total of 187 cereals and its product samples were investigated for ZEN contamination from China in 2019. The results of the ZEN levels using the quantitative ICA are shown in Table 3. The ZEN-positive rate was 62.6% in 117 out of the 187 total samples, which detected positivity in 28 out of 40 corn samples, 7 out of 19 wheat samples, 17 out of 39 wheat flour samples, 39 out of 49 cereal product samples, and 26 out of 40 feed samples. From the entire sample, the ZEN-positive levels ranged from 2.7 to 867.0 ng/g, with an average ZEN-positive level being 85.0 ng/g.

Table 3. The investigation of the ZEN in the original samples using the quantitative ICA from China in 2019.

Item	All	Corn	Wheat	Wheat Flour	Cereal Product	Feed
Total samples	187	40	19	39	49	40
ZEN-positive samples	117	28	7	17	39	26
ZEN-positive rate (%)	62.6	70.0	36.8	43.6	79.6	65.0
Average ZEN-positive (ng/g)	85.0	46.8	27.1	12.4	95.9	173.0
ZEN-positive range (ng/g)	2.7–867.0	3.2–743.2	3.0–117.5	6.8–21.3	2.7–677.7	7.2–867.0

The highest ZEN-positive rate was found in the cereal product samples (79.6%), which, along with the corn samples (70.0%) and the feed samples (65.0%), had above the average ZEN-positive rate (62.6%; Figure 6A). The average ZEN-positive levels for the cereal product and feed samples reached 95.9 ng/g and 173.0 ng/g, which were higher than the average ZEN-positive level of all of the samples (85.0 ng/g; Figure 6B). The lowest average ZEN-positive level was detected in the wheat flour samples, thus demonstrating the lower ZEN-exposure risk for wheat flour. To make a further assessment of the ZEN-exposure levels from China in 2019, the distribution of the ZEN-positive sample and ZEN-positive rate at different levels is shown in Figure 6C. The ZEN-positive levels were classified into five levels, namely, <10 ng/g, 10–60 ng/g, 60–100 ng/g, 100–350 ng/g, and >350 ng/g, according to the MLs from European Commission and the China National Standard. The ZEN-positive levels of 68.4% with 80 out of 117 ZEN-positive samples were lower than the ML value of China (60 ng/g), while 94% of samples had ZEN-positive levels less than the European Commission regulation of 350 ng/g. However, it is noteworthy that 6.0% in 117 of the ZEN-positive samples showed ZEN levels more than 350 ng/g, exceeding the ML value of China and the European Commission. In addition, the highest ZEN-positive level was detected as 867.0 ng/g in a feed sample, which had been 14-fold and 2.5-fold higher than the ML values for China and the European Commission, respectively. These results indicate that ZEN contamination was a commonly occurring problem for the authentic cereals and their product samples from China in 2019, and most of the contamination levels were within the bounds of the control, but some samples were seriously ZEN-positive and needed to be focused on in order to ensure food safety and human health.

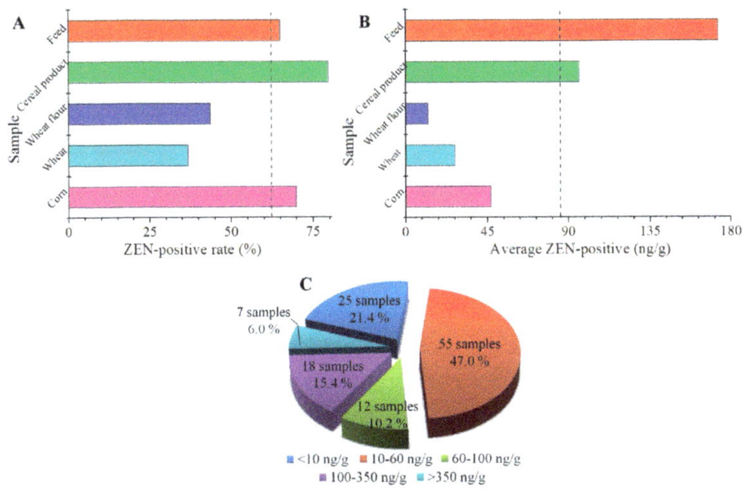

Figure 6. The distribution of ZEN-positive rate (**A**), average ZEN-positive level (**B**) and ZEN-positive sample at different levels (**C**) from China in 2019.

The LC-MS/MS detection showed that ZEN contaminations in the positive samples ranged from 3.2 to 761.7 ng/g (Table S2). The large correlation of results from the quantitative ICA method and LC-MS/MS method for detecting the ZEN in the authentic samples was $R^2 = 0.9794$ ($Y = 0.94X + 4.8442$; Figure S3), which further verified the reliability and accuracy of the proposed ICA method.

3. Conclusions

To further enhance the detection ability and ensure food safety, a semi-quantitative and a quantitative detection method of ICA were successfully established and used for detecting ZEN in cereal and feed samples. The results of the ICA could be judged visually by the naked eye or with a digitized strip reader within 5 min. The visual LOD for ZEN was 0.50 ng/mL using the semi-quantitative ICA. The quantitative ICA had a lower LOD of 0.25 ng/mL, and a wider detection range, which could obtain accurate ZEN levels. The powerful detection capability of the developed ICA was demonstrated by the evaluation of its sensitivity, specificity, stability, accuracy, and precision. The ICA could dramatically shorten the analytical procedure and the overall detection time when compared with the micro-well based ELISA or chromatographic-based HPLC method. A total of 187 samples of authentic cereals and their products from China in 2019 were investigated for ZEN contamination by both the developed ICA and the referenced LC-MS/MS, in order to demonstrate the reliability of the proposed ZEN detection method. The ZEN-positive rate was 62.6%, and the ZEN-positive levels ranged from 2.7 to 867.0 ng/g, with an average ZEN-positive level being 85.0 ng/g. The highest ZEN-positive level was detected as 867.0 ng/g in a feed sample. It is noteworthy that the ZEN contamination levels of 6.0% in 117 ZEN-positive samples exceeded the ML value of the China and European Commission. The results of this investigation suggest that ZEN contamination in China occurred widely and had a high detection rate. The efficiency and accuracy of the ZEN detection could have further improvement, and the study could provide an alternative approach and valuable information about ZEN contamination in China.

4. Materials and Methods

4.1. Reagents and Materials

ZEN, its analogs, and common mycotoxin standards were purchased from Pribolab Pte. Ltd. (Biopolis, Singapore). The BSA and GAM-IgG were supplied by Sigma-Aldrich (St. Louis, MO, USA). The trisodium citrate, $HAuCl_4$, and K_2CO_3 were bought from Aladdin (Shanghai, China). The NC membranes were from Millipore (Bedford, MA, USA). The absorbent pad, sample pad, polyvinyl chloride (PVC) sheet, and glass-fiber conjugate pad were provided by Jiening Bio. Tech. Co., Ltd. (Shanghai, China). The ZEN antigen (ZEN-BSA) and McAb against ZEN (ZEN-McAb) were supplied by Biosco Biological Tech. Co., Ltd. (Dalian, China). The other reagents were all of analytical grade.

The XYZ3030 dispense platform (Kinbio Tech. Co., Ltd., Shanghai, China) and the HGS201 automatic programmable cutter (Hangzhou Autokun Tech. Co., Ltd., Hangzhou, China) were used in the preparation of the ICA strips. The Tecnai 12 transmission electron microscope (Philips, Eindhoven, Netherlands) was used to scan the diameter and shape of the nanoparticles. The Milli-Q purification system (Millipore, Bedford, MA, USA) was used for preparing the purified water. The TG16-WS high-speed centrifuge was used for centrifuging (Cence Co., Changsha, China). The vortex mixer was provided by North TZ-Biotech Develop. Co., Ltd. (Beijing, China). The results of the ICA for detecting ZEN were confirmed with an Agilent 1200-6460 LC-MS/MS equipped with electrospray ionization (Agilent, Wilmington, DE, USA). The results of the ICA were digitized on an HG-1721 strip reader (Vict Tech. Co., Ltd., Suzhou, China).

4.2. Preparation of the Nanoparticles

The GNPs were prepared using the reducing $HAuCl_4$ method with trisodium citrate. Briefly, one milliliter of 1% $HAuCl_4$ aqueous solution was added to 99 mL of boiling threefold-distilled water. Then, 1.5 mL of 1% trisodium citrate was quickly added to the above solution under the boiling and stirring condition. The mixed reaction solution was kept boiling for 5 min in order to let the GNPs develop after the color of solution changed from deep black into brilliant wine red. Then, the prepared GNPs were cooled down and stored at 4 °C. Finally, the diameters and dispersion of the GNPs were evaluated through a transmission electron microscope.

4.3. Preparation of the ZEN-McAb-GNP Probes

The ZEN-McAb-GNP probes were the essential biosensor materials for reflecting the detection results, which were prepared as per the previously reported method [30,31]. Firstly, appropriate dosages of 0.2 mol/L K_2CO_3 of an aqueous solution were used to adjust the GNPs solution (1 mL) to pH 8.5. Next, the 3.0 μL diluted ZEN-McAb was coupled with the GNPs at room temperature for 0.5 h after mixing well. Then, the BSA in purified water was used to block the nonspecific surface for 0.5 h. After removing the weak particles at low speed centrifugation, the reaction solution was separated through centrifugation in 8000 r/min at 4 °C for 15 min. Finally, the sediment resulted in the ZEN-McAb-GNP probes, which were redissolved in a 10 mmol/L Tris buffer containing a stabilizer and were stored at 4 °C.

4.4. Preparation of the ICA Strip

To prepare the ICA strip for ZEN detection, the various biochemical reagents were immobilized on the corresponding locations. Then, 30 μL of the diluted GAM-IgG or ZEN-antigen were uniformly sprayed on the NC membrane with 5 mm parallel spacing, and were dried for 5 h at 60 °C, which were the control line (C line) and test line (T line), respectively. The prepared ZEN-McAb-GNP probes were also uniformly sprayed on the glass-fiber conjugate pad and dried for 1 h at 37 °C. The assembly of the

ICA strip for ZEN was the same as the previous technological process. The immobilized sample pad, glass-fiber conjugate pad, NC membrane, and absorbent pad were pasted onto a 30 cm PVC sheet. Then, the assembled pad was cut into a 4 mm wide ICA strip and stored in dry and dark conditions.

4.5. Procedure and Judgment of the ICA

When using the prepared ICA to detect ZEN, the ZEN standard solution or sample extraction solution (100 µL) was put onto the sample pad of the ICA strip. The ICA detection could be finished after 5 min under the capillary action and immune reaction. At that moment, the detection results of the red color intensity would be reflected on the T line and C line, which were inversely proportional with the ZEN levels in the detection solutions. The depth of the red color intensity could be evaluated by two models—the naked eye and strip reader—as shown in Figure 1. Using model I, the ICA results could be judged as the following four cases: (1) negative (−), the ZEN was less than the LOD or free; (2) weakly positive (±), the ZEN was more than the LOD but lower than the concentration of the complete inhibition of the immune reaction; (3) positive (+), the ZEN was above the concentration of the complete inhibition of the immune reaction; and (4) ineffective, the ICA strip was not working. Using model II, the ICA results could be detected digitally. The gray values of the T lines could be obtained, and the inhibition ratios could be used to establish a linear relationship with the concentrations of ZEN levels using model II.

4.6. Optimization of the ICA

The optimizations of the biochemical reagent dosages were essential for developing a sensitive and desirable ICA method for ZEN detection [32]. The key parameters were evaluated to improve the performance and sensitivity of the ICA, such as the dosage of K_2CO_3 to adjust the pH value; BSA to block the nonspecific site at the surface of the GNPs; and the concentrations of ZEN-McAb, ZEN-antigen, and GAM-IgG to prepare the ICA strips. During the optimization process, the sensitivity of detection and clear judgment of the red color intensity by the naked eye should be focused on and selected.

4.7. Evaluation of the ICA

Under the optimized vital parameters, the sensitivity of the ICA for ZEN would be improved. The two models for judging the results of the ICA could obtain semi-quantitative and quantitative detection. The corresponding minimum ZEN concentration, which could significantly reduce the red color in the T line compared with the reference, was defined as the semi-quantitative LOD. In the digitized ICA, the minimum ZEN concentration of the linear range was determined as the quantitative LOD. In addition, the cross-reactivity test and storage test were carried out to evaluate the specificity and stability of the developed ICA.

4.8. Detection of Spiked and Authentic Samples

The recovery test of the spiked samples and the verification of the authentic samples by LC-MS/MS were used to evaluate the accuracy and precision of the developed ICA for ZEN detection. A total of 187 authentic cereal and its product samples, including 40 corn, 19 wheat, 39 wheat flour, 49 cereal product, and 40 feed samples, were collected from China in 2019 (Tables S3 and S4). The cereal product samples were mainly corn gluten meal, corn gluten, corn germ, soybean meal, peanut meal, and rice bran meal. All of the authentic samples were homogenized and stored at −20 °C before the detection procedure.

For the recovery test, the cereal and feed samples (5.0 g), which had been confirmed to be free of ZEN contamination, were completely crushed and spiked the ZEN standard at 0, 3, 20, and 50 ng/g, and were stored for 2 h at room temperature. The methanol/water (12.5 mL, 1:1, *v/v*) was used as an extracting solution. Vortex blending was performed on the mixture for 5 min, and was then centrifuged for 10 min at 4000 rpm/min. The supernatant solution was diluted four-fold with a working buffer and was adjusted to pH 6–8 and detected by the developed ICA method. After 5 min, the strip of

ICA flowed over the absorption pad, and the result was judged by the naked eye and the strip reader, respectively. The pretreatment and detection for the authentic cereal and its product samples were carried out in the same way as the above procedures. The correlation between the results of the ICA and LC-MS/MS was also evaluated.

To verify the reliability of the ICA method using the LC-MS/MS method [33,34], the completely crushed corn, wheat, or feed sample (5.0 g) was added in an extract solution of acetonitrile/water/formic acid (10 mL, 80: 19: 1, *v/v/v*). The mixture was mixed with an ultrasonic bath for 30 min, and then centrifuged at 4000 rpm/min for 5 min. The supernatant (10 mL) was transferred into another tube, followed by adding C_{18} (100 mg) and $MgSO_4$ (200 mg). The mixture was vortexed for 3 min and centrifuged at 4000 rpm/min for 5 min. The supernatant (1 mL) was concentrated to dryness by nitrogen gas. The residue was redissolved in methanol/water (400 μL, 50:50, *v/v*) and filtered with a 0.22 μm nylon filter, then analyzed by the LC-MS/MS with an Agilent Zorbax SB C18 reverse-phase column (3.5 μm, 150 mm × 2.1 mm) with a column temperature of 40 °C. The mobile phase was the different volume ratio of the acetonitrile/water, and flowed on the gradient elution program at a flow rate of 0.2 mL/min. The analysis was performed by multi-reaction monitoring (MRM) technology, and the ion source was an electrospray ion (ESI). The capillary voltage was at 3.0 kV, and the argon collision pressure was 2.60×10^{-4} Pa. The mass-to-charge ratios of the ZEN parent ion, quantitative ion, and qualitative ion were 317.1, 174.9, and 273.9, respectively.

Supplementary Materials: The following are available online at http://www.mdpi.com/2072-6651/12/8/521/s1, Table S1: The parameters of ELISA and ICA for ZEN, Table S2: The ZEN-positive levels in the authentic samples from China in 2019, Table S3: The information of the authentic samples from China in 2019, Table S4: The information of the authentic samples from China in 2019, Figure S1: The standard curve of ELISA for ZEN, Figure S2. The stability and sensitivity tests of the ICA strip for ZEN in the storage of 3 months, Figure S3. The correlation between the quantitative ICA and LC-MS/MS for ZEN in the authentic samples.

Author Contributions: X.H. executed and validated the experiments; Y.M. interpreted the results and analyzed the data; C.Y. collected the data; Z.L. drafted the manuscript; M.L. conceived and designed the experiments, and D.D. modified and validated the manuscript. All authors have read and agreed to the published version of the manuscript.

Funding: This study was supported by the National Natural Science Foundation of China (31701687), the Natural Science Foundation of Jiangsu Province (BK20170537), the National Key Research and Development Program of China (2017YFC1200100), the Senior Talent Scientific Research Initial Funding Project of Jiangsu University (16JDG035), the Project Funded by the Priority Academic Program Development of Jiangsu Higher Education Institutions (PAPD-2018-87) and the Jiangsu Collaborative Innovation Center of Technology and Material of Water Treatment.

Conflicts of Interest: The authors declare no conflict of interest.

References

1. Juan, C.; Oueslati, S.; Mañes, J.; Berrada, H. Multimycotoxin Determination in Tunisian Farm Animal Feed. *J. Food Sci.* **2019**, *84*, 3885–3893. [CrossRef] [PubMed]

2. Li, M.; Yang, C.; Mao, Y.; Hong, X.; Du, D.-L. Zearalenone Contamination in Corn, Corn Products, and Swine Feed in China in 2016-2018 as Assessed by Magnetic Bead Immunoassay. *Toxins* **2019**, *11*, 451. [CrossRef]

3. Hao, K.; Suryoprabowo, S.; Song, S.; Liu, L.; Kuang, H. Rapid detection of zearalenone and its metabolite in corn flour with the immunochromatographic test strip. *Food Agric. Immunol.* **2017**, *29*, 498–510. [CrossRef]

4. Buszewska-Forajta, M. Mycotoxins, invisible danger of feedstuff with toxic effect on animals. *Toxicon* **2020**, *182*, 34–53. [CrossRef] [PubMed]

5. Gadzała-Kopciuch, R.; Cendrowski, K.; Cesarz, A.; Kiełbasa, P.; Buszewski, B. Determination of zearalenone and its metabolites in endometrial cancer by coupled separation techniques. *Anal. Bioanal. Chem.* **2011**, *401*, 2069–2078. [CrossRef] [PubMed]

6. Ibrahim, O.O.; Menkovska, M. The Nature, Sources, Detections and Regulations of Mycotoxins That Contaminate Foods and Feeds Causing Health Hazards for Both Human and Animals. *J. Agric. Chem. Environ.* **2019**, *8*, 33–57. [CrossRef]

7. Rogowska, A.; Pomastowski, P.; Sagandykova, G.; Buszewski, B. Zearalenone and its metabolites: Effect on human health, metabolism and neutralisation methods. *Toxicon* **2019**, *162*, 46–56. [CrossRef]

8. Złoch, M.; Rogowska, A.; Pomastowski, P.; Railean-Plugaru, V.; Walczak-Skierska, J.; Rudnicka, J.; Buszewski, B. Use of Lactobacillus paracasei strain for zearalenone binding and metabolization. *Toxicon* **2020**, *181*, 9–18. [CrossRef]

9. Rogowska, A.; Pomastowski, P.; Walczak-Skierska, J.; Railean-Plugaru, V.; Rudnicka, J.; Buszewski, B.; Plugaru, R. Investigation of Zearalenone Adsorption and Biotransformation by Microorganisms Cultured under Cellular Stress Conditions. *Toxins* **2019**, *11*, 463. [CrossRef]

10. Ji, J.; Zhu, P.; Li, J.; Pi, F.; Zhang, Y.; Li, Y.; Wang, J.-S.; Sun, X. The Antagonistic Effect of Mycotoxins Deoxynivalenol and Zearalenone on Metabolic Profiling in Serum and Liver of Mice. *Toxins* **2017**, *9*, 28. [CrossRef]

11. *European Commission Regulation, 2007*; EC Regulation No. 1126/2007; Official Journal of the European Union: Brussels, Belgium, 2007.

12. Zinedine, A.; Soriano, J.M.; Moltó, J.C.; Mañes, J. Review on the toxicity, occurrence, metabolism, detoxification, regulations and intake of zearalenone: An oestrogenic mycotoxin. *Food Chem. Toxicol.* **2007**, *45*, 1–18. [CrossRef] [PubMed]

13. *China National Standard, 2017*; No. GB 2761-2017; Ministry of Health of P. R. China: Beijing, China, 2017.

14. Drzymala, S.S.; Weiz, S.; Heinze, J.; Marten, S.; Prinz, C.; Zimathies, A.; Garbe, L.-A.; Koch, M. Automated solid-phase extraction coupled online with HPLC-FLD for the quantification of zearalenone in edible oil. *Anal. Bioanal. Chem.* **2015**, *407*, 3489–3497. [CrossRef] [PubMed]

15. Li, C.; Deng, C.; Zhou, S.; Zhao, Y.; Wang, D.; Wang, X.; Gong, Y.Y.; Wu, Y. High-throughput and sensitive determination of urinary zearalenone and metabolites by UPLC-MS/MS and its application to a human exposure study. *Anal. Bioanal. Chem.* **2018**, *410*, 5301–5312. [CrossRef]

16. Pei, S.-C.; Lee, W.-J.; Zhang, G.; Hu, X.-F.; Eremin, S.A.; Zhang, L.-J. Development of anti-zearalenone monoclonal antibody and detection of zearalenone in corn products from China by ELISA. *Food Control.* **2013**, *31*, 65–70. [CrossRef]

17. Wang, D.; Zhang, Z.; Zhang, Q.; Wang, Z.; Zhang, W.; Yu, L.; Li, H.; Jiang, J.; Li, P. Rapid and sensitive double-label based immunochromatographic assay for zearalenone detection in cereals. *Electrophoresis* **2018**, *39*, 2125–2130. [CrossRef] [PubMed]

18. Liu, Z.; Zhang, Z.; Zhu, G.; Sun, J.; Zou, B.; Li, M.; Wang, J. Rapid screening of flonicamid residues in environmental and agricultural samples by a sensitive enzyme immunoassay. *Sci. Total Environ.* **2016**, *551*, 484–488. [CrossRef]

19. Zhan, S.; Huang, X.; Chen, R.; Li, J.; Xiong, Y. Novel fluorescent ELISA for the sensitive detection of zearalenone based on H2O2-sensitive quantum dots for signal transduction. *Talanta* **2016**, *158*, 51–56. [CrossRef]

20. Li, M.; Zhang, Y.; Xue, Y.; Hong, X.; Cui, Y.; Liu, Z.; Du, D.-L. Simultaneous determination of β 2-agonists clenbuterol and salbutamol in water and swine feed samples by dual-labeled time-resolved fluoroimmunoassay. *Food Control.* **2017**, *73*, 1039–1044. [CrossRef]

21. Liu, R.; Shi, R.; Zou, W.; Chen, W.; Yin, X.; Zhao, F.; Yang, Z. Highly sensitive phage-magnetic-chemiluminescent enzyme immunoassay for determination of zearalenone. *Food Chem.* **2020**, *325*, 126905. [CrossRef]

22. Zhang, Y.; Li, M.; Cui, Y.; Hong, X.; Du, D. Using of Tyramine Signal Amplification to Improve the Sensitivity of ELISA for Aflatoxin B1 in Edible Oil Samples. *Food Anal. Methods* **2018**, *11*, 2553–2560. [CrossRef]

23. Li, M.; Cui, Y.; Liu, Z.; Xue, Y.; Zhao, R.; Li, Y.; Du, D.-L. Sensitive and selective determination of butyl benzyl phthalate from environmental samples using an enzyme immunoassay. *Sci. Total Environ.* **2019**, *687*, 849–857. [CrossRef] [PubMed]

24. Ji, F.; Mokoena, M.P.; Zhao, H.; Olaniran, A.O.; Shi, J. Development of an immunochromatographic strip test for the rapid detection of zearalenone in wheat from Jiangsu province, China. *PLoS ONE* **2017**, *12*, e0175282. [CrossRef] [PubMed]

25. Shao, Y.; Duan, H.; Guo, L.; Leng, Y.; Lai, W.; Xiong, Y. Quantum dot nanobead-based multiplexed immunochromatographic assay for simultaneous detection of aflatoxin B1 and zearalenone. *Anal. Chim. Acta* **2018**, *1025*, 163–171. [CrossRef] [PubMed]

26. Li, R.; Meng, C.; Wen, Y.; Fu, W.; He, P. Fluorometric lateral flow immunoassay for simultaneous determination of three mycotoxins (aflatoxin B1, zearalenone and deoxynivalenol) using quantum dot microbeads. *Microchim. Acta* **2019**, *186*, 748. [CrossRef] [PubMed]

27. Zhang, X.; Yu, X.; Wen, K.; Li, C.; Mari, G.M.; Jiang, H.; Shi, W.; Shen, J.; Wang, Z. Multiplex Lateral Flow Immunoassays Based on Amorphous Carbon Nanoparticles for Detecting Three Fusarium Mycotoxins in Maize. *J. Agric. Food Chem.* **2017**, *65*, 8063–8071. [CrossRef]

28. Wu, S.; Liu, L.; Duan, N.; Li, Q.; Zhou, Y.; Wang, Z. Aptamer-Based Lateral Flow Test Strip for Rapid Detection of Zearalenone in Corn Samples. *J. Agric. Food Chem.* **2018**, *66*, 1949–1954. [CrossRef]

29. Li, X.M.; Wang, J.; Yi, C.Q.; Jiang, L.L.; Wu, J.X.; Chen, X.M.; Shen, X.; Sun, Y.M.; Lei, H.T. A smartphone-based quantitative detection device integrated with latex microsphere immunochromatographic for on-site detection of zearalenone in cereals and feed. *Sensor. Actuat. B-Chem.* **2019**, *290*, 170–179. [CrossRef]

30. Zeng, K.; Wei, W.; Jiang, L.; Zhu, F.; Du, D. Use of Carbon Nanotubes as a Solid Support To Establish Quantitative (Centrifugation) and Qualitative (Filtration) Immunoassays To Detect Gentamicin Contamination in Commercial Milk. *J. Agric. Food Chem.* **2016**, *64*, 7874–7881. [CrossRef]

31. Jiang, L.; Wei, D.; Zenga, K.; Shao, J.; Zhu, F.; Du, D. An Enhanced Direct Competitive Immunoassay for the Detection of Kanamycin and Tobramycin in Milk Using Multienzyme-Particle Amplification. *Food Anal. Methods* **2018**, *11*, 2066–2075. [CrossRef]

32. Zhu, F.; Mao, C.; Du, D. Time-resolved immunoassay based on magnetic particles for the detection of diethyl phthalate in environmental water samples. *Sci. Total Environ.* **2017**, *601*, 723–731. [CrossRef]

33. *China National Standard, 2016*; No. GB 5009-209; Ministry of Health of P. R. China: Beijing, China, 2016.

34. Wu, A.-B.; Liu, N.; Yang, L.; Deng, Y.; Wang, J.; Song, S.; Lin, S.; Wu, A.; Zhou, Z.; Hou, J. Multi-mycotoxin analysis of animal feed and animal-derived food using LC–MS/MS system with timed and highly selective reaction monitoring. *Anal. Bioanal. Chem.* **2015**, *407*, 7359–7368. [CrossRef]

Article

Aflatoxin B1 and Sterigmatocystin Binding Potential of Lactobacilli

Judit Kosztik [1],*, Mária Mörtl [2] **, András Székács [2], József Kukolya [1] and Ildikó Bata-Vidács [1]**

[1] Department of Environmental and Applied Microbiology, Agro-Environmental Research Institute, National Agricultural Research and Innovation Centre, 1022 Budapest, Hungary; kukolya.jozsef@akk.naik.hu (J.K.); batane.vidacs.ildiko@akk.naik.hu (I.B.-V.)

[2] Department of Environmental Analysis, Agro-Environmental Research Institute, National Agricultural Research and Innovation Centre, 1022 Budapest, Hungary; mortl.maria@akk.naik.hu (M.M.); szekacs.andras@akk.naik.hu (A.S.)

* Correspondence: kosztik.judit@akk.naik.hu

Received: 9 November 2020; Accepted: 26 November 2020; Published: 30 November 2020

Abstract: Due to global climate change, mould strains causing problems with their mycotoxin production in the tropical–subtropical climate zone have also appeared in countries belonging to the temperate zone. Biodetoxification of crops and raw materials for food and feed industries including the aflatoxin B1 (AFB1) binding abilities of lactobacilli is of growing interest. Despite the massive quantities of papers dealing with AFB1-binding of lactobacilli, there are no data for microbial binding of the structurally similar mycotoxin sterigmatocystin (ST). In addition, previous works focused on the detection of AFB1 in extracts, while in this case, analytical determination was necessary for the microbial biomass as well. To test binding capacities, a rapid instrumental analytical method using high-performance liquid chromatography was developed and applied for measurement of AFB1 and ST in the biomass of the cultured bacteria and its supernatant, containing the mycotoxin fraction bound by the bacteria and the fraction that remained unbound, respectively. For our AFB1 and ST adsorption studies, 80 strains of the genus *Lactobacillus* were selected. Broths containing 0.2 µg/mL AFB1and ST were inoculated with the *Lactobacillus* test strains. Before screening the strains for binding capacities, optimisation of the experiment parameters was carried out. Mycotoxin binding was detectable from a germ count of 10^7 cells/mL. By studying the incubation time of the cells with the mycotoxins needed for mycotoxin-binding, co-incubation for 10 min was found sufficient. The presence of mycotoxins did not affect the growth of bacterial strains. Three strains of *L. plantarum* had the best AFB1 adsorption capacities, binding nearly 10% of the mycotoxin present, and in the case of ST, the degree of binding was over 20%.

Keywords: aflatoxin B1; sterigmatocystin; lactobacilli; mycotoxin binding; detoxification

Key Contribution: Eighty strains belonging to species of *Lactobacillus* were screened for aflatoxin B1 and sterigmatocystin binding abilities using a rapid instrumental analytical method developed for both the bacterial biomass and its supernatant. It is the first time in the literature that sterigmatocystin binding of lactobacilli is presented. Aflatoxin B1 and sterigmatocystin binding abilities of strains belonging to the same *Lactobacillus* species vary highly.

1. Introduction

Mycotoxins are secondary metabolic products produced by moulds common in the food chain, causing major economic losses and becoming also sources of public health threats. These mycotoxins have a number of adverse health effects in humans and animals. They can be carcinogenic, immune-damaging, teratogenic, neurotoxic, kidney and liver-damaging depending on the species, age, and sex of the consumer. A mould can produce a variety of mycotoxins, and these compounds can amplify the harmful effects of each other. Due to global climate change, mould strains so far only causing problems with their mycotoxin production in the tropical climate zone have also appeared in Hungary [1]. Some 300 compounds have been recognised as mycotoxins of which around thirty are considered as a threat to human or animal health [2].

An example of a mycotoxin producing mould is *Aspergillus flavus*, a species of several strains able to produce mycotoxins. By infecting fodder plants like corn, wheat, and oily seeds as for example peanuts and walnuts, the mycotoxin formed enters the food chain [3]. The four most important aflatoxins produced by *A. flavus* are AFB1, AFB2, AFG1, and AFG2 [4].

Aflatoxin B1 (AFB1) is one of the most dangerous mycotoxins, primarily carcinogenic and genotoxic, harmful to the liver. The IARC classifies AFB1 in Group 1 (Carcinogenic to humans). It is a relatively heat-stable compound, up to 250 °C it is unchanged in roasted nuts, but in aqueous environments, it almost completely decomposes at 160 °C [5]. In accordance with Regulation (EU) No 574/2011, the maximum permitted level for AFB1 in feed is 0.02 mg/kg [6].

Sterigmatocystin (ST) is a precursor of aflatoxin. It is also produced by fungal species like *A. flavus*, *A. parasiticus*, *A. versicolor*, and *A. nidulans*. *A. flavus* and *A. parasiticus* are able to convert ST into aflatoxin, while *A. versicolor* and *A. nidulans* are not capable of this, resulting in elevated levels of ST in crops infected by them [7,8]. Rice and oats are typically the most contaminated with ST [9]. It is possible to reduce the level of ST by roasting [10]. Although experiments have shown genotoxicity and carcinogenicity of ST, limited data are available on the tumorigenic effect of the mycotoxin, which is why IARC has classified it as a potential human carcinogen (Group 2B).

Co-occurrence of aflatoxin and sterigmatocystin is recently gaining attention, as researches are being conducted and published on the sterigmatocystin contaminations as well for example in wheat and wheat products in the supermarkets in China [11] or corn, soybean meal, and formula feed in Japan [12].

Physical, chemical, and biological methods exist to prevent mycotoxins from entering the food chain. Microbes are used in biological detoxification. They may be capable of either inhibiting the growth of mycotoxin-producing fungi or of binding the mycotoxin to their surface, or, in rare cases, of degrading the mycotoxin itself [13]. The most detailed model of microbial mycotoxin binding has been described for zearalenone binding of *Saccharomyces* spp. In the adsorption of the mycotoxin, the beta-1,3/1,6-glucan moieties play a crucial role [14]. For AFB1-binding, glucomannans and mannanoligosaccharides have been proposed to be responsible for yeast cell walls. Similar to yeast, polysaccharides have been proposed to be the most crucial elements responsible for AFB1 binding in lactic acid bacteria (LAB) [15]. These polysaccharides are present in three main forms in the cell wall of lactobacilli: exopolysaccharides (EPS), peptidoglycan, and teichoic or lipoteichoic acids [16,17]. Lahtinen et al. [18] reported the ability of peptidoglucan to bind AFB1 in *L. rhamnosus*, and stated that the other glucan fractions, like EPS, lacked the mycotoxin-binding ability. However, the prominent role of peptidoglycan in binding is questionable, because, in 2010, Chapot-Chartier et al. described a new non-EPS cell wall polysaccharide, WPS, in *L. lactis*, which covalently binds to peptidoglycan forming a layer over it [19]. WPS appear as omnipresent components of the cell surface of LAB and exhibit most probably high structural diversity between strains even belonging to the same species.

Lactic acid bacteria (LAB) are found in both the animal and the human body. They got their name from the fact that glucose is fermented into lactic acid by them. They are Gram-positive, non-sporulating, oxidase and catalase-negative, anaerobic aerotolerant microorganisms. The most important genera belonging here are *Lactobacillus*, *Lactococcus*, *Leuconostoc*, *Enterococcus* and *Pediococcus*. Three hundred and three known species belong to the genus *Lactobacillus*, 17 species to the genus *Lactococcus*, 69 species to the genus *Enterococcus*, 15 species to the genus *Pediococcus*, and 27 species to the genus *Leuconostoc*.

As a member of the gut microbiota, they inhibit the growth of harmful microbes. Furthermore, they produce vitamins (e.g., vitamin B1, vitamin B2, vitamin B12, and vitamin K) [20] and stimulate the immune system [21]. In addition, numerous studies have shown that certain strains of some LAB species can bind mycotoxins, for example, AFB1, to their surface [22–24].

At our department, microbes with colony morphology of lactic acid bacteria were isolated on LAB selective MRS (de Man, Rogosa and Sharpe) plates from 14 exotic animals of the Budapest Zoo and Botanical Garden. The molecular taxonomical identification of the strains was carried out by 16S rDNA sequencing and analysis. At present, the collection comprises nearly 1000 strains and is constantly expanding. Most of our strains belong to the genera *Lactobacillus* and *Enterococcus*, but we also managed to isolate strains belonging to the other LAB genera.

Our goal was to screen strains of the genus *Lactobacillus* from our collection for AFB1 and ST binding capacities. For this purpose, a rapid high-performance liquid chromatography method was developed and used for analytical determination of AFB1 and ST in both the bacterial biomass and its supernatant.

2. Results

For screening lactobacilli for AFB1 and ST binding capacities, several parameters had to be optimised. The effects of cell concentration, incubation time with the mycotoxins on the mycotoxin binding capacities, and the effect of the mycotoxin itself on the cell counts of the lactobacilli had to be considered before the screening.

2.1. Analytical Determination of the Mycotoxins

Instrumental analysis of AFB1 by high-performance liquid chromatography is well described in the literature, and recent work [25] presents a robust method for simultaneous quantification of several aflatoxins from fungal cultures, therefore, AFB1 was found to be sufficient to be detected at a single wavelength of 365 nm. Peak purities for ST, as a relatively novel analyte for HPLC detection, were systematically checked in all analytical determinations by recording absorption at two wavelengths of 240 and 325 nm, and peak area ratios at those wavelengths were compared to the ratios characteristic to standard solutions of the given analyte (ST). As blank microbial biomasses did not contain interfering matrix components, the limits of detection were found to be 0.010 µg/mL for both AFB1 and ST, and it was the same for both matrices, namely in spiked supernatants and in liquid matrices extracted from blank biomass. Therefore, quantisation in the analytical determination was based on instrumental (external) calibration with standard solutions in the range between 0.010 and 2.00 µg/mL.

2.2. Optimisation for Mycotoxin Binding Experiments

2.2.1. Study of the Effect of Bacterial Count on Mycotoxin Binding of *Lactobacillus* Strains

According to the method described in Section 4.4.1, the effect of bacterial concentration on the mycotoxin binding capacity of *Lactobacillus* strains was determined for AFB1 and ST. The result shown in Figure 1 indicates that detectable mycotoxin binding could only be found above 10^7 cells/mL for both AFB1 and ST.

This result is in agreement with the findings of Ma et al. [26], who only found one strain that was able to bind mycotoxin at 10^6 cell/mL concentration (4.27%), and they obtained cut-off values at 10^9 cell/mL.

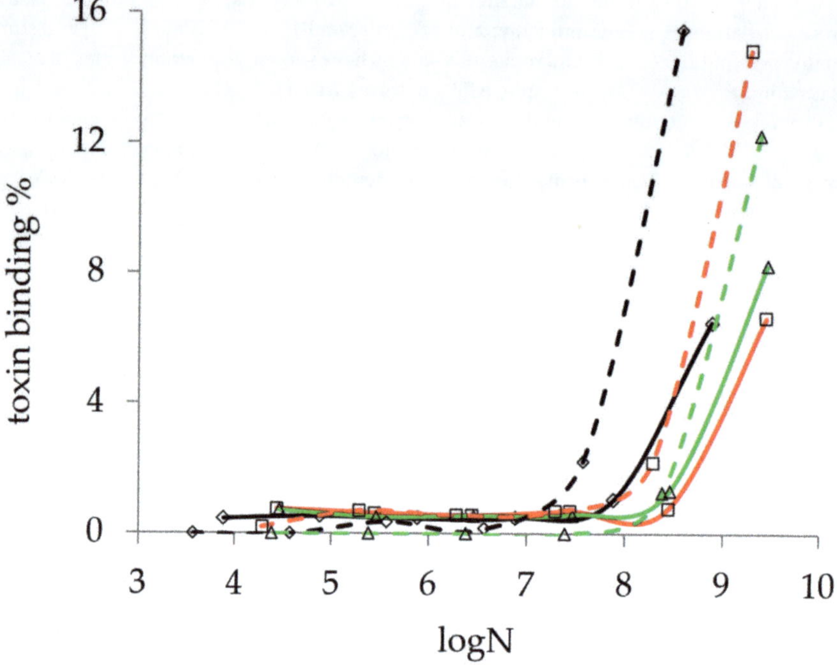

Figure 1. The effect of bacterial concentration on the aflatoxin B1 (continuous line) and sterigmatocystin (dashed line) binding of *L. pentosus* TV3 (black), *L. paracasei* MA2 (red), and *L. plantarum* TS23 (green). (logN means the logarithm of the number of colony-forming units per mL of bacterial cell suspension).

2.2.2. Study of the Effect of Incubation Time on Mycotoxin Binding of *Lactobacillus* Strains

The effect of incubation time on the AFB1 mycotoxin binding capacity of *Lactobacillus* strains was also examined. For this experiment, five strains of different genera were selected. The strains were incubated with AFB1 mycotoxin for 10 min or 48 h, according to the method described in Section 4.4.2. The two-time values were selected according to the literature data available on AFB1 binding (see at the end of the paragraph), 10 min was the lowest with satisfactory results and 48 h is the incubation period in which lactobacilli reached the highest cell count under the study parameters. Our aim was to determine whether it is necessary to add the mycotoxin at the beginning of culturing or it is enough to add it after the bacteria reached their final cell concentrations. On the studied strains, very diverse results were obtained (Figure 2). For strain TV3, a significantly ($p < 0.005$) higher mycotoxin binding was found for the shorter incubation time, on the other hand, for TS23, significantly ($p < 0.00005$) better binding could be observed for the longer incubation in the presence of the mycotoxin. For strains MA2, TV24, and SK63 the incubation times had no significant ($p > 0.4$, 0.6 and 0.5, respectively) effect on the mycotoxin binding capacity. Regarding data in the literature, contradictory results can also be found. Studying 1, 10, 30 and 60 min, no effect on the incubation time on mycotoxin binding was found by Bueno et al. [22]. In another paper of Kasmani et al. [27], however, AFB1 binding was assayed at 0, 0.5, 4, 12, 24, and 72 h, with the lowest mycotoxin binding obtained at 0.5 h and the highest at 12 h, with a twofold difference. El-Nezami et al. and Peltonen et al. studied the binding of AFB1 by different species for 0, 4, 24, 48 and 72 h, and suggested that mycotoxin elimination is a rapid process [24,28]. As no satisfactory conclusions could be drawn regarding the optimal incubation time with the mycotoxin, a practical decision was made, the mycotoxin binding experiments were performed with 10 min incubation with the mycotoxin for our experiments.

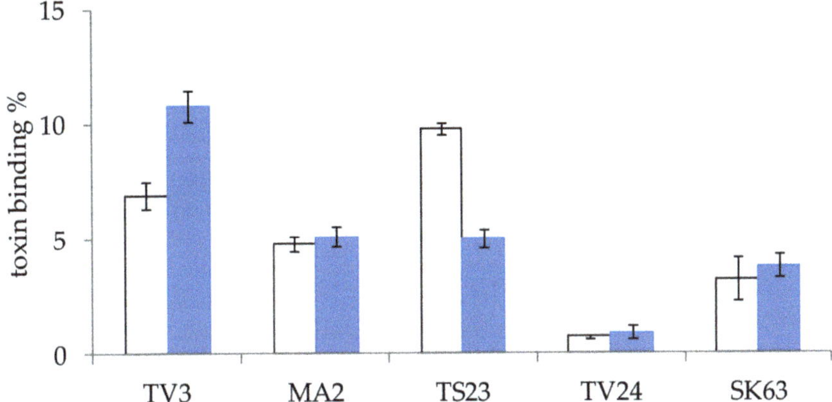

Figure 2. The effect of incubation time of 10 min (blue) and 48 h (white) on the aflatoxin B1 binding of *L. pentosus* TV3, *L. paracasei* MA2, *L. plantarum* TS23, *L. graminis* TV24, and *L. salivarius* SK63. (means ± standard deviation, $N = 5$).

2.2.3. Study of the Effect of Mycotoxins on *Lactobacillus* Cell Count

As mycotoxins cause serious health damage to higher organisms, the question arises, whether they have a negative effect on bacteria, too. So the possible changes in bacterial counts were also studied in the presence of mycotoxins. For the experiment, three *Lactobacillus* strains from different genera were selected. It was observed that neither AFB1 nor ST at the studied concentration of 0.2 µg/mL caused a significant reduction ($p > 0.5$ for AFB1 and $p > 0.4$ for ST) in the bacterial count compared to the control Figure 3).

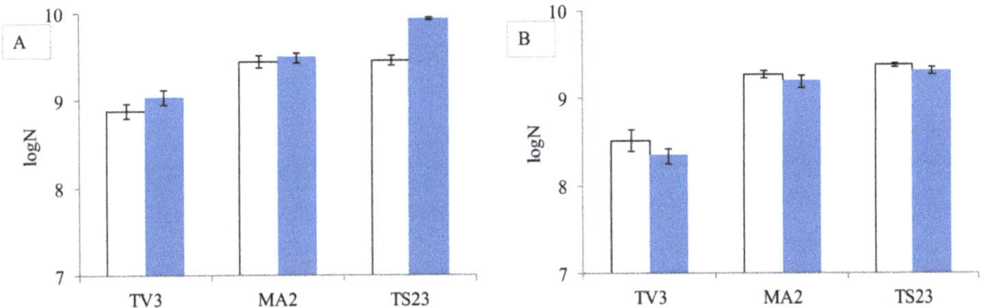

Figure 3. The effect of aflatoxin B1 (**A**) and sterigmatocystin (**B**) at concentration 0.2 µg/mL on the cell count *L. pentosus* TV3, *L. paracasei* MA2, *L. plantarum* TS23 (control-white, with mycotoxin-blue). (means±standard deviation, N = 3) (logN means the logarithm of the number of colonies forming units per ml of bacterial cell suspension).

2.3. Screening Lactobacillus Strains for Mycotoxin Binding Capacities

For screening AFB1 and ST binding abilities of lactobacilli, 80 strains from our collection were selected. A phylogenetic tree was prepared with all known *Lactobacillus* strains by 16S rDNA sequences. In the case of larger clades, where our strain collection was missing the species, those missing strains were obtained from the BCCM strain collection (see Section 4.1). With these 25 strains ordered from the BCCM, a comprehensive study was conducted on the mycotoxin binding abilities of lactic acid bacteria, with emphasis on the genus *Lactobacillus*.

The mycotoxin binding experiments were performed according to the method described in Section 4.5. The cell concentration of the lactobacilli during the test was set to 10^8 cfu/mL based on the findings in Section 2.2.1. The incubation time with the mycotoxins was 10 min based on our results presented in Section 2.2.2.

2.3.1. Aflatoxin B1 Binding Capacities of Lactobacilli

Figure 4 shows the lactic acid bacteria that could bind aflatoxin the best from the studied 105 strains. Only 14 strains were able to bind AFB1 above 5% at the studied mycotoxin concentration. The best AFB1 binding capacities in MRS broth were obtained for *L. pentosus* TV3 with 11.5% and *L. plantarum* AT26, AT3, and AT1 with 8–9% (Figure 4).

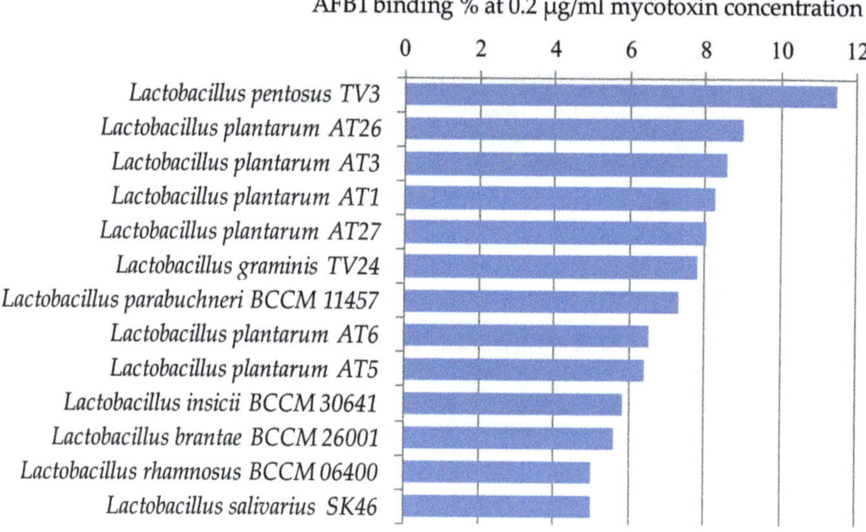

Figure 4. *Lactobacillus* strains with AFB1 binding capacities above 5% at 0.2 µg/mL mycotoxin concentration in MRS broth.

Thirty-three more strains were found with AFB1 binding capacities of 3–4% (Figure 5). For the remaining 58 strains only a smaller, less than 3%, the percentage of AFB1 binding could be observed. These results are significantly below the binding values generally presented between 17% and 83% in the literature [22–24].

The highest AFB1 binding abilities were found in our study for strains of *L. pentosus*, *L. plantarum*, and *L. graminis*. In the study of Huang et al., *L. plantarum* C88 presented the highest binding ability with AFB1 using AFB1 binding assay in vitro compared with other strains [29]. Though not aflatoxin, a high percentage of OTA reduction was obtained by *L. plantarum* and *L. graminis* in the studies of Belkacem-Hanf et al. [30]. It can be seen in Figure 6 that there is a close phylogenetic relationship among the lactobacilli strains with the best mycotoxin binding abilities, which might be an explanation for their good mycotoxin binding abilities.

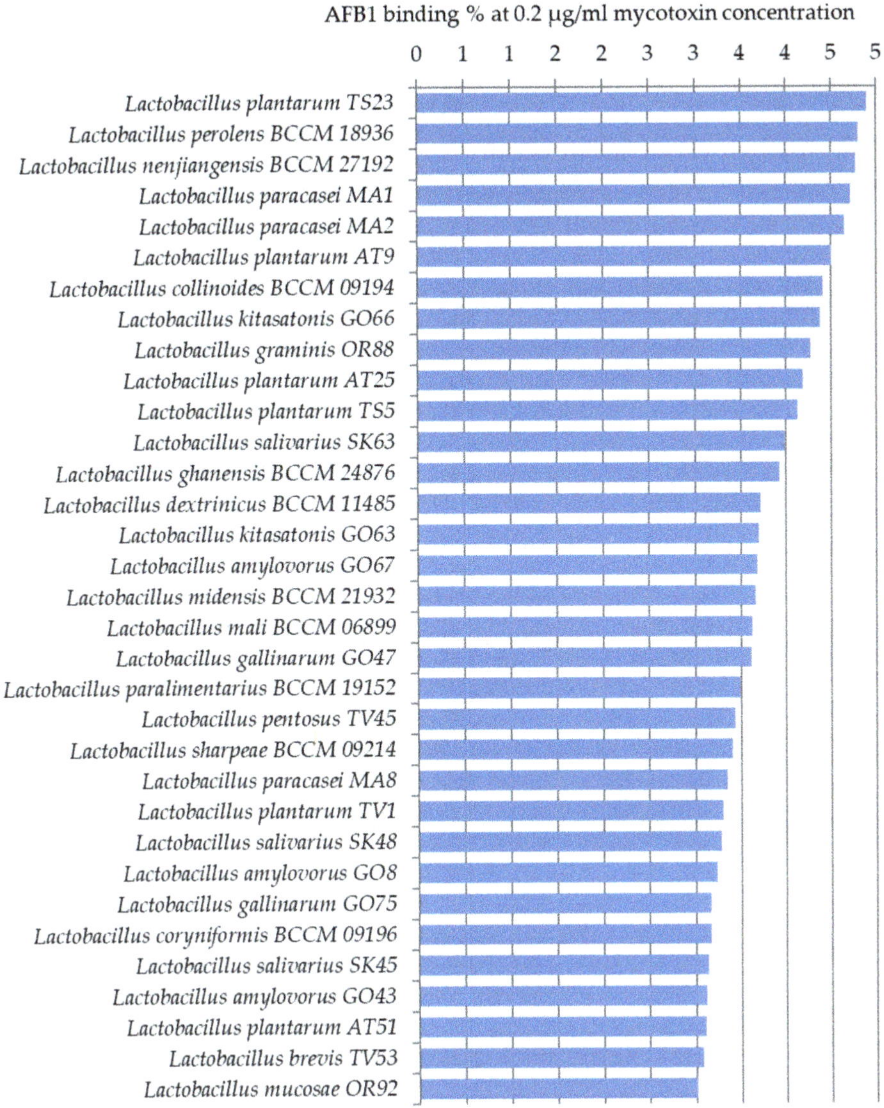

Figure 5. *Lactobacillus* strains with AFB1 binding capacities of 3–5% at 0.2 µg/mL mycotoxin concentration in MRS broth.

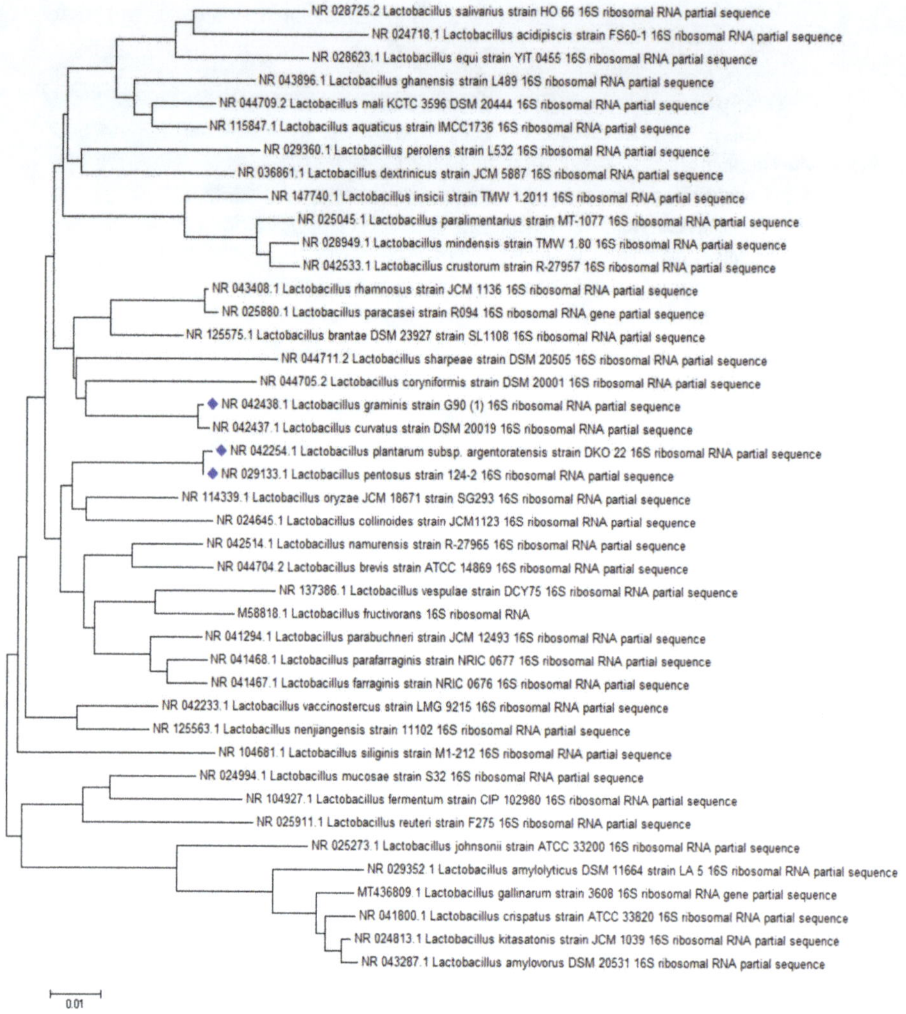

Figure 6. Phylogenetic relationship of the best aflatoxin B1 binding *Lactobacillus* strains.

2.3.2. Sterigmatocystin Binding Capacities of Lactobacilli

ST binding abilities of 14 *Lactobacillus* strains from our collection and 25 strains ordered from BCCM were studied. Based on our experiments, *L. plantarum* TV1, AT1, AT3, AT5, *L. paracasei* MA8, and *L. pentosus* TV3 proved to be the strains with the best adsorption abilities, able to bind more than 20% of ST under the studied parameters (Figure 7). Similar to aflatoxin binding, it can be seen in Figure 8 that there is a close phylogenetic relationship among these strains as well, furthermore, good AFB1 and ST binding ability seems to be related (Figures 6 and 8). So far, no results have been published in the literature that address the ST-binding ability of lactobacilli.

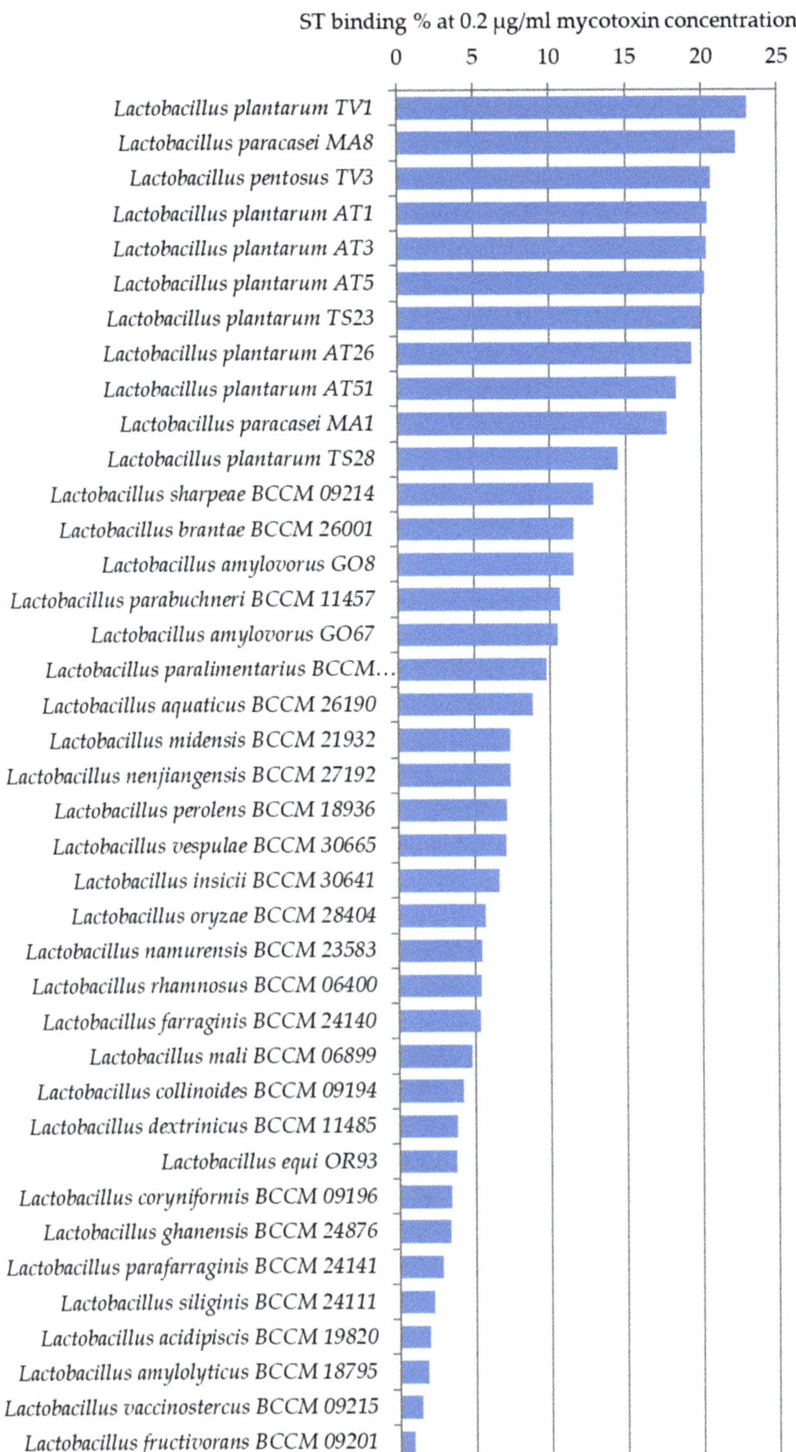

Figure 7. Sterigmatocystin binding capacities (%) of *Lactobacillus* strains at 0.2 µg/mL mycotoxin concentration in MRS broth.

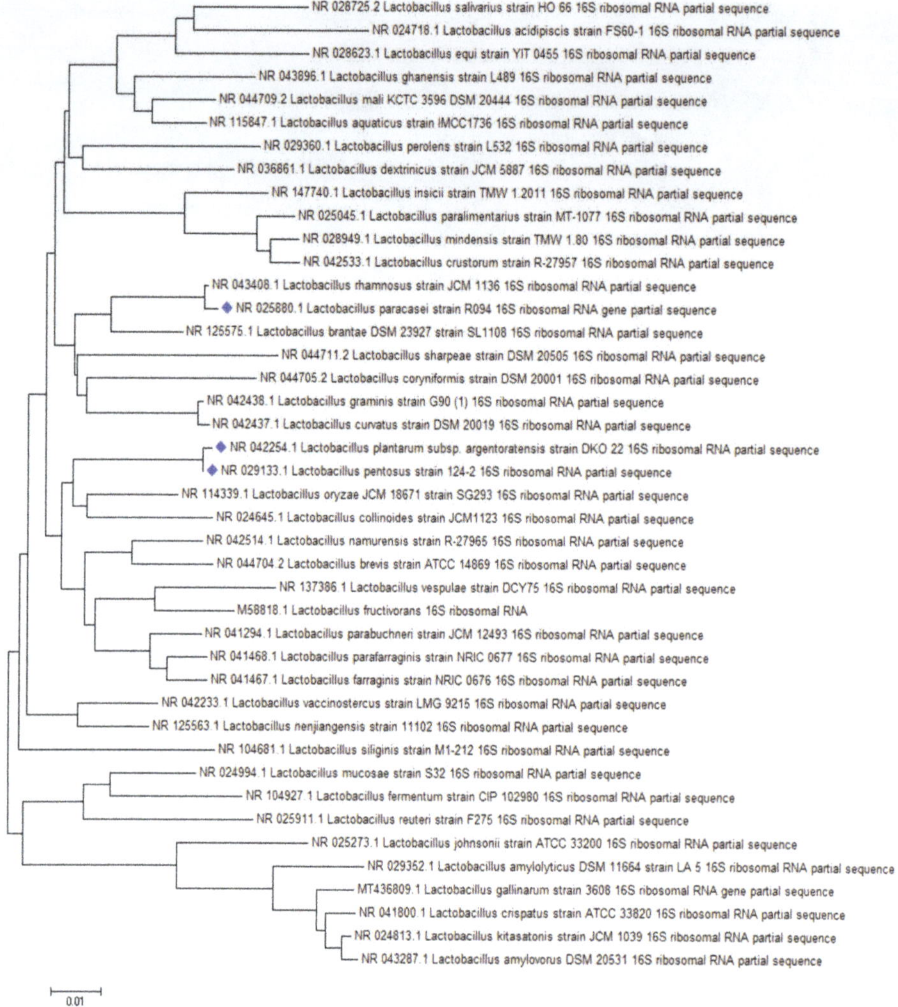

Figure 8. Phylogenetic relationship of the best sterigmatocystin binding *Lactobacillus* strains.

3. Conclusions

For mycotoxin binding abilities, broths containing 0.2 µg/mL AFB1 or ST were inoculated with the *Lactobacillus* test strains. Before screening the strains for binding capacities, optimisation of the experiment parameters was carried out. Mycotoxin binding was detectable from a germ count of 10^7 cells/mL at 0.2 µg/mL mycotoxin concentration in MRS broth, so for the screening, a cell concentration of 10^8 cells/mL was chosen. The incubation time of the cells with the mycotoxins was studied from 10 min to 48 h. It was found that 2 days of co-incubation was not required for mycotoxin binding, after 10 min of incubation nearly the same binding values were obtained for the majority of the tested strains, though some anomalies could be observed as for *L. pentosus* TV3 shorter incubation time, while in the case of *L. plantarum* TS23 longer incubation time was slightly more efficient. Based on our experiments, it can be said that neither AFB1 nor ST affected the growth of bacterial strains at the studied concentration.

One hundred and five strains were tested for AFB1 binding; the highest capacities were obtained for *L. pentosus* TV3 with 11.5% and *L. plantarum* AT26, AT3, and AT1 with 8–9%. Interestingly, in the case of ST with a very similar structure, the degree of binding was more than 20%. ST binding ability was examined in 39 *Lactobacillus* strains. *L. plantarum* TV1, AT1, AT3, AT5, *L. paracasei* MA8, and *L. pentosus* TV3 proved to be the strains with the best adsorption abilities. The results found in the literature on the mycotoxin binding abilities of lactobacilli are diverse due to the different methodologies used.

Toxin binding of lactobacilli was measured in the MRS medium, the optimal medium for LAB. The highest mycotoxin binding values found in the literature for lactobacilli were measured in vitro in PBS buffer, 87% for AFB1 by *L. acidophilus* [31], 96% binding was found by Liew and co-workers by *L. casei* Shirota at AFB1 concentration of 2 µg/mL [32], nevertheless, Hernandez-Mendoza et al. showed that the percentage of AFB1 bound by the same species was approximately 30% at AFB1 concentration of 4.6 µg/mL after 4 h of incubation at 37 °C [33]. These latter findings underline that even in the same medium the same *Lactobacillus* species might present very different mycotoxin binding abilities in different experiments. Though the most results for AFB1 binding in the literature is measures in PBS buffer, however, MRS medium represents better the possible environment for LAB to be used for mycotoxin binding purposes. Thus, the results of our AFB1 binding assay could not be directly compared to values in the literature.

The same location of AFB1 and ST binding is assumed by our result that the most efficient mycotoxin binding species were representatives of *L. plantarum* and *L. pentosus* species for both mycotoxins (Figures 6 and 8). This is consistent with literature data for AFB1 binding, where these strains are among the most effective within the genus *Lactobacillus* [34].

In our studies, we consistently found that the ST binding potential of *Lactobacillus* strains was approximately twice that of AFB1 binding. This phenomenon may be due to the higher ST affinity of binding-critical cell wall polysaccharide fragments, but this may be explained by the nature of ST in aqueous media: ST in aqueous media may form a unique type of aggregate [35].

An interesting result of our studies is that we also found a large difference in AFB1 and ST binding potential between *Lactobacillus* strains belonging to a given species. This may be explained by the strain-specific, different polysaccharide composition of the WPS fraction of cell surface polysaccharides, as peptidoglucan has too conservative a structure to account for differences between strains [16].

Our work is the first report on microbial ST binding. The investigated LAB type strains had different ST and AFB1 binding abilities. These data, especially the altered binding potential of the *Lactobacillus* strains belonging to the same species, would be very useful in the future for investigating the molecular mechanism of bacterial mycotoxin adsorption and developing aflatoxin bio-binders.

4. Materials and Methods

4.1. Bacterial Strains

Eighty *Lactobacillus* strains of our collection isolated from faeces samples of zoo animals were used for the studies. The strains were identified by the 16S rDNA sequence extracted from pure bacterial cultures and sequenced by BaseClear (Table 1). In addition, 25 other *Lactobacillus* strains have been obtained from BCCM (Belgian Coordinated Collections of Microorganisms) (Table 2). The *Lactobacillus* strains stored at −80 °C in 43.5% glycerine were thawed on ice before culturing.

4.2. Mycotoxins

AFB1 and ST were purchased from Sigma-Aldrich (Merck, Darmstadt, Germany). Standard solutions were made by diluting the mycotoxin powder with methanol (puriss., MOLAR Chemicals Ltd., Halásztelek, Hungary) to make stock solutions of 50 µg/mL. The concentrations of the stock solutions were verified by HPLC measurement. The mycotoxin concentrations for our experiments were set at 0.2 µg/mL, which is a tenth of the maximum permitted level for AFB1 by EU Regulation No.574/2011.

Table 1. *Lactobacillus* species of our collection with the strains used in this study.

Species	Strains
Lactobacillus amylovorus	GO5, GO8, GO43, GO45, GO67
L. brevis	AT70, TV23, TV50, TV53
L. crispatus	GO48
L. crustorum	TV19
L. curvatus	TS4
L. equi	OR7, OR25, OR86, OR93
L. fermentum	SK64
L. gallinarum	GO47, GO75, GO78
L. graminis	OR12, OR81, OR88, TV24, TV35
L. johnsonii	GO76
L. kitasatonis	GO6, GO13, GO16, GO17, GO63, GO66, GO73, GO95, GO98
L. mucosae	OR2, OR13, OR17, OR23, OR28, OR48, OR63, OR66, OR80, OR92
L. paracasei	MA1, MA2, MA4, MA8, MA99
L. pentosus	TV3, TV45
L. plantarum	AT1, AT3, AT5, AT6, AT9, AT25, AT26, AT27, AT51, TS5, TS16, TS23, TS62, TV1
L. reuterii	VO12, VO26
L. salivarius	SK6, SK12, SK17, SK20, SK29, SK41, SK42, SK45, SK46, SK48, SK63, VO20

Table 2. Type strains of the *Lactobacillus* species ordered from BCCM for this study.

Strains:
Lactobacillus farraginis BCCM 24140
L. acidipiscis BCCM 19820
L. fructivorans BCCM 09201
L. oryzae BCCM 28404
L. vaccinostercus BCCM 09215
L. siliginis BCCM 24111
L. parafarraginis BCCM 24141
L. amylolyticus BCCM 18795
L. namurensis BCCM 23583
L. aquaticus BCCM 26190
L. vespulae BCCM 30665
L. coryniformis BCCM 09196
L. sharpeae BCCM 09214
L. paralimentarius BCCM 19152
L. mali BCCM 06899
L. midensis BCCM 21932
L. dextrinicus BCCM 11485
L. ghanensis BCCM 24876
L. collinoides BCCM 09194
L. nenjiangensis BCCM 27192
L. perolens BCCM 18936
L. rhamnosus BCCM 06400
L. brantae BCCM 26001
L. insicii BCCM 30641
L. parabuchneri BCCM 11457

4.3. Mycotoxin Extraction and Analytical Determination

The mycotoxin content of the samples was determined by UV detection after high performance liquid chromatographic separation (HPLC-UV) on the basis of literature methods for AFB1 [19]) and ST [36,37]. First, the mycotoxin was extracted as follows. The cultures in the Falcon tubes were centrifuged for 40 min at room temperature at 4000 rpm. The supernatant contains the remaining unbound mycotoxin and the residue, referred to as the biomass hereinafter, contains the mycotoxin bound by the bacteria. One millilitre of the supernatant transferred to an empty Falcon tube was shaken with 1 mL of dichloromethane for 20 min in a horizontal shaker in the dark. From the dichloromethane phase, 0.5 mL was taken out and concentrated in a clean Eppendorf tube at 45 °C under a fume hood. For the extraction of the mycotoxin from the biomass, 1.8 mL of dichloromethane and 0.2 mL of methanol were added to the Falcon tube containing the biomass. The mixture was pipetted into Eppendorf tubes. The tubes were vortexed (Vortex Genie 2, MO BIO Laboratories, Carlsbad, CA, USA) for 20 min in the dark and then centrifuged at 3000 rpm for 10 min. One ml of the supernatant was evaporated as before. The residues of the extracts were resolved in 1.0 mL eluent, and determined by HPLC on a Younglin YL9100 HPLC system equipped with a YL9150 autosampler (YL Instruments Co., Anyang, Korea). For the analysis, 30 μL of the extracts were applied onto a Brisa (Technochrome) C18 column (5 μm, 15 cm × 0.46 cm) at 30 °C. The separation was carried out at a flow rate of 1 mL/min using isocratic elution, containing 60:20:20 or 40:30:30 ($v/v\%$) of water, methanol and acetonitrile for AFB1 and ST, respectively. The detector wavelengths were 365 nm or 325 and 240 nm for AFB1 and ST, respectively. All determinations were performed in triplicates from three parallel samples. Relative standard deviations established for binding capacities for three parallel samples ranged between 0.53% and 1.35%.

4.4. Optimisation for Mycotoxin Binding Experiments

4.4.1. Study of the Effect of Bacterial Count on Mycotoxin Binding of LAB Strains

Three strains (TV3, MA2, TS23) with good mycotoxin binding capacities, selected by the results obtained in preliminary experiments (results not shown), were used for the study. The strains were grown in 9 mL of MRS broth (de Man Rogosa and Sharpe Broth, VWR) for 48 h at 37 °C. From these cultures of 10^8 cfu/mL, ten-fold dilutions were performed in MRS broth until the concentration of 10^3 cfu/mL. From the five dilutions, 15–15 mL was transferred to 15 mL plastic Falcon tubes. Bacterial concentrations were checked by plating on MRS agar. To each sample a uniform amount of mycotoxin equal to 0.2 μg/mL was added, the samples were mixed and incubated for 10 min at room temperature. The tubes were then centrifuged at 4000 rpm for 40 min (Centrifuge 5810 R, Eppendorf, Wien, Austria). The supernatant was decanted and the mycotoxin was extracted from the biomass (see Section 4.3).

4.4.2. Study of the Effect of Incubation Time on Mycotoxin Binding of *Lactobacillus* Strains

The effect of incubation time was studied with 5 efficient AFB1 binding strains (TV3, TV24, MA2, TS23, SK63) selected by the results obtained in preliminary experiments (results not shown). In one case, the *Lactobacillus* strains were grown in 15 mL of MRS broth in the presence of 0.2 μg/mL AFB1 mycotoxin for 48 h at 37 °C. In the other case, the strains were cultivated in the same manner, but without the presence of the mycotoxin for 48 h at 37 °C, then the mycotoxin was added to the culture broths. The tubes were vortexed and incubated for 10 min at room temperature. The samples were then centrifuged at 4000 rpm for 40 min. The supernatant was decanted and the mycotoxin was extracted from the biomass (see Section 4.3).

4.4.3. Study of the Effect of Mycotoxin on *Lactobacillus* Cell Count

In addition to *Lactobacillus* strains grown in 15 mL MRS broth in the presence of 0.2 μg/mL mycotoxin, the number of bacterial cells was also determined under the same conditions but in mycotoxin free MRS broth by plating on MRS agar to determine the effect of the mycotoxin on the bacterial growth.

4.5. Screening LAB Strains for Mycotoxin Binding Capacities

Lactobacillus strains were taken from −20 °C storage, thawed on ice, and 20 μl of the suspension was transferred to 9 mL MRA broth. The tubes were incubated at 37 °C for 24 h. Falcon tubes containing 15 mL of MRS broth were inoculated with 50 μl of the cultures. The tubes were incubated at 37 °C for two days. Three replicates were prepared with each strain.

After the incubation, 0.2 µg/mL of AFB1 or ST were added to the tubes. Pure MRS broth was used as a negative control, and mycotoxin-only MRS broth without bacteria was used as a positive control. The tubes were mixed by shaking and the tubes were incubated with the mycotoxin for 10 min at room temperature. The tubes were centrifuged at 4000 rpm for 40 min to separate the biomass from the supernatant. The supernatant was transferred to an empty sterile Falcon tube and stored at −80 °C until further analysis. The AFB1 and ST contents of the biomasses were determined by the HPLC method described in Section 4.3.

4.6. Statistical Analyses

Statistical calculations of F- and *t*-Tests were performed in Microsoft Excel 2007 program.

Author Contributions: Conceptualization, J.K. (József Kukolya); methodology, J.K. (Judit Kosztik), I.B.-V., M.M. and A.S.; validation, M.M.; investigation, J.K. (Judit Kosztik) and I.B.-V.; writing—Original draft preparation, J.K. (Judit Kosztik), I.B.-V. and M.M.; writing—Review and editing, I.B.-V. and A.S.; supervision, J.K. (József Kukolya) and A.S.; funding acquisition, J.K. (József Kukolya). All authors have read and agreed to the published version of the manuscript.

Funding: This research was funded by the National Research, Development and Innovation Office in Hungary, projects NVKP-16-1-2016-0009, NVKP-16-1-2016-0049 and OTKA K116631.

Conflicts of Interest: The authors declare no conflict of interest.

References

1. Dobolyi, C.; Sebők, F.; Varga, J.; Kocsubé, S.; Szigeti, G.; Baranyi, N.; Szécsi, Á.; Tóth, B.; Varga, M.; Kriszt, B.; et al. Occurrence of aflatoxin producing Aspergillus flavus isolates in maize kernel in Hungary. *Acta Aliment.* **2013**, *42*, 451–459. [CrossRef]

2. Alassane-Kpembi, I.; Schatzmayr, G.; Taranu, I.; Marin, D.; Puel, O.; Oswald, I.P. Mycotoxins co-contamination: Methodological aspects and biological relevance of combined toxicity studies. *Crit. Rev. Food Sci. Nutr.* **2017**, *57*, 3489–3507. [CrossRef] [PubMed]

3. Jelinek, C.F.; E Pohland, A.; E Wood, G. Worldwide occurrence of mycotoxins in foods and feeds—An update. *J. Assoc. Off. Anal. Chem.* **1989**, *72*, 223–230. [CrossRef] [PubMed]

4. Pitt, J.I. Toxigenic fungi and mycotoxins. *Br. Med. Bull.* **2000**, *56*, 184–192. [CrossRef] [PubMed]

5. Raters, M.; Matissek, R. Thermal stability of aflatoxin B1 and ochratoxin A. *Mycotoxin Res.* **2008**, *24*, 130–134. [CrossRef] [PubMed]

6. Commission Regulation (EU) No 574/2011. Annex I to Directive 2002/32/EC of the European Parliament and of the Council as Regards Maximum Levels for Nitrite, Melamine, *Ambrosia* spp. and Carry-over of Certain Coccidiostats and Histomonostats and Consolidating Annexes I and II. 2011, L 159/7. Available online: https://eur-lex.europa.eu/eli/reg/2011/574/oj?locale=en (accessed on 1 October 2020).

7. Sweeney, M.J.; Dobson, A.D. Molecular biology of mycotoxin biosynthesis. *FEMS Microbiol. Lett.* **1999**, *175*, 149–163. [CrossRef] [PubMed]

8. Cole, R.J.; Cox, R.H. *Handbook of Toxic Fungal Metabolites*; Academic Press: London, UK, 1981; pp. 67–93.

9. Mol, J.G.J.; Pietri, A.; MacDonald, S.J.; Anagnostopoulos, C.; Spanjer, M. *Survey on Sterigmatocystin in Food*; Research report EFSA supporting publication 2015 EN-774; RIKILT-Wageningen UR: Wageningen, The Netherlands, 2015.

10. Bokhari, F.M.; Aly, M.M. Evolution of traditional means of roasting and mycotoxins contaminated coffee beans in Saudi Arabia. *Adv. Biol. Res.* **2009**, *3*, 71–78.

11. Zhao, Y.; Wang, Q.; Huang, J.; Ma, L.; Chen, Z.; Wang, F. Aflatoxin B1 and sterigmatocystin in wheat and wheat products from supermarkets in China. *Food Addit. Contam. Part B* **2018**, *11*, 9–14. [CrossRef]

12. Nomura, M.; Aoyama, K.; Ishibashi, T. Sterigmatocystin and aflatoxin B1 contamination of corn, soybean meal, and formula feed in Japan. *Mycotoxin Res.* **2017**, *34*, 21–27. [CrossRef]

13. Bianchini, A.; Bullerman, L.B. Biological control of molds and mycotoxins in foods. In *Mycotoxin Prevention and Control in Agriculture*; Appell, M., Kendra, D., Trucksess, M., Eds.; American Chemical Society: Washington, DC, USA, 2010; Volume 1031, pp. 1–16.

14. Yiannikouris, A.; François, J.; Poughon, L.; Dussap, C.G.; Bertin, G.; Jeminet, G.; Jouany, J.-P. Adsorption of Zearalenone by beta-D-glucans in the Saccharomyces cerevisiae cell wall. *J. Food Prot.* **2004**, *67*, 1195–1200. [CrossRef]

15. Guan, S.; Gong, M.; Yin, Y.; Huang, R.; Ruan, Z.; Zhou, T.; Xie, M. Occurrence of mycotoxins in feeds and feed ingredients in China. *J. Food Agric. Environ.* **2011**, *9*, 163–167.

16. Chapot-Chartier, M.-P. Interactions of the cell-wall glycopolymers of lactic acid bacteria with their bacteriophages. *Front. Microbiol.* **2014**, *5*, 236. [CrossRef] [PubMed]

17. Fochesato, A.; Cuello, D.; Poloni, V.; Galvagno, M.A.; Dogi, C.A.; Cavaglieri, L.R. Aflatoxin B1 adsorption/desorption dynamics in the presence of *Lactobacillus rhamnosus* RC007 in a gastrointestinal tract-simulated model. *J. Appl. Microbiol.* **2018**, *126*, 223–229. [CrossRef] [PubMed]

18. Lahtinen, S.J.; Haskard, C.A.; Ouwehand, A.C.; Salminen, S.J.; Ahokas, J.T. Binding of aflatoxin B1 to cell wall components of *Lactobacillus rhamnosus* strain GG. *Food Addit. Contam.* **2004**, *21*, 158–164. [CrossRef]

19. Chapot-Chartier, M.P.; Vinogradov, E.; Sadovskaya, I.; Andre, G.; Mistou, M.Y.; Trieu-Cuot, P.; Furlan, S.; Bidnenko, E.; Courtin, P.; Péchoux, C.; et al. The cell surface of *Lactococcus lactis* is covered by a protective polysaccharide pellicle. *J. Biol. Chem.* **2010**, *285*, 10464–10471. [CrossRef]

20. Linares, D.M.; Fitzgerald, G.; Hill, C.; Stanton, C.; Ross, P. Production of vitamins, exopolysaccharides and bacteriocins by probiotic bacteria. In *Probiotic Dairy Products*, 2nd ed.; Tamime, A., Thomas, L., Eds.; John Wiley & Sons Ltd.: Hoboken, NJ, USA, 2018; pp. 359–388.

21. Tsai, Y.T.; Cheng, P.C.; Pan, T.M. The immunomodulatory effects of lactic acid bacteria for improving immune functions and benefits. *Appl. Microbiol. Biotechnol.* **2012**, *96*, 853–862. [CrossRef]

22. Bueno, D.J.; Casale, C.H.; Pizzolitto, R.P.; Salvano, M.A.; Oliver, G. Physical adsorption of aflatoxin B1 by lactic acid bacteria and *Saccharomyces cerevisiae*: A theoretical model. *J. Food. Prot.* **2007**, *70*, 2148–2154. [CrossRef]

23. Wacoo, A.P.; Mukisa, I.M.; Meeme, R.; Byakika, S.; Wendiro, D.; Sybesma, W.; Kort, R. Probiotic enrichment and reduction of aflatoxins in a traditional African maize-based fermented food. *Nutrients* **2019**, *11*, 265. [CrossRef]

24. Peltonen, K.; El-Nezami, H.; Haskard, C.; Ahokas, J.; Salminen, S. Aflatoxin B1 binding by dairy strains of lactic acid bacteria and bifidobacteria. *J. Dairy Sci.* **2001**, *84*, 2152–2156. [CrossRef]

25. Alshannaq, A.F.; Yu, J.-H. A liquid chromatographic method for rapid and sensitive analysis of aflatoxins in laboratory fungal cultures. *Toxins* **2020**, *12*, 93. [CrossRef]

26. Ma, Z.X.; Amaro, F.X.; Romero, J.J.; Pereira, O.G.; Jeong, K.C.; Adesogan, A.T. The capacity of silage inoculant bacteria to bind aflatoxin B1 in vitro and in artificially contaminated corn silage. *J. Dairy Sci.* **2017**, *100*, 7198–7210. [CrossRef] [PubMed]

27. Kasmani, F.B.; Karimi Torshizi, M.A.; Allameh, A.A.; Shariatmadari, F. Aflatoxin detoxification potential of lactic acid bacteria isolated from Iranian poultry. *Iran. J. Vet. Res.* **2012**, *13*, 152–155.

28. El-Nezami, H.; Kankaanpaa, P.; Salminen, S.; Ahokas, J. Ability of dairy strains of lactic acid bacteria to bind a common food carcinogen, aflatoxin B1. *Food Chem. Toxicol.* **1998**, *36*, 321–326. [CrossRef]

29. Huang, L.; Cuicui Duan, C.; Zhao, Y.; Gao, L.; Niu, C.; Xu, J.; Li, S. Reduction of aflatoxin B1 toxicity by *Lactobacillus plantarum* C88: A potential probiotic strain isolated from Chinese traditional fermented food "tofu". *PLoS ONE* **2017**, *12*, e0170109. [CrossRef]

30. Belkacem-Hanfi, N.; Fhoula, I.; Semmar, N.; Guesmi, A.; Perraud-Gaime, I.; Ouzari, H.-I.; Boudabous, A.; Roussos, S. Lactic acid bacteria against post-harvest moulds and ochratoxin A isolated from stored wheat. *Biol. Control.* **2014**, *76*, 52–59. [CrossRef]

31. Apás, A.L.; González, S.N.; Arena, M.E. Potential of goat probiotic to bind mutagens. *Anaerobe* **2014**, *28*, 8–12. [CrossRef]

32. Liew, W.-P.-P.; Nurul-Adilah, Z.; Than, L.T.L.; Mohd-Redzwan, S. The binding efficiency and interaction of *Lactobacillus casei* Shirota toward aflatoxin B1. *Front. Microbiol.* **2018**, *9*, 1503. [CrossRef]

33. Hernandez-Mendoza, A.; Garcia, H.; Steele, J. Screening of *Lactobacillus casei* strains for their ability to bind aflatoxin B1. *Food Chem. Toxicol.* **2009**, *47*, 1064–1068. [CrossRef]

34. Oluwafemi, F.; Da-Silva, F.A. Removal of aflatoxins by viable and heat-killed *Lactobacillus* species isolated from fermented maize. *J. Appl. Biosci.* **2009**, *16*, 871–876.

35. Jakšić, D.; Klarić, M. Šegvić; Crnolatac, I.; Vujičić, N. Šijaković; Smrečki, V.; Górecki, M.; Pescitelli, G.; Piantanida, I. Unique Aggregation of Sterigmatocystin in Water Yields Strong and Specific Circular Dichroism Response Allowing Highly Sensitive and Selective Monitoring of Bio-Relevant Interactions. *Mar. Drugs* **2019**, *17*, 629.

36. Despot, D.J.; Kocsubé, S.; Bencsik, O.; Kecskeméti, A.; Szekeres, A.; Vágvölgyi, C.; Varga, J.; Klarić, M. Šegvić Species diversity and cytotoxic potency of airborne sterigmatocystin-producing Aspergilli from the section Versicolores. *Sci. Total. Environ.* **2016**, *562*, 296–304. [CrossRef] [PubMed]

37. Liu, S.; Fan, J.; Huang, X.; Jin, Q.; Zhu, G. Determination of Sterigmatocystin in Infant Cereals from Hangzhou, China. *J. AOAC Int.* **2016**, *99*, 1273–1278. [CrossRef] [PubMed]

Publisher's Note: MDPI stays neutral with regard to jurisdictional claims in published maps and institutional affiliations.

 *toxins*

Article

Aflatoxin B1 and Sterigmatocystin Binding Potential of Non-*Lactobacillus* LAB Strains

Ildikó Bata-Vidács [1],*, Judit Kosztik [1], Mária Mörtl [2] , András Székács [2] and József Kukolya [1]

[1] Department of Environmental and Applied Microbiology, Agro-Environmental Research Institute, National Agricultural Research and Innovation Centre, 1022 Budapest, Hungary; kosztik.judit@akk.naik.hu (J.K.); kukolya.jozsef@gmail.com (J.K.)

[2] Department of Environmental Analysis, Agro-Environmental Research Institute, National Agricultural Research and Innovation Centre, 1022 Budapest, Hungary; mortl.maria@akk.naik.hu (M.M.); szekacs.andras@akk.naik.hu (A.S.)

* Correspondence: batane.vidacs.ildiko@akk.naik.hu

Received: 17 November 2020; Accepted: 11 December 2020; Published: 14 December 2020

Abstract: Research on the ability of lactic acid bacteria (LAB) to bind aflatoxin B1 (AFB1) has mostly been focusing on lactobacilli and bifidobacteria. In this study, the AFB1 binding capacities of 20 *Enterococcus* strains belonging to *E. casseliflavus*, *E. faecalis*, *E. faecium*, *E. hirae*, *E. lactis*, and *E. mundtii*, 24 *Pediococcus* strains belonging to species *P. acidilactici*, *P. lolii*, *P. pentosaceus*, and *P. stilesii*, one strain of *Lactococcus formosensis* and *L. garviae*, and 3 strains of *Weissella soli* were investigated in MRS broth at 37 °C at 0.2 μg/mL mycotoxin concentration. According to our results, among non-lactobacilli LAB, the genera with the best AFB1 binding abilities were genus *Pediococcus*, with a maximum binding percentage of 7.6% by *P. acidilactici* OR83, followed by genus *Lactococcus*. For AFB1 bio-detoxification purposes, beside lactobacilli, pediococci can also be chosen, but it is important to select a strain with better binding properties than the average value of its genus. Five *Pediococcus* strains have been selected to compare their sterigmatocystin (ST) binding abilities to AFB1 binding, and a 2–3-fold difference was obtained similar to previous findings for lactobacilli. The best strain was *P. acidilactici* OR83 with 18% ST binding capacity. This is the first report on ST binding capabilities of non-*Lactobacillus* LAB strains.

Keywords: aflatoxin B1; sterigmatocystin; lactic acid bacteria; mycotoxin binding; detoxification

Key Contribution: Forty-eight strains belonging to the genera *Enterococcus*, *Pediococcus*, *Weissella*, and *Lactococcus* were screened for aflatoxin B1 binding abilities, and also 5 *Pediococcus* strains have been tested for sterigmatocystin binding potential. It is the first time in the literature that sterigmatocystin binding of pediococci is presented. AFB1 and ST binding abilities of strains belonging to the same species vary highly.

1. Introduction

Mycotoxins are secondary metabolites produced by microfungi that are capable of causing disease and death in humans and other animals [1]. The effects of some food-borne mycotoxins are acute with symptoms of severe illness appearing rapidly after consumption of food products contaminated with mycotoxins [2]. Of the several hundred mycotoxins identified so far, about a dozen have gained the most attention due to their severe effects on human health and their occurrences in food; and among the most commonly observed mycotoxins that present a concern to human health and livestock are aflatoxins [2].

Aflatoxins are amongst the most poisonous mycotoxins produced by species within *Aspergillus* section *Flavi*, which grow in soil, decaying vegetation, hay, grains, and various other substances [3,4].

Crops that are frequently affected by *Aspergillus* spp. include cereals (corn, sorghum, wheat, and rice), oilseeds (soybean, peanut, sunflower, and cotton seeds), spices (chili peppers, black pepper, coriander, turmeric, and ginger), and tree nuts (pistachio, almond, walnut, coconut, and Brazil nut) [1,5–9]. The four major aflatoxins are called aflatoxins B1, B2, G1, and G2 based on their fluorescence under UV light (blue or green). Among them, aflatoxin B1 (AFB1) is one of the most hazardous mycotoxins, primarily carcinogenic and genotoxic [10] and harmful to the liver [11]. The International Agency for Research on Cancer (IARC) classifies AFB1 as Group 1 carcinogen (carcinogenic to humans) [12]. In accordance with Regulation (EU) No 574/2011, the maximum permitted level for AFB1 in feed is 0.02 mg/kg.

Sterigmatocystin (ST) is a late metabolite in the aflatoxin pathway and is also produced as a final biosynthetic product by a number of species such as *Aspergillus versicolor* and *Aspergillus nidulans* [1,13]. ST is both mutagenic and tumorigenic but is less potent than aflatoxin [13]. Although experiments have shown genotoxicity and carcinogenicity of ST, limited data are available on the tumorigenic effect of the mycotoxin, which is why IARC classified it as a potential human carcinogen (Group 2B) in 1987 and has not revised this opinion ever since [14].

Lactic acid bacteria (LAB), Gram-positive, nonsporulating, oxidase and catalase negative, anaerobic aerotolerant microorganisms, are found in both the animal and the human body [15]. They ferment glucose to lactic acid. The most important genera are *Lactobacillus*, *Lactococcus*, *Leuconostoc*, *Enterococcus*, and *Pediococcus*, with 294, 21, 15, 59, and 11 species belonging to each genus, respectively [16].

Food-borne lactic acid bacteria able to bind mycotoxins can prevent their biotransformation to more toxic metabolites in the digestive tract, as the mycotoxin-microorganism adduct can pass through the body and be excreted in the feces, similarly as in the case of industrial mycotoxin binders, like aluminosilicates or glucomannan [17,18]. Numerous studies have shown that certain strains of some LAB species can bind mycotoxins, among them AFB1, to their surface [19–21]. The published results indicate that the adsorption of AFB1 to microorganisms is a rapid process. The binding involves the formation of a reversible complex between the chemically unmodified mycotoxin molecule and the microorganism surface, and the yield of AFB1 removal is dependent of the concentration of both the mycotoxin and the bacteria [22]. The binding mechanism is not yet elucidated, but the binding of AFB1 to the glycan components of the cell wall of probiotic bacteria has been suggested as a key momentum in the process [23–25].

According to literature data, the binding of aflatoxins by LAB strains is highly strain specific. The ratio of AFB1 bound by 10^9 cfu/mL of 8 strains of *L. casei* varied from 14% to 49% from the available 4.6 µg/mL in the studies of Hernandez-Mendoza et al. [26]. Reasons for the strain-specificity of AFB1 binding are yet unknown, but differences in cell wall components, particularly in the peptidoglycan content, may be implicated [27].

There has been quite a bit of research done on the ability of LAB to mitigate the detrimental effects of aflatoxin-producing fungal strains and their AFB1 binding capacity [20,22,28–30], though focusing mostly on lactobacilli and bifidobacteria. In addition, data on aflatoxin binding abilities of different strains belonging to the same species are scarce.

At our department, microbes with colony morphology of lactic acid bacteria were isolated on LAB selective MRS (de Man, Rogosa and Sharpe, VWR) plates from 14 exotic animals of the Budapest Zoo and Botanical Garden [31,32]. The identification of the strains was done by sequencing. At present, the collection comprises of nearly 1000 strains and is constantly expanding. Most of our strains belong to the genera *Lactobacillus* and *Enterococcus*, but we also managed to isolate strains belonging to the other LAB genera.

Our goal was to screen strains of the genus *Enterococcus*, *Lactococcus*, *Pediococcus*, and *Weissella* from our collection for AFB1 and ST binding capacities. In the literature, PBS medium (phosphate buffer solution) is most commonly used in mycotoxin binding assays of LAB strains. As our work was carried out as part of a probiotic development project, we chose a medium, MRS medium (Lactobacillus Agar according to DeMan, Rogosa and Sharpe), for our experiments, which is closer to real conditions due to its much higher organic matter content and provides optimal conditions for the microbes. Mycotoxin concentrations were monitored in rapid analysis by an instrumental method, high-performance liquid chromatography (HPLC) coupled with UV detection of the target analytes, AFB1 and ST, upon solvent extraction. Separation of the mycotoxins was achieved on hydrophobic linear alkylsilane stationary phase.

2. Results and Discussion

2.1. Instrumental Analysis of Mycotoxins

Analysis of 54 biomasses as well as the corresponding 6 MRS broth samples was carried out by high-performance liquid chromatography (HPLC) coupled with UV detection upon solvent extraction on the basis of a method optimization. Recent methods use mainly acetonitrile for the extraction of mycotoxins from foodstuffs, followed by cleanup with different modes of solid phase extraction (e.g., immunoaffinity, dispersive, etc.) to eliminate matrix effects. For cultivated bacteria or fungal strains, methanol [33] or chloroform [34,35] are the most frequently applied solvents, then the extracts are either subjected to a cleanup [33] or only filtered [35] prior to the analysis by a liquid chromatographic method. For complex matrices, gradient elution is applied [33], but for the bacteria investigated in the present study, a simple isocratic method gave sufficient separation. Typical chromatograms of target compounds are shown in Figure 1. The retention times were 6.07 and 7.13 min for AFB1 and ST, respectively. Although mycotoxins could be determined directly from the MRS broth, removal of the most polar matrix components resulted in better baseline and less interference (see Figure 2). For the extraction of MRS broth, the more volatile solvent, dichloromethane, allowed good recoveries (93.4 ± 3.1 and 97.6 ± 4.8% for AFB1 and ST, respectively), while for biomass, addition of 10% methanol to dichloromethane significantly increased the extraction efficiency by enhancing the solvent penetration to the cells. Ultrasound agitation seemed to be less effective than shaking of samples. Both analytes were determined from HPLC peak areas at the corresponding retention times with excellent linear calibration characteristics. For quantification of target compounds, peak areas determined for AFB1 at 365 nm and for ST at 240 nm were used. Peak purity for ST was checked by the ratios of signal intensities (peak areas) recorded at 240 and 325 nm, which was found 2.03 for standard solutions. The linear regression values of external calibration curves were 0.9992 and 0.9997, and the slopes were 110.7 and 145.3 for AFB1 and ST, respectively. The limits of detection, determined with standard solutions, were 0.010 µg/mL for both mycotoxins, and they were the same in spiked liquid matrices extracted from blank.

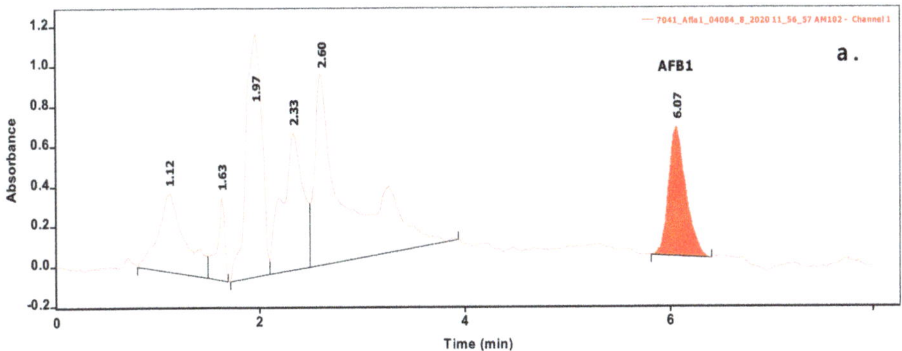

Figure 1. *Cont.*

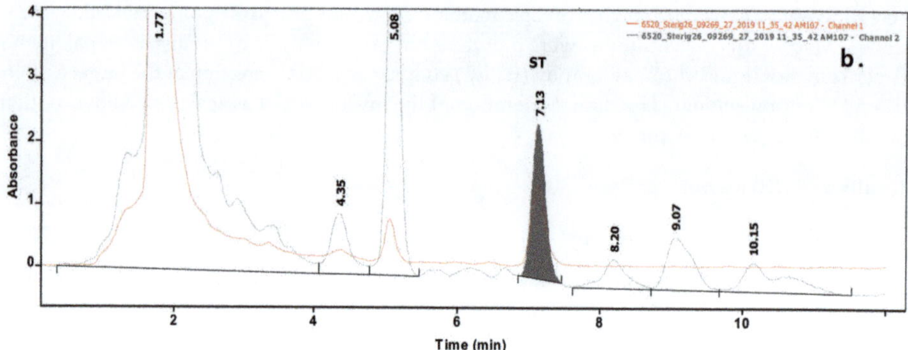

Figure 1. Chromatograms of samples extracted from biomass containing Aflatoxin B1 (AFB1) (**a**) or sterigmatocystin (ST) (**b**) at levels of 0.070 and 0.236 µg/mL, respectively.

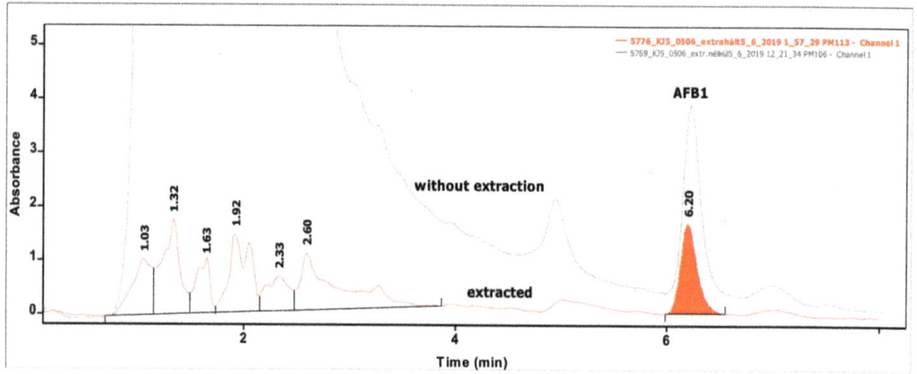

Figure 2. Chromatograms of a sample measured directly without any extraction (upper line) and that of the same sample extracted from MRS broth (lower line) containing AFB1 at the level of 0.50 µg/mL.

2.2. Aflatoxin B1 Binding Capacities of LAB Other Then Lactobacilli

2.2.1. Aflatoxin B1 Binding Capacities of *Enterococcus* Strains

Twenty *Enterococcus* strains from our lactic acid bacterium culture collection were selected for this study. One strain belonged to *E. casseliflavus*, 4 to *E. faecalis*, 1 to *E. faecium*, 6 to *E. hirae*, 3 to *E. lactis*, and 4 to *E. mundtii*. Two strains had higher AFB1 binding ability, *E. hirae* AT12 and *E. lactis* SK34 with 4.62% and 3.40%, respectively, for the other strains, the binding was below 1.61% (Figure 3). Regarding species, the best average AFB1 binding capacities were also obtained for species *E. lactis* and *E. hirae*, though for these two species, the standard deviations were higher than for the other species studied (Table 1). Juri et al. [36] found much higher AFB1 binding percentages for *Enterococcus faecium* GJ40 with 24–27% and 17–24%, and *E. faecium* MF4 with 36–42% and 27–32% at 0.05 and 0.10 µg/mL, respectively. The stability of those bacteria-AFB1 complexes formed was found to be high, up to 50% of AFB1 remained bound in bacterial cell after three washes with phosphate buffered saline. These differences in the results might be explained by the different strains or cultivation parameters and methods used in the studies; for example, in most studies, the bound mycotoxin concentration is calculated from the mycotoxin content remaining in the supernatant of the culture suspension, while in our investigations, the mycotoxin content of the biomass was determined directly.

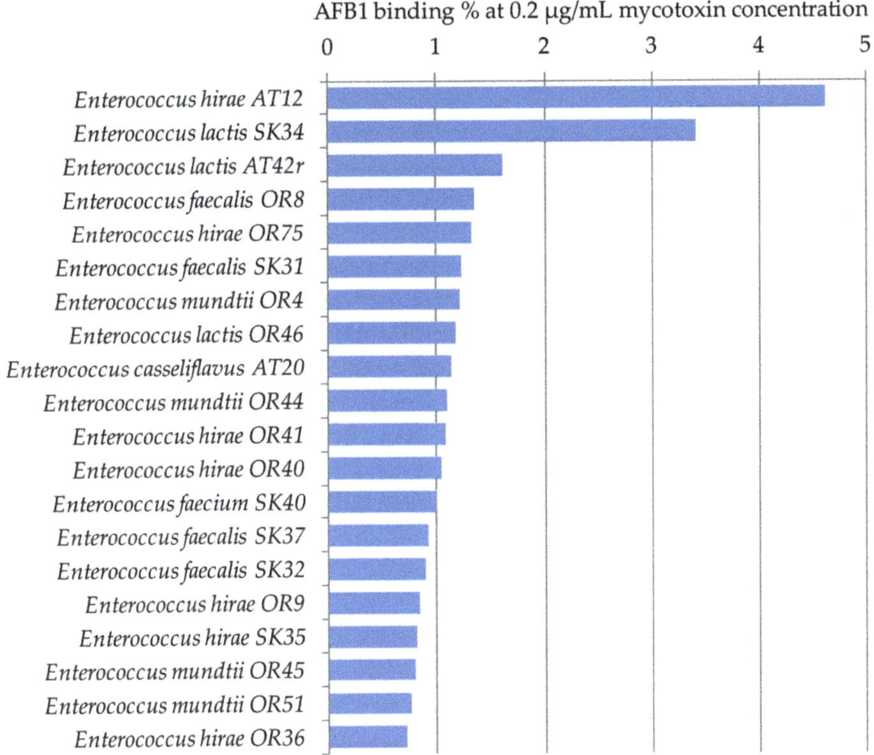

Figure 3. AFB1 binding capacities of *Enterococcus* strains at 0.2 µg/mL mycotoxin concentration in MRS broth.

Table 1. Percentage AFB1 binding capacities of *Enterococcus* species at 0.2 µg/mL mycotoxin concentration in MRS broth.

Species	Number of Strains	Average Binding %	STD	Min Binding %	Max Binding %
Enterococcus lactis	3	2.06	1.18	1.17	3.40
Enterococcus hirae	7	1.49	1.39	0.72	4.62
Enterococcus casseliflavus	1	1.14		1.14	1.14
Enterococcus faecalis	4	1.10	0.23	0.89	1.35
Enterococcus faecium	1	1.00		1.00	1.00
Enterococcus mundtii	4	0.97	0.22	0.77	1.21

2.2.2. Aflatoxin B1 Binding Capacities of *Pediococcus* Strains

From the genus *Pediococcus*, the AFB1 binding capacities of 24 strains were studied. The strains belonged to species *P. acidilactici* (8 strains), *P. lolii* (3 strains), *P. pentosaceus* (12 strains), and *P. stilesii* (1 strain). According to the results shown in Figure 4, the best AFB1 binding ability was found for strain *P. acidilactici* OR83. For the other strains, the AFB1 binding percentages were around or below 4% (Table 2). The average binding capacities of the species were around 3% with the exception of *P. pentosaceus* at 2%. The highest standard deviation of the AFB1 binding abilities of the strains belonging to one species was obtained for *P. acidilactici*. These results are in agreement with data presented in the literature, where Zinedine et al. [37] found that *Pediococcus acidilactici* strain P55 removed 1.80% AFB1.

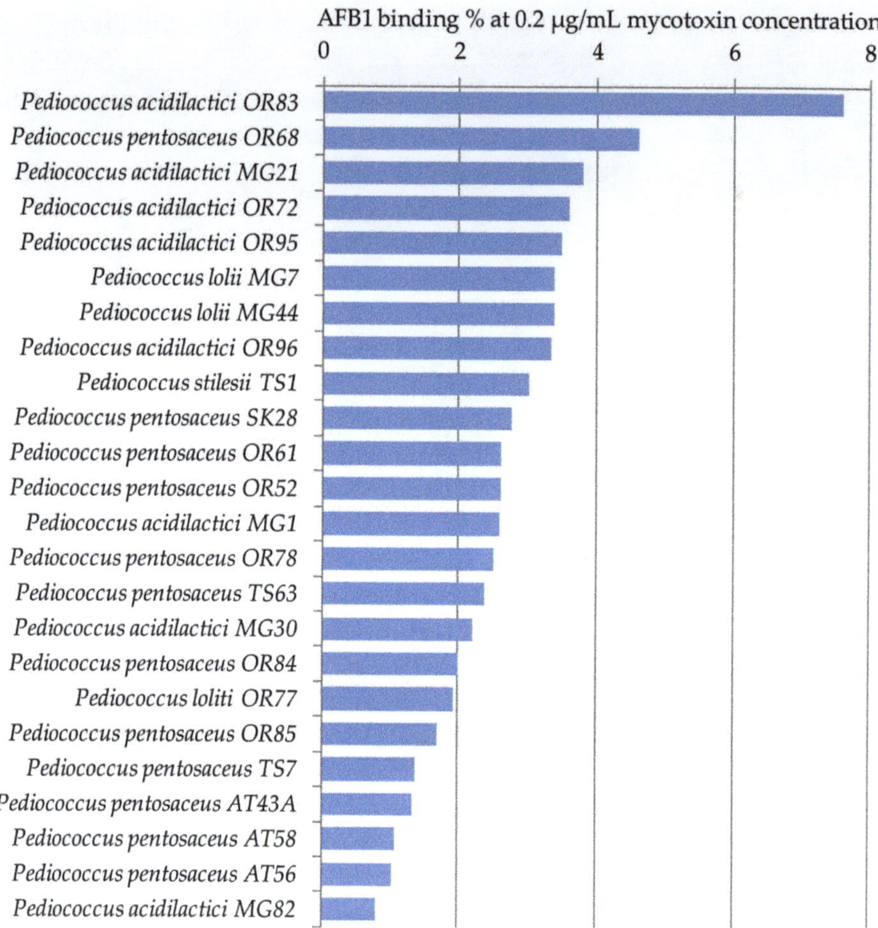

Figure 4. *Pediococcus* strains with percentage AFB1 binding capacities at 0.2 µg/mL mycotoxin concentration in MRS broth.

Table 2. Percentage AFB1 binding capacities of *Pediococcus* species at 0.2 µg/mL mycotoxin concentration in MRS broth.

Species	Number of Strains	Average Binding %	STD	Min Binding %	Max Binding %
Pediococcus acidilactici	8	3.43	1.95	0.80	7.60
Pediococcus stilesii	1	3.03		3.03	3.03
Pediococcus lolii	3	2.90	0.84	1.93	3.39
Pediococcus pentosaceus	12	2.18	0.99	1.05	4.60

2.2.3. Aflatoxin B1 Binding Capacities of *Lactococcus* and *Weissella* Strains

For the study of AFB1 binding capacities of the genera *Lactococcus* and *Weissella*, only a limited number of strains has been used, one strain of *Lactococcus formosensis*, 1 strain of *L. garviae*, and 3 strains of *Weissella soli*. All studied strains have low mycotoxin binding capacities at the parameter setup of the experiment (Figure 5). Peltonen et al. [29] found that the three *Lactococcus lactis* strains studied bound 5.6 to 41.1% AFB1, which shows the wide range of binding capacities depending on the strains of a species. For aflatoxin binding of *Weissella* spp., only a few papers can be found in the literature. Binding with AFB1 was found to be 43.7% for *Weissella cibaria* NN20 by Nduti et al. [38] in skim milk at 10 ng/mL AFB1 concentration, also the EPS produced by *Weissella confusa* was proved to have aflatoxin binding capacity up to 34.79% at 100 mg/mL concentration of EPS, though no binding could be observed under 20 mg/mL EPS concentration [39]. Differences between our findings and the results might be explained by the different strains or methods used in the studies.

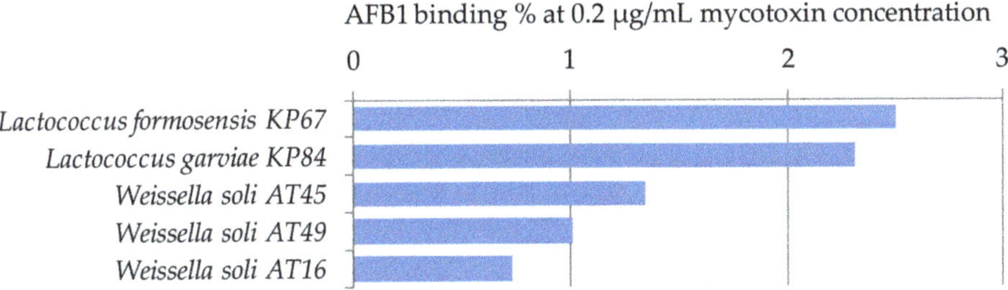

Figure 5. *Lactococcus* and *Weissella* strains with percentage AFB1 binding capacities at 0.2 µg/mL mycotoxin concentration in MRS broth.

2.2.4. AFB1 Binding Capacities of Lactic Acid Bacteria, Regarding Genus

Among the major genera belonging to lactic acid bacteria are *Enterococcus*, *Lactobacillus*, *Lactococcus*, *Pediococcus*, and *Weissella*. To compare the AFB1 binding capacities of the genera, averages, standard deviations, minimum and maximum values were calculated from the data obtained for the strains belonging to the same genus. Data obtained from our previous studies [32] were used to calculate the values for genus *Lactobacillus*. Results are presented in Table 3. The genus with the best AFB1 binding ability was the genus *Lactobacillus*, with an average binding of 3.16%. In addition, the standard deviation of the data for the abilities of its strains was the highest among genera, presenting a 20-fold difference between minimum and maximum values. The second best genus was genus *Pediococcus*, with average binding percentage of 2.72%, and the third place was taken by the genus *Lactococcus*. It can be concluded that for AFB1 bio-detoxification purposes, lactobacilli, pediococci, or lactococci should be chosen, but it is important to select a strain with better binding properties than the average value of their genera.

Table 3. Percentage AFB1 binding capacities of lactic acid bacteria, regarding genus, at 0.2 µg/mL mycotoxin concentration in MRS broth (*Lactobacillus* results are from previous studies [32]).

Genus	Number of Strains	Average Binding %	STD	Min Binding %	Max Binding %
Lactobacillus	105	3.16	1.98	0.55	11.50
Pediococcus	24	2.72	1.42	0.80	7.60
Lactococcus	2	2.40	0.14	2.31	2.50
Enterococcus	20	1.35	0.96	0.72	4.62
Weissella	3	1.03	0.31	0.73	1.35

2.3. Sterigmatocystin Binding Capacities of Pediococcus Strains

ST binding abilities of 5 *Pediococcus* strains of our LAB collection were determined according to Section 4.3. The results are summarized in Figure 6. Mycotoxin binding values were between 9–18%. These results are in agreement with our previous findings for *Lactobacillus* strains, that ST binding is 2–3 times the AFB1 binding capacity [32]. So far, no other results have been published in the literature that addressed the ST binding ability of lactobacilli.

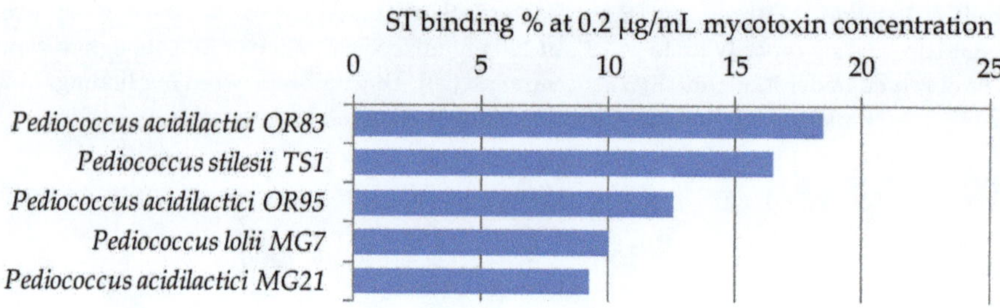

Figure 6. Sterigmatocystin binding capacities (%) of *Pediococcus* strains at 0.2 µg/mL mycotoxin concentration in MRS broth.

3. Conclusions

For the analytical determination of target components, a simple isocratic separation was suitable after the appropriate sample preparation. Extraction of MRS broth by dichloromethane removed the polar matrix components, resulting in lower baseline and lower detection limit. Extraction of biomass required addition of 10% of methanol to dichloromethane to facilitate better release of target components from the surface of the cells. Accurate determination of mycotoxins in biomass is especially important in those cases when the binding capacity is low (e.g., 1%) and RSD values for the remaining mycotoxin in the MRS broth are comparable to those bound by the cells.

For the study of their AFB1 binding abilities, 49 lactic acid bacteria other than lactobacilli were selected from our culture collection. The results of our mycotoxin binding assays in MRS medium cannot be compared directly with PBS-based binding assays. However, it is a perfectly suitable method for determining the binding potential among our strains.

From the 20 *Enterococcus* strains belonging to 6 species, two had higher AFB1 binding ability, *E. hirae* AT12 and *E. lactis* SK34 with 4.62% and 3.40%, respectively. From the genus *Pediococcus*, the AFB1 binding capacities of 24 strains belonging to 4 species were studied, strain *P. acidilactici* OR83 stood out with a value of 7.60%, for the other pediococci, binding values of around 3% were obtained. For the genera *Lactococcus* and *Weissella*, low AFB1 binding capacities were found, though there was only limited number of strains studied. It can be concluded that for AFB1 bio-detoxification purposes, beside lactobacilli, pediococci can also be chosen, but it is important to select a strain with better binding properties than the average value of its genus. On the ST binding ability of strains belonging to the genus *Pediococcus*, the results are in agreement with our previous findings for *Lactobacillus* strains, that ST binding is 2–3 times the AFB1 binding capacity.

It should be noted that the best aflatoxin binding *Pediococcus* strain was the best ST binding, as well. This can be explained by the fact that the two structurally similar mycotoxins bind to the same cell wall polysaccharide (WPS) receptor of the bacterium. The binding strength may be stronger for ST than for AFB1. The different mycotoxin binding ability of strains of the same species, which can also be seen in the literature, may be due to their highly variable WPS cell wall components [40] rather than to the much more conserved peptidoglycan cell wall.

In this work, we report strong ST binding of non-lactobacillus LAB strains for the first time in the literature. The detection of different AFB1 and ST binding of LAB strains belonging to the same species with different binding activity may represent a model system that will allow the exploration of the exact molecular mechanism of the binding of these mycotoxins in the future.

4. Materials and Methods

4.1. Bacterial Strains

Forty-nine lactic acid bacterium strains of our collection isolated from feces samples of exotic herbivorous zoo animals were used for the studies (Table 4). The strains were identified by the 16S rDNA sequence extracted from pure bacterial cultures and sequenced by BaseClear (Leiden, The Netherlands). The LAB strains stored at −80 °C in 43.5% glycerin were thawed on ice before culturing.

Table 4. Strains of lactic acid bacterium species of our collection used in the current study.

Species	Strains
Enterococcus	
E. casseliflavus	AT20
E. faecalis	OR8, SK31, SK32, SK37
E. faecium	SK40
E. hirae	AT12, OR9, OR36, OR40, OR41, OR75, SK35
E. lactis	AT42r, OR46, SK34
E. mundtii	OR4, OR44, OR45, OR51
Lactococcus	
L. formosensis	KP67
L. garviae	KP84
Pediococcus	
P. acidilactici	MG1, MG21, MG31, MG82, OR72, OR83, OR95, OR96
P. lolii	MG7, MG44, OR77
P. pentosaceus	AT43A, AT56, AT58, OR52, OR61, OR68, OR78, OR84, OR85, SK28, TS7, TS63
P. stilesii	TS1
Weissella	
W. soli	AT16, AT45, AT49

4.2. Mycotoxins

AFB1 and ST were purchased from Sigma-Aldrich (Budapest, Hungary). Standard solutions were made by diluting the mycotoxin powder with methanol (puriss., MOLAR Chemicals Ltd., Halásztelek, Hungary) to make stock solutions of 50 μg/mL. The complete dissolution of the mycotoxins was ensured by mild heating and sonication (Ultrasonic Cleaning Instrument, Falc Instruments, Treviglio, Italy). The concentrations of the stock solutions were verified by HPLC measurement. These stock solutions were used in all experiments. The mycotoxin concentrations for our experiments were set at 0.2 μg/mL, which is the tenfold value of the maximum permitted level for AFB1 by EU Regulation No 574/2011.

4.3. Screening LAB Strains for Mycotoxin Binding Capacities

LAB strains were taken from −20 °C storage, thawed on ice, and 20 μl of the suspension was transferred to 9 mL lactic acid bacterium selective MRS (de Man, Rogosa and Sharpe, VWR) broth. The tubes were incubated at 37 °C for 24 h. Falcon tubes containing 15 mL of MRS broth were inoculated with 50 μl of the cultures. The tubes were incubated at 37 °C for two days. Three replicates were prepared with each strain.

After the incubation, the cell concentrations of the cultures were set at 10^8 cfu/mL, and then 0.2 μg/mL of AFB1 or ST was added to the tubes. Pure MRS broth was used as negative control, and mycotoxin-only MRS broth without bacteria was used as positive control. The tubes were mixed by shaking and the tubes were incubated with the mycotoxin for 10 min at room temperature. The tubes were centrifuged at 4000 rpm for 40 min to separate the biomass from the supernatant. The supernatant was discarded [20]. The AFB1 and ST contents of the biomasses were determined by HPLC method described in Section 4.4.

4.4. Mycotoxin Extraction and HPLC Measurements

The amount of mycotoxin was determined by high performance liquid chromatography (HPLC) analysis using a YL9100 HPLC system equipped with a YL9150 autosampler (YL Instruments, Gyeonggi, Korea). For the measurement, the mycotoxin was extracted from the samples by the following steps. For the extraction of the mycotoxin from the biomass, 1.8 mL of dichloromethane and 0.2 mL of methanol were added to the Falcon tube containing the biomass, using the ratio that gave best results in preliminary experiments. The mixture was pipetted into Eppendorf tubes. The tubes were vortexed in a horizontal shaker for 20 min in the dark and then centrifuged at 3000 rpm for 10 min. One ml of the supernatant was taken out and the solvent was evaporated to the dryness in a clean Eppendorf tube at 45 °C under a fume hood in a Thermo Shaker (TS-100, Biosan, Riga, Latvia). MRS broth was extracted similarly, but 1 mL of dichloromethane was shaken with one milliliter of supernatant for 20 min. From the dichloromethane phase, 0.5 mL was taken out and concentrated in a clean Eppendorf tube at 45 °C under a fume hood. The residues were solved in 1 mL of eluent (see below) and the sample was filtered through a 0.45 μm hydrophilic polytetrafluoroethylene (PTFE) syringe filter (Labex Ltd., Budapest, Hungary) prior to HPLC determination.

The mycotoxin content of the samples was determined by UV detection (HPLC-UV) after an isocratic liquid chromatographic separation. UV detector signals were recorded at $\lambda = 365$ nm or $\lambda = 240$ and 325 nm for AFB1 and ST, respectively. The separation was performed on a Brisa (Technochroma, Barcelona, Spain) C18 column (5 μm, 15 cm × 0.46 cm) at 30 °C. The eluent flow rate was set to 1.0 mL/min and 30 μl of samples were injected. The eluent consisted of 60:20:20 = A:B:C eluents, and 40:30:30 = A:B:C eluents (A = 90% water: 10% MeOH, B = MeOH, C = Acetonitrile), held till 8 and 12 min for AFB1 and ST, respectively. Extracts of blank non-spiked control biomass did not contain interfering matrix components, therefore quantitation was based on instrumental (external) calibration with standard solutions in the range between 0.010 and 2.00 μg/mL. Recoveries at concentration of 0.2 μg/mL in the spiked samples were determined by adding a known concentration of AFB1 or ST to the liquid of blank samples. Peak purities were systematically checked by recording absorption at two wavelengths, and peak area ratios at those wavelengths were compared to the ratios characteristic to standard solutions of the analyte (ST). Binding capacities (%) were calculated on the basis of analyte concentrations in the extracted biomass samples related to the initial MRS broth levels considering the corresponding concentration factor applied (see sample preparation, above). RSD values calculated from the three parallel injections of standard solutions ranged between 0.2 and 1.4%.

Author Contributions: Conceptualization, J.K. (József Kukolya); methodology, J.K. (Judit Kosztik), I.B.-V., M.M. and A.S.; validation, M.M.; investigation, J.K. (Judit Kosztik) and I.B-V.; writing—original draft preparation, J.K. (Judit Kosztik), I.B.-V and M.M.; writing—review and editing, I.B.-V. and A.S.; supervision, J.K. (József Kukolya) and A.S.; funding acquisition, J.K. (József Kukolya). All authors have read and agreed to the published version of the manuscript.

Funding: This research was funded by the National Research, Development and Innovation Office in Hungary, projects NVKP-16-1-2016-0009, NVKP-16-1-2016-0049 and the Hungarian National Research Fund, OTKA K116631.

Conflicts of Interest: The authors declare no conflict of interest.

References

1. Bennett, J.W.; Klich, M. Mycotoxins. *Clin. Microbiol. Rev.* **2003**, *16*, 497–516. [CrossRef] [PubMed]
2. World Health Organization (WHO). Available online: https://www.who.int/news-room/fact-sheets/detail/mycotoxins (accessed on 1 December 2020).
3. Frisvad, J.C.; Hubka, V.; Ezekiel, C.N.; Hong, S.-B.; Nováková, A.; Chen, A.J.; Arzanlou, M.; Larsen, T.O.; Sklenář, F.; Mahakarnchanakul, W.; et al. Taxonomy of *Aspergillus* section *Flavi* and their production of aflatoxins, ochratoxins and other mycotoxins. *Stud. Mycol.* **2019**, *93*, 1–63. [CrossRef] [PubMed]
4. Singh, P.; Callicott, K.A.; Orbach, M.J.; Cotty, P.J. Molecular Analysis of S-morphology Aflatoxin Producers From the United States Reveals Previously Unknown Diversity and Two New Taxa. *Front. Microbiol.* **2020**, *11*, 1236. [CrossRef] [PubMed]

5. Forgacs, J. Mycotoxicoses—The neglected diseases. *Feedstuffs* **1962**, *34*, 124–134.

6. Diener, U.L.; Cole, R.J.; Sanders, T.H.; Payne, G.A.; Lee, L.S.; Klich, M.A. Epidemiology of aflatoxin formation by *Aspergillus flavus*. *Annu. Rev. Phytopathol.* **1987**, *25*, 249–270. [CrossRef]

7. Detroy, R.W.; Lillehoj, E.B.; Ciegler, A. Aflatoxin and related compounds. In *Microbial Toxins, Volume VI: Fungal Toxins*; Ciegler, A., Kadis, S., Ajl, S.J., Eds.; Academic Press: New York, NY, USA, 1971; pp. 3–178.

8. Moral, J.; Garcia-Lopez, M.T.; Camiletti, B.X.; Jaime, R.; Michailides, T.J.; Bandyopadhyay, R.; Ortega-Beltran, A. Present Status and Perspective on the Future Use of Aflatoxin Biocontrol Products. *Agronomy* **2020**, *10*, 491. [CrossRef]

9. Singh, P.; Cotty, P.J. Aflatoxin contamination of dried red chilies: Contrasts between the United States and Nigeria, two markets differing in regulation enforcement. *Food Control* **2017**, *80*, 374–379. [CrossRef]

10. Marchese, S.; Polo, A.; Ariano, A.; Velotto, S.; Costantini, S.; Severino, L. Aflatoxin B1 and M1: Biological Properties and Their Involvement in Cancer Development. *Toxins* **2018**, *10*, 214. [CrossRef]

11. Henry, S.H.; Bosch, F.X.; Bowers, J.C. Aflatoxin, hepatitis and worldwide liver cancer risks. In *Mycotoxins and Food Safety*; Springer: Boston, MA, USA, 2002; pp. 229–233.

12. International Agency for Research on Cancer (IARC). Aflatoxins. *IARC Monogr. Eval. Carcinog. Risks Hum.* **2012**, *100F*, 225–248.

13. Berry, C.L. The pathology of mycotoxins. *J. Pathol.* **1988**, *154*, 301–311. [CrossRef]

14. International Agency for Research on Cancer (IARC). Overall evaluations of carcinogenicity. *IARC Monogr. Eval. Carcinog. Risks Hum.* **1987**, *7*, 72.

15. Mokoena, M.P. Lactic Acid Bacteria and Their Bacteriocins: Classification, Biosynthesis and Applications against Uropathogens: A Mini-Review. *Molecules* **2017**, *22*, 1255. [CrossRef] [PubMed]

16. LPSN. Available online: https://lpsn.dsmz.de/ (accessed on 2 December 2020).

17. Gimeno, A.; Martins, M.L. *Micotoxinas y Micotoxicosis en Animales y Humanos*, 3rd ed.; Special Nutrients: Miami, FL, USA, 2011; pp. 92–93.

18. Kumara, S.S.; Gayathri, D.; Hariprasad, P.; Venkateswaran, G.; Swamy, C.T. In vivo AFB1 detoxification by Lactobacillus fermentum LC5/a with chlorophyll and immunopotentiating activity in albino mice. *Toxicon* **2020**, *187*, 214–222. [CrossRef] [PubMed]

19. Byakika, S.; Mukisa, I.M.; Wacoo, A.P.; Kort, R.; Byaruhanga, Y.B.; Muyanja, C. Potential application of lactic acid starters in the reduction of aflatoxin contamination in fermented sorghum-millet beverages. *Int. J. Food Contam.* **2019**, *6*, 4. [CrossRef]

20. El-Nezami, H.; Kankaanpaa, P.; Salminen, P.; Ahokas, J. Ability of dairy strains of lactic acid bacteria to bind a common food carcinogen, aflatoxin B1. *Food Chem. Toxicol.* **1998**, *36*, 321–326. [CrossRef]

21. Liew, W.-P.; Nurul-Adilah, Z.; Than, L.T.L.; Mohd-Redzwan, S. The Binding Efficiency and Interaction of Lactobacillus casei Shirota toward Aflatoxin B1. *Front. Microbiol.* **2018**, *9*, 1503. [CrossRef]

22. Bueno, D.J.; Casale, C.H.; Pizzolitto, R.P.; Salvano, M.A.; Oliver, G. Physical Adsorption of Aflatoxin B1 by Lactic Acid Bacteria and Saccharomyces cerevisiae: A Theoretical Model. *J. Food Prot.* **2007**, *70*, 2148–2154. [CrossRef]

23. Ahlberg, S.H.; Joutsjoki, V.; Korhonen, H.J. Potential of lactic acid bacteria in aflatoxin risk mitigation. *Int. J. Food Microbiol.* **2015**, *207*, 87–102. [CrossRef]

24. El-Nezami, H.S.; Kankaanpää, P.E.; Salminen, S.; Ahokas, J. Physicochemical Alterations Enhance the Ability of Dairy Strains of Lactic Acid Bacteria to Remove Aflatoxin from Contaminated Media. *J. Food Prot.* **1998**, *61*, 466–468. [CrossRef]

25. Haskard, C.A.; El-Nezami, H.S.; Kankaanpaa, P.E.; Salminen, S.; Ahokas, J. Surface Binding of Aflatoxin B1 by Lactic Acid Bacteria. *Appl. Environ. Microbiol.* **2001**, *67*, 3086–3091. [CrossRef]

26. Hernández-Mendoza, A.; Garcia, H.S.; Steele, J.L. Screening of Lactobacillus casei strains for their ability to bind aflatoxin B1. *Food Chem. Toxicol.* **2009**, *47*, 1064–1068. [CrossRef] [PubMed]

27. Lahtinen, S.J.; Haskard, C.A.; Ouwehand, A.C.; Salminen, S.J.; Ahokas, J.T. Binding of aflatoxin B1 to cell wall components of *Lactobacillus rhamnosus* strain GG. *Food Addit. Contam.* **2004**, *21*, 158–164. [CrossRef] [PubMed]

28. Dalié, D.; Deschamps, A.; Richard-Forget, F. Lactic acid bacteria—Potential for control of mold growth and mycotoxins: A review. *Food Control* **2010**, *21*, 370–380. [CrossRef]

29. Peltonen, K.; El-Nezami, H.; Haskard, C.; Ahokas, J.; Salminen, S. Aflatoxin B1 Binding by Dairy Strains of Lactic Acid Bacteria and Bifidobacteria. *J. Dairy Sci.* **2001**, *84*, 2152–2156. [CrossRef]

30. Wacoo, A.P.; Mukisa, I.M.; Meeme, R.; Byakika, S.; Wendiro, D.; Sybesma, W.; Kort, R. Probiotic Enrichment and Reduction of Aflatoxins in a Traditional African Maize-Based Fermented Food. *Nutrients* **2019**, *11*, 265. [CrossRef]

31. Tóth, Á.; Baka, E.; Bata-Vidács, I.; Luzics, S.; Kosztik, J.; Tóth, E.; Kéki, Z.; Schumann, P.; Táncsics, A.; Nagy, I.; et al. Micrococcoides hystricis gen. nov., sp. nov., a novel member of the family Micrococcaceae, phylum Actinobacteria. *Int. J. Syst. Evol. Microbiol.* **2017**, *67*, 2758–2765. [CrossRef]

32. Kosztik, J.; Mörtl, M.; Székács, A.; Kukolya, J.; Bata-Vidács, I. Aflatoxin B1 and Sterigmatocystin Binding Potential of Lactobacilli. *Toxins* **2020**, *12*, 756. [CrossRef]

33. Saldan, N.C.; Almeida, R.T.R.; Avíncola, A.; Porto, C.; Galuch, M.B.; Magon, T.F.S.; Pilau, E.J.; Svidzinski, T.I.E.; Oliveira, C.C. Development of an analytical method for identification of Aspergillus flavus based on chemical markers using HPLC-MS. *Food Chem.* **2018**, *241*, 113. [CrossRef]

34. Pereyra, M.L.G.; Martínez, M.P.; Petroselli, G.; Balsells, R.E.; Cavaglieri, L.R. Antifungal and aflatoxin-reducing activity of extracellular compounds produced by soil Bacillus strains with potential application in agriculture. *Food Control* **2018**, *85*, 392–399. [CrossRef]

35. Alshannaq, A.F.; Yu, J.-H. A Liquid Chromatographic Method for Rapid and Sensitive Analysis of Aflatoxins in Laboratory Fungal Cultures. *Toxins* **2020**, *12*, 93. [CrossRef]

36. Juri, F.M.G.; Dalcero, A.M.; Magnoli, C.E. In vitro aflatoxin B1 binding capacity by two Enterococcus faecium strains isolated from healthy dog faeces. *J. Appl. Microbiol.* **2015**, *118*, 574–582. [CrossRef] [PubMed]

37. Zinedine, A.; Faid, M.; Benlemlih, M. In vitro reduction of aflatoxin B1 by strains of lactic acid bacteria isolated from Moroccan sourdough bread. *Int. J. Agric. Biol.* **2005**, *7*, 67–70.

38. Nduti, N.N.; Reid, G.; Sumarah, M.; Hekmat, S.; Mwaniki, M.; Njeru, P.N. *Weissella cibaria* Nn20 isolated from fermented kimere shows ability to sequester AFB1 in vitro and ferment milk with good viscosity and phin comparison to yogurt. *Food Sci. Nutr. Technol.* **2018**, *3*, 000137. [CrossRef]

39. Kavitake, D.; Singh, S.P.; Kandasamy, S.; Devi, P.B.; Shetty, P.H. Report on aflatoxin-binding activity of galactan exopolysaccharide produced by Weissella confusa KR780676. *3 Biotech* **2020**, *10*, 181. [CrossRef]

40. Chapot-Chartier, M.P.; Kulakauskas, S. Cell wall structure and function in lactic acid bacteria. *Microb. Cell Factories* **2014**, *13*, S9. [CrossRef]

Publisher's Note: MDPI stays neutral with regard to jurisdictional claims in published maps and institutional affiliations.

MDPI

St. Alban-Anlage 66

4052 Basel

Switzerland

Tel. +41 61 683 77 34

Fax +41 61 302 89 18

www.mdpi.com

Toxins Editorial Office

E-mail: toxins@mdpi.com

www.mdpi.com/journal/toxins